ADVANCES IN VIBRATION ANALYSIS

Advances in Vibration Analysis
Research Collection
http://dx.doi.org/10.5772/ 59847
Chapters from books edited by: Francisco Beltran-Carbajal, Mauricio Zapateiro, Farzad Ebrahimi and Natalie Baddour

CBS, Edition **2018**

Published by InTech
Janeza Trdine 9, 51000 Rijeka, Croatia

Additional hard copies can be obtained from orders@intechopen.com

Advances in Vibration Analysis - Research Collection

p. cm.

ISBN 978-953-51-1833-6

INTECH

open science | open minds

Contents

Preface

Vibration analysis plays a key role in the monitoring of machine systems to ensure safety and efficiency, and is routinely performed in a huge variety of fields. This book brings the reader the latest research from around the world. In the first part, new trends in vibration analysis and control are presented. Topics include smart materials for seismic response control and design for vibration mitigation in wind turbine systems. In part two, new advances in analysis, both in theory and practice, are considered. Topics include vibration control for a cable-stayed bridge. In part three, advances in vibration analysis research are showcased. The chapters cover topics including analysis of long space joist floors and recent trends in the torsional vibration analysis of internal combustion engines. In part four, advances in engineering and structural dynamics are explored. Subjects include modelling and control for rotors on active magnetic bearings and recent advances in force identification. Finally, part five offers the very latest developments and directions for future work. Contributors present research on topics such as beam structural modelling in the hydroelastic modelling of container ships. The book is ideal for readers seeking the state of the art in the field as well as new ideas for their own research.

VIBRATION ANALYSIS AND CONTROL – NEW TRENDS AND DEVELOPMENTS

Edited by **Francisco Beltrán-Carbajal**

Seismic Response Control Using Smart Materials

Sreekala R, Muthumani K, Nagesh R Iyer
CSIR/Structural Engineering Research Centre
India

1. Introduction

Earthquakes are highly destructive natural phenomena resulting in the massive deterioration of civil infrastructure, which becomes highly significant wih increasing urban population. Performance of structural systems needs to be improved recalling the huge loss of life and destruction to constructed facilities caused by various earthquakes. Protection of lifelines and infrastructural facilities are of utmost importance during a seismic event. There have been considerable research efforts in seismic vibration control for the past several decades. Developments of new techniques and new materials, which are not traditionally used in civil engineering structures, offer significant promise in reducing the seismic risk. Smart materials may be described as materials that can sense an external stimulus (e.g:-stress, pressure, temperature change, magnetic field, etc.) and initiate a response. They may belong to one of the four classes namely metals or alloys, polymers, ceramics or composites. Metals and alloys of different metals are considered as classical materials with lot of research activities around the globe . Shape Memory Alloys (SMA) belongs to the class of smart materials which are well known for its peculiar characteristics, which can be stress or temperature, induced. The research area which deals with structural applications of this variety of smart materials are promising (Sreekala& Muthumani, 2009) for structural health monitoring and vibration control.

Nickel Titanium (NiTi) Alloys are well known for its super elastic and shape memory properties and they belong to the class of Shape Memory Alloys (SMA). Presently SMA's are mainly applied in medical sciences, electrical, aerospace and mechanical engineering and also can open new applications in civil engineering specifically in seismic protection of buildings. Super elastic nitinol is found to be very effective for passive vibration control, as it can sustain large amounts of inelastic deformation and recover that deformation at the end of the process with good energy dissipation compared to regular metallic materials.

Various tests are conducted on NiTi wires in CSIR-SERC to evaluate the possibility of using SMA as an energy dissipating material with re-centering capabilities. Ability to sustain and recover large amounts of inelastic deformation during reversed cyclic loading with extra ordinary fatigue resistance make super elastic NiTi suitable candidate for seismic risk reduction. Quasi static and dynamic tests conducted with various parameters like pre strain, amplitudes and frequency for a number of cycles establishes the behavior which is suitable for seismic applications. It is interesting to find that from the quasi-static behavior of the material an optimum value of pre strain, which is material dependent, can be selected and adopted in vibration control applications. Mathematical models were developed to predict the maximum energy dissipation capability of the material under study. The tests verify the potential of pre-strained NiTi wires as kernel components in seismic protection devices. It also highlights the effects of repeated cycling on pre strained wires, as special devices for passive protection during earthquake are meant for repeated cycling. Various types of vibration control devices can be made with the material used.

The technologies using smart materials are useful for both new and existing constructions. The chapter highlights various structural applications of this class of smart materials available in addition to the suitability of this material for vibration control applications. Protection of structures from damage during earthquakes can be addressed using passive protection devices designed to have maximum energy dissipation capacity with re centering capabilities. Innovative means of structural control scheme can be developed through passive systems, consisting of materials and devices useful in the context of seismic design and retrofit of structures

Ductility and energy dissipation are important parameters to be evaluated and quantified for seismic resistant design of structural components. Equivalent viscous damping is a significant quantity usually employed in earthquake engineering to characterize the dissipation capability of the material under investigation. This parameter expresses the effectiveness of the material in vibration damping. The relevance of the study is made clear with the associated terms in earthquake engineering

2. Properties, which enable SMA for civil engineering application

1. Repeated absorption of large amounts of strain energy under loading without permanent deformation
2. Possibility to obtain a wide range of cyclic behaviour from supplemental and fully re centering to highly dissipating by simply varying the characteristics of SMA components.
3. Strain range of 2 to 10%
4. Extraordinary fatigue resistance under large strain cycles.
5. Their greater durability and reliability in the long run.

2.1 Shape memory alloys in structural control

The existence of SMA property in certain alloys has been known from 1932(when it was first observed in Gold-Cadmium alloy exhibits a rubber like behavior). SMA's have unique properties, which are not present in many materials traditionally used in engineering applications; especially shape memory effect and super elasticity. While shape memory alloys have been commercially available since the 1960s, their application has been limited (Pelton et al., 2000). Early literatures on SMA materials shows successful uses of these for the design of biomedical and clinical devices, particularly for fixing body fractures internally with various 'NiTi' SMA interlocking intramedullary nails(Da et al,2001). Among other designs, research related to the use of SMA materials as micro-pumps, micro-grippers, flexible SMA active catheters, micro nano-tribological devices and SMA thin films for electromechanical system have been increasingly revealed in recent years (Stockel & Melzer, 1995). Those achievements exhibited an excellent reliability and confidence in using these materials for other primarily structural elements. The SMA materials possess many unique mechanical and geometrical properties such as shape memory effects which provide a large recovery strain and force simultaneously, pseudo-elastic behavior upon heating and super-elastic properties in ambient conditions for specified SMA materials. Birman (1997) and Humbeeck (2001) gave comprehensive reviews on the properties of SMA materials. Based on the studies under MANSIDE project, 1999(Italian Department for National technical services) Ni–Ti SMA's appear to be the best candidates for use in seismic applications. But selection of the SMA wires for energy dissipation with or without pre strain had to be addressed separately (Sreekala et al., 2010). So the need arose for finding the effect of pre strain as well as repeated cycling in the energy dissipation of SMA wires.

2.2 Mechanism behind super elasticity

Super elasticity is related to the isothermal response of SMA specimens to applied mechanical loads. The phase transformations are solid to solid diffusionless processes between a crystallographically more ordered phase, the austenite (A) and a crystallographically less ordered phase the martensite (M). Fig. 1 shows a schematic stress–strain cyclic curve of a super elastic SMA. It is characterized by five branches. Branches 1 and 4 correspond to the elastic deformation of the two stable phases of SMA, respectively, austenite and martensite. Branches 2 and 3 correspond, respectively, to the forward (from austenite to detwinned martensite) and inverse (from detwinned martensite to austenite) phase transformation. Branch 5 corresponds to the onset of plastic deformation of detwinned martensite. In the figure σ_{Fs} and σ_{Ff} represent the critical stresses at which the forward transformation, respectively, starts and finishes, while σ_{Is} and σ_{If} are those at which the inverse transformation, respectively, starts and finishes. Since loading and unloading paths are different from each other, a certain amount of energy is dissipated over the cycle.

3. Research significance

Most of the applications of SMA have focused on developing various types of actuators. However, most of the commercially successful applications - such as couplings, electrical connectors, cardiovascular stents and eyeglass frames - use the full shape- memory effect or the full super elastic cycle only a few times. This indicates that SMA components can be used successfully for several non-actuator applications as well. One such area is the use of SMA components for damage control in structures. The SMA component can undergo finite deformations either due to super-elasticity or due to rearrangement of martensite variants at relatively small stresses. This can be used to prevent permanent plastic deformation due to unexpected loads in the rest of the structure. The ability of the SMA component to recover its original dimensions may be an additional advantage, during the seismic events.

Shape Memory Alloys (SMA) show the potential to eliminate some limitations involved in present technologies, allowing a broader application. The limitations of existing technologies for passive protection devices can be summarized as follows

1. Problems related to ageing and durability (e.g., rubber components)
2. Difficulty in maintenance(those based on fluid viscosity)
3. Installation complexity or replacement and geometry restoration after strong events (those based on steel yielding or lead extrusion)
4. Variable performances depending on temperature(polymer based devices)

The usable strain range in the order of 10% in shape memory alloys provide them very high energy dissipation per unit mass of material. Comparing the well known rubber isolators and steel hysteretic dampers which belong to the class of quasi-elastic devices and elasto-plastic devices respectively, re-centering devices gain the best mechanical characteristics of both. The shortcomings of the traditional restrainers can potentially be addressed with the use of SMAs (Fig.2). The comparative behaviour of steel, rubber and SMA is given in Fig.2.

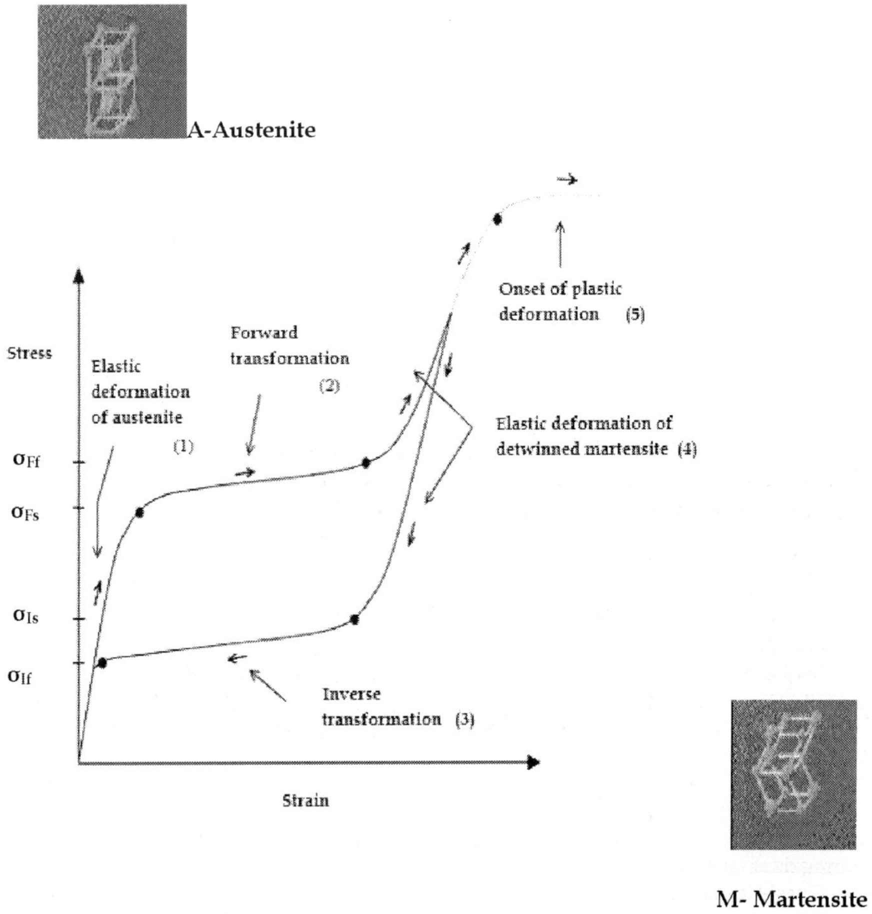

Fig. 1. Stress-strain curve showing super elasticity of SMA-Schematic sketch describing the phase change with the deformation process and the associated crystal structure

On the one hand, they recover the initial position of the structure, with a good control of the displacements; on the other hand they put a threshold to the force transmitted to the super-structure. The full possibility of designing the mechanical behavior combining the self-centering and the energy dissipation capability, permit to calibrate the desired features and fit the specific needs. The availability of such features opens considerable room for improvement of the structural system design. But the mechanical behavior of SMAs is strongly dependent upon the alloy composition and the thermo-mechanical treatment. Special care should be taken while alloy selection is made.

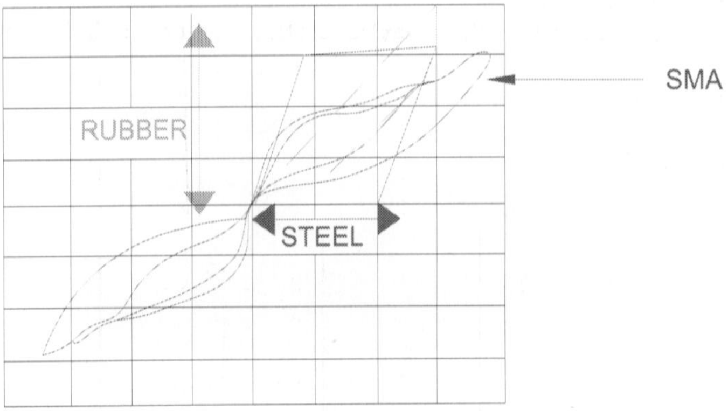

Fig. 2. Comparative behaviour of Steel, Rubber and SMA

4. Experimental investigation and inferences

The experimental tests were carried out on austenite wire samples of 0.4mm diameter and 1.2mm. The wires were available in spool form. The typical photograph of the material – wires in spool form-is shown in fig.3. The compositions of wires were NiTi alloy with 55 % Ni and balance titanium spooled NiTi. The wire selection is made in such a way that Ni-Ti alloy wire with equi-atomic composition possesses better dissipation property and higher resistance to corrosion and fatigue. The material is straight annealed super elastic NiTi wires and has its latent heat and specific gravity 14500 J/kg and 6479.85 kg/m^3 (0.234 lbs/in^3) respectively. The test was carried out under ambient temperature conditions around 27^0 C. Quasi-Static and dynamic tests were conducted to evaluate the super-elastic properties and energy dissipation capabilities under ambient temperature. Dynamic tests were carried out by applying sinusoidal cyclic deformation to wire samples. A displacement controlled test set up was used to perform the tests. The typical test sequence is characterized in terms of number of cycles, strain amplitude and frequency as they could provide very important information regarding earthquake applications.

The test programme on SMA wires of two diameters namely 0.4 and 1.2 mm can be summarized as follows:

1. Quasi-static tests
2. Dynamic tests -frequency of operation 0.5 Hz (sinusoidal cyclic deformation)
    a. Tests to evaluate the number of cycles to failure for a constant amplitude
    b. Variable amplitude tests to evaluate the energy dissipation capability for increased amplitudes.
3. Tests on pre-strained wires –frequency of operation 0.5 Hz .
4. Tests on wires for varying frequencies.

The quasi-static (slow rate) tests were performed till the failure of the specimen. The test was run in a position controlled test set up at constant rate of loading 0.025mm/sec. Fig.4 shows the stress strain curve for quasi-static loading. It is compared with the behavior of the 1mm diameter steel wire which is used for binding purposes in construction A sinusoidal cyclic test on SMA wires was performed for maximum amplitude of 5 mm.

Fig. 3. The photograph of the super elastic SMA wire samples (Available in spool form)

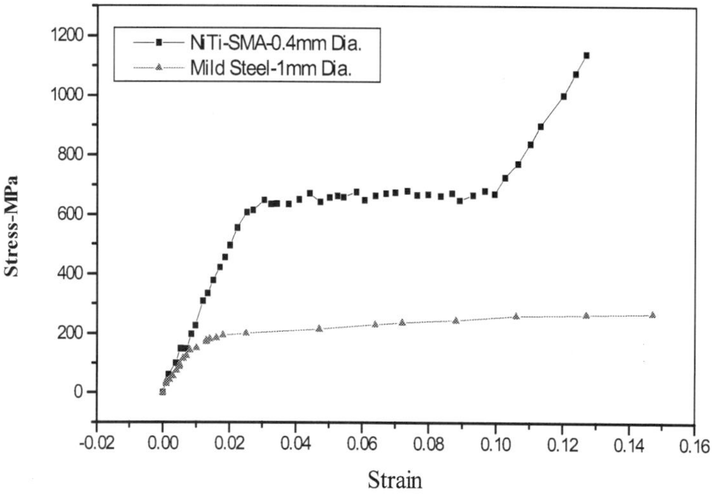

Fig. 4. Behaviour of Ni-Ti wires – quasi static tests- Comparison with 1 mm mild- steel binding wire used for reinforcement stirrup tying in structural construction.

The mechanical behavior such as the transformation stress levels, fatigue behaviour and the frequency dependence were found to be similar in both diameter wires. The tests were also carried out with increasing number of cycles to obtain the fatigue characteristics.

Typical load deformation characteristics of 1.2mm diameter wire with 11 mm amplitude are shown in Fig.6. As the displacement amplitude increased as observed in the force deformation behaviour, presence of the martensitic phase -after completion of the phase transformation- provides additional stiffness. Strain hardening is observed towards higher amplitudes which correspond to 8-9% strain in both the wires seen in Fig.4 & Fig.6. The increase in stress levels is quite evident in Fig.6. Here cycling up to 9% strain variation fetches a stress of 830 Mpa.

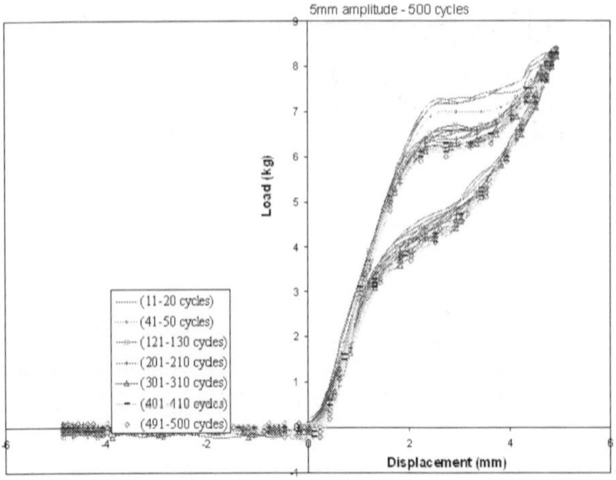

Fig. 5. Load deformation curves for 0.4mm diameter wire with 5mm amplitude (strain variation 3%, maximum stress 650 Mpa)- frequency of loading 0.5 Hz-0.4mm diameter wires

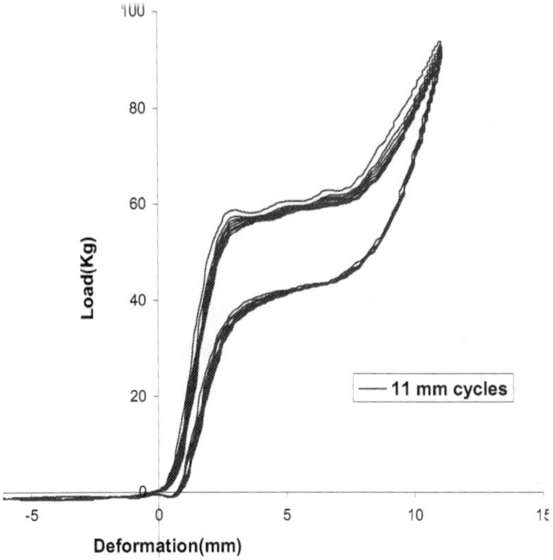

Fig. 6. Load-deformation curves for 1.2 mm diameter wire with 11mm amplitude cycles (strain range 9%, max. stress 830MPa)- frequency of loading 0.5 Hz Cycles (strain range 9%, max. stress 830MPa)

The various parameters and the testing scheme are selected in a way to understand the behavior of the material during earthquakes. Initially a sinusoidal cyclic test on SMA wires was performed for constant amplitude of 5 mm [Fig.7(a)- 7(d)]. Usually earthquake vibrations are dynamic- reversed cyclic in nature which consists many number of cycles. The tests were also carried out with increasing number of cycles to obtain the fatigue characteristics. Further tests were carried out with increasing amplitude from 5 mm to 11 mm [i.e increasing the total strains from 4% to 9%, Fig.7 (e)]. Strain measurements were based on the actuator displacement as the wires were having very smooth oxide finish surface

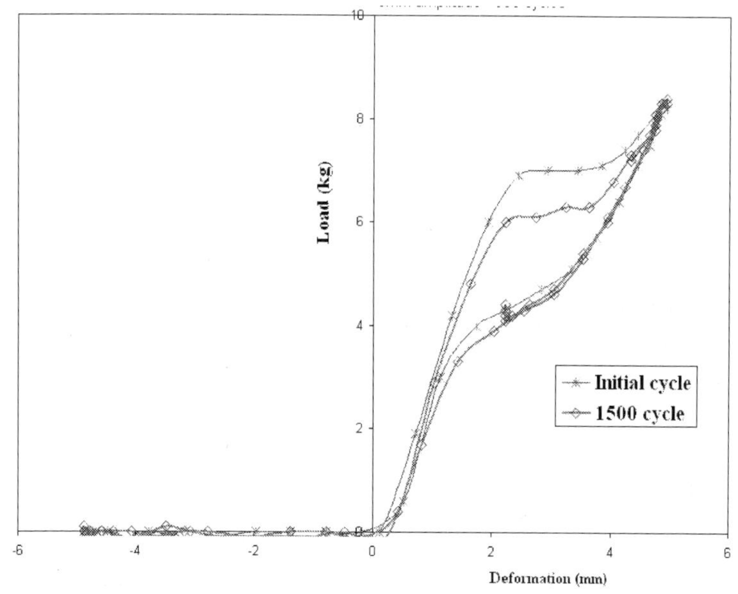

(a)

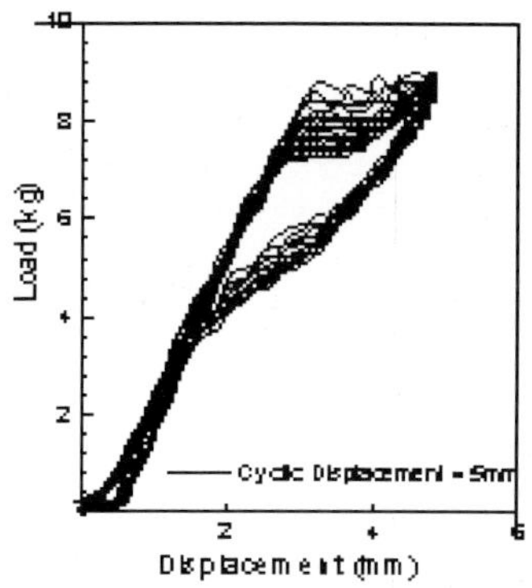

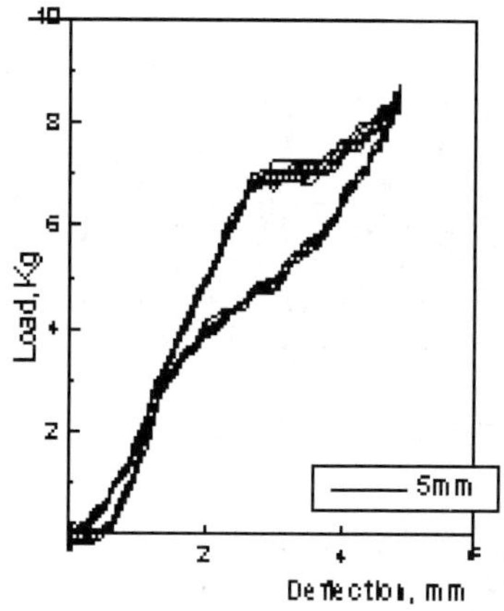

(b) Initial 28 cycles

(c) Subsequent cycles

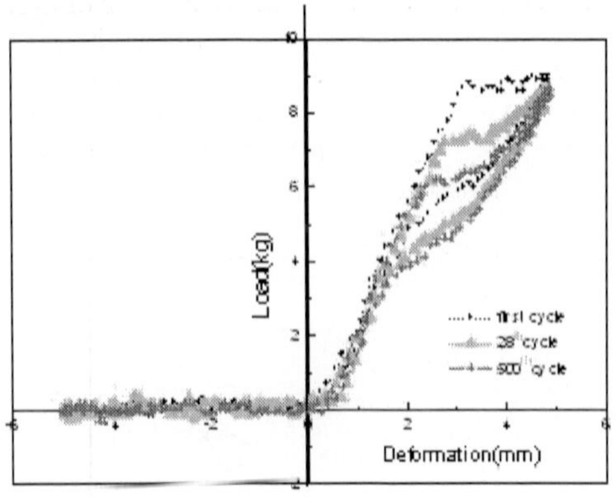

(d)

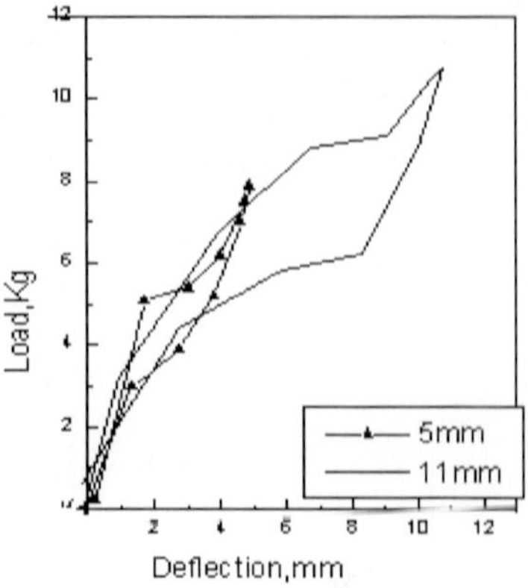

(e)

Fig. 7. (a) to (d) Cyclic tests on non pre-strained wires cycling for +/- 3 percentage strain variation (5mm amplitude). (e) Effect of increasing the amplitude beyond 1500 cycles.

Due to the repeated cyclic deformation of the same amplitude it is observed that the hysteresis loops translate downwards and narrows (Fig.7a-7d). Consequently the stress levels for the transformation decreases. Energy dissipation capacity under constant amplitude loading remained same up to 500 cycles, 20% reduction was observed in the next 500 cycles and around 30% reduction towards the failure from the initial value. Effect of increasing the amplitude beyond 1500 cycles can be observed in figure.7e.

4.1. Effect of pre-strain

The effect of pre-strain was found out by imposing pre-strain amplitude in the wire. The wire is held at that point of pre-strain and reversed-cycling performed for various amplitudes (Fig.8b). Like wise for various pre strain values the dynamic tests were performed by giving sinusoidal deformation to the wire samples (Fig.8a). These tests help to simulate the characteristics of earthquake loading and understand the behaviour of the material. The frequency dependence on loading is also investigated during the experiment (Fig.9). The frequency of loading was varied from 0.5 Hz to 3 Hz, at intervals of 0.5 Hz, since most of the earthquake applications come within this range. Table.1 gives the test programme for the nitinol wire and the parameters evaluated in the study are tabulated. Temperature variation during cycling was observed using a thermocouple for quite a few cycles.

5. Discussion of test results

Quasi-static tests of the wire clearly shows the super elastic property of the SMA wire. Fig.4. presents the stress strain curves obtained under quasi-static loading using a hydraulically driven test system (instron make). The shape of the stress strain curve exhibits some of the fundamental characteristics of SMA. Test result shows the super elastic property of the SMA wires in the strain range 2% to 10 %. The initial elastic to plateau transition, that is the transition from linearly elastic to super-elastic occurs at a stress of approximately 597 MPa. The yield stress of 597 Mpa and ultimate stress of 1100 Mpa is comparable with the idealized behavior of SMA (Dolce, 1994).

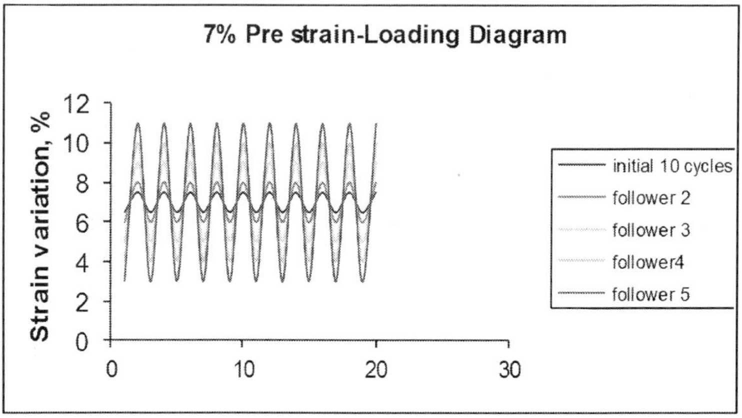

(a)

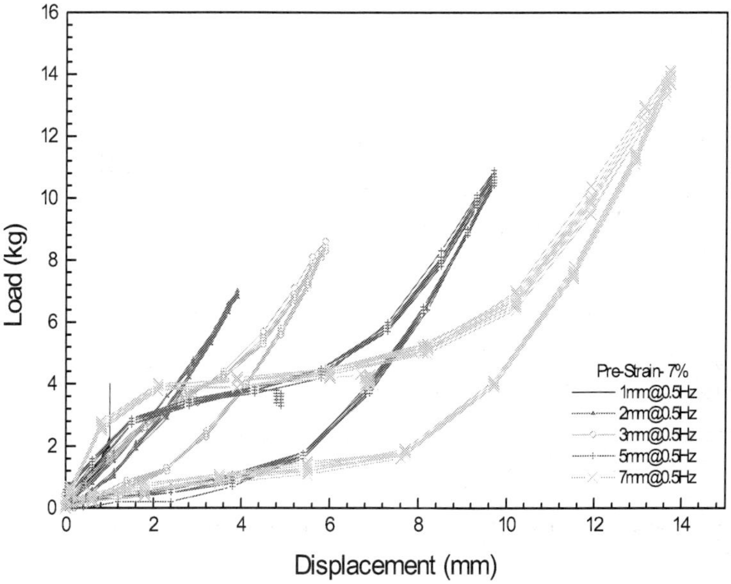

(b)

Fig. 8. Typical sketch showing the loading and the hysteretic curves obtained at 7% pre-strain -variable amplitudes.

If the material is unloaded after being loaded into the super elastic region, the unloading portion of the stress strain curve does not follow the loading portion, but follows a lower path back to the origin which indicates the transformation of stress induced martensite back to the austenite (Fig.5,6). This makes the hysteretic energy dissipation which is an important parameter to be assessed for earthquake applications and hence in passive control devices. The super elasticity based applications take advantage of the following features 1) the possibility of recovering large deformations 2) the existence of a transformation stress plateau, which guarantees constant stress over non-negligible strain intervals. The frequency dependence on loading is also investigated during the experiment (Fig.9). The frequency of loading was varied from 0.5 Hz to 3 Hz, at intervals of 0.5 Hz, since most of the earthquake applications come within this range. It is interesting to note that the number of cycles for the same amplitude is not sensitive to the frequencies in the range 0.5-3 Hz as observed in both the diameter wires.

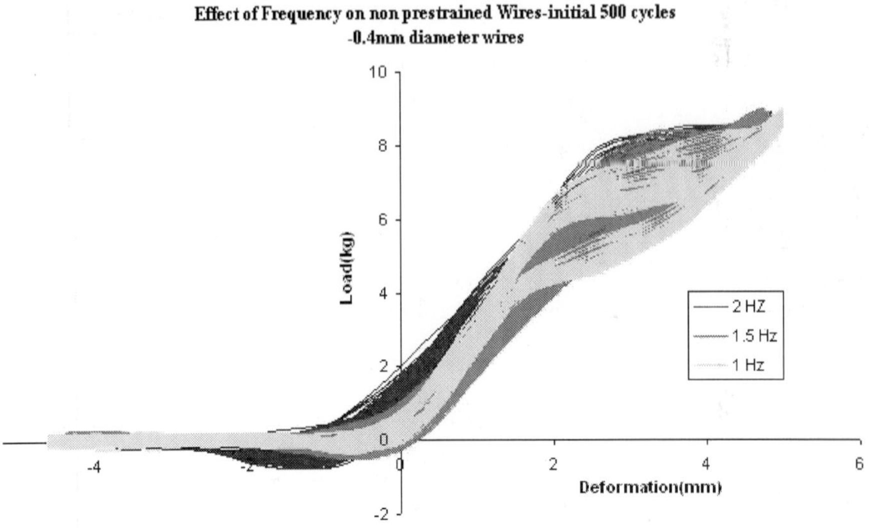

Fig. 9. Typical Cyclic behavior (tension compression tests) of non pre-strained wires for varying frequencies.

It is reported in the literature that the effect of frequency is negligible in the energy dissipation in the range of interest 0.5 - 3 Hz for seismic applications (Cardone et al,1999) . But it is observed from the test that the dissipated energy during the sinusoidal loading at 0.5 Hz is 25 % more than the energy dissipated at 1Hz to 3Hz, which is almost a constant value. Effect of increasing the amplitude beyond 1500 cycles can be observed in figure.7e. The high fatigue resistance combined with negligible residual deformation, observed in SMA material help to reduce the post earthquake repair costs, if dampers are made out using these.

It is found from the experiment that the hysteresis loops narrow and translate upwards when there is an increase in strain amplitude, while the branches of the curve relevant to the phase transformations harden, thus yielding an increase in the stress levels (Sreekala et al,2008). This trend is observed in the case of cycling around 7% pre-strain, as the strain reaches to 10% the end of the transformation range (Fig.8). This phenomenon is related to the elastic deformation of the de-twinned martensite found at the end of the phase transformation. This is true for pre strained wires as well as without pre strained. It is a favorable aspect in seismic applications as the system stiffens rather than softening, if the expected design seismic action is exceeded, thus ensuring a good control of displacements.

Constant amplitude tests for many cycles for varying frequencies indicated that there is slight variation in the shape of the loops. Different heat exchange modalities with ambient temperature, at different strain rates are the causes of these effects from quasi-static to dynamic range (0.5-3 Hz). As strain rate increases, the heat-exchange condition increasingly digresses from the isothermal one. The latent heat of transformation causes the specimen to self-heat and then increase its average temperature during test. At the same time, during each loading cycle, the specimen temperature oscillates around an average value, according to the sinusoidal variation of strain. The temperature variation is found to be 9^0c.While the instantaneous temperature of the wire oscillates around the ambient temperature (27^0c) with excursions of the order of 9^0c, the average temperature remains practically unchanged during all the loading history equal to the ambient temperature.

As a result of the self heating (cooling) there occurs instantaneous increase (decrease) of the stress required for transformation. It is observed from the test that the dissipated energy during the sinusoidal loading at 0.5 Hz is 25 % more than the energy dissipated at 1Hz to 3Hz, which is almost a constant value. The hardening of the transformation branches, the narrowing of the cycle, apparent change in the shape of the loop resulting in a reduced energy loss all of which are associated with slight temperature variation due to different heat exchange modalities are caused by the stabilization of the material due to repeated cyclic loading(Fig.10). Possible increase of average temperature during cycling at higher strain rates produces an upward translation [Miyazaki et al,1986b].

Stabilization of the material behavior due to repeated cyclic loading also can be observed from the shape change of the loops in the tests on non-pre strained wires (Fig.10). The effect of pre-strain is found at 2.5% pre strain applied in the form of deformation to the sample, on two cyclic strains namely 4 % and 6 % (Fig.11.).

The figure reveals the fact that the hysteresis loops translate upwards as it near the transformation stress level. The comparison between the energy dissipation capability at various pre strain under different amplitudes, namely 3mm and 7mm is shown in Fig.12.

It is interesting to find that the quasi-static curve serves as the envelope curve in this process and as and when the stress reaches the transformation stress level, as marked in fig.12, further stiffening happens in the material. This is an added advantage for seismic applications that the material gets stiffen, rather softening to have good control of displacements. Fig. 12 also shows the hysteresis curve for the wire pre strained at 6% and cycled between 2 % and 10% along with other strain variations.

Tests on wires pre strained to different levels, indicated a particular value of pre-strain, from which cycling in the pseudo-elastic range fetches maximum energy dissipation (Fig.8.).

The equivalent viscous damping expresses the effectiveness of the material in vibration damping. It is calculated based on the average energy dissipated per cycle and the damping values observed are also found to be significant. Looking into 6% pre strain one can see the higher equivalent damping values. Cyclic variation of 4% strain on either direction at 6% pre strain gives higher equivalent damping due to higher energy dissipation (Table.1). An

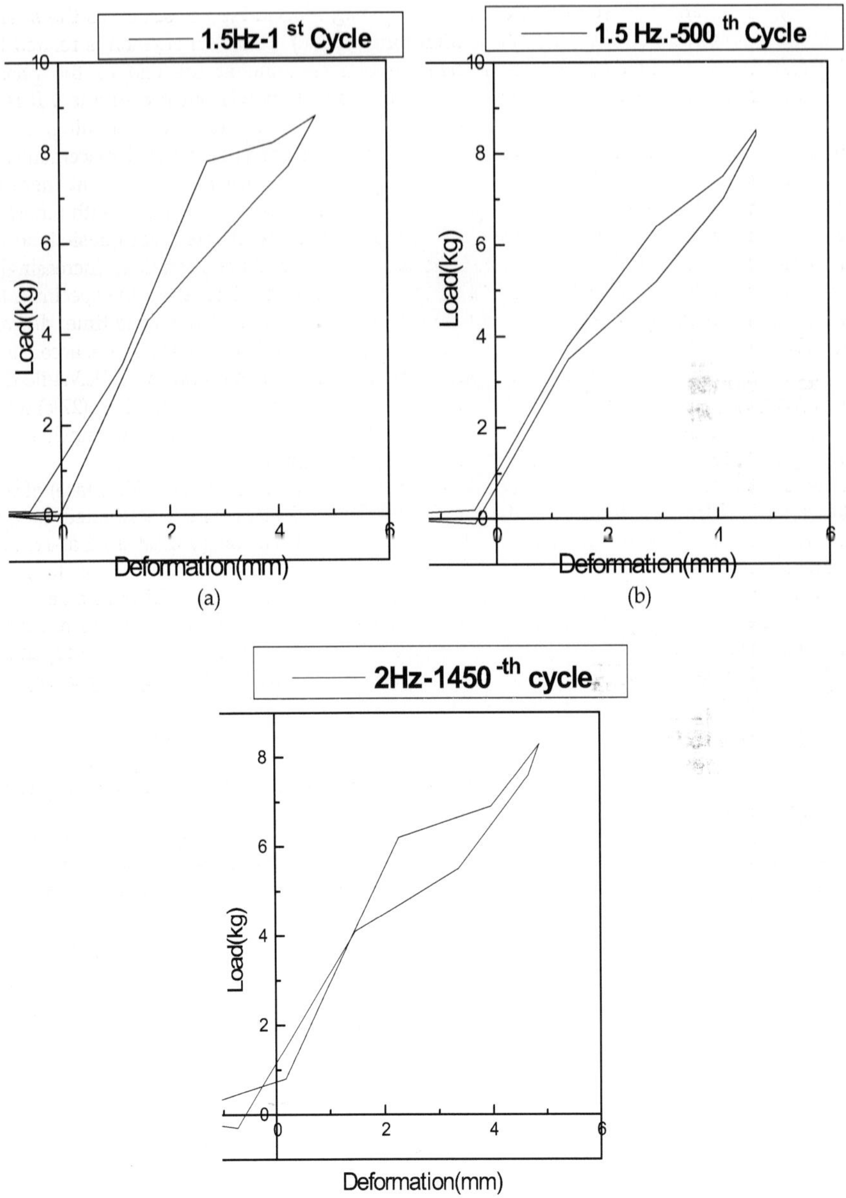

(c)

Fig. 10. (a) to (c) Changes in the shape of the curve during varying frequencies.

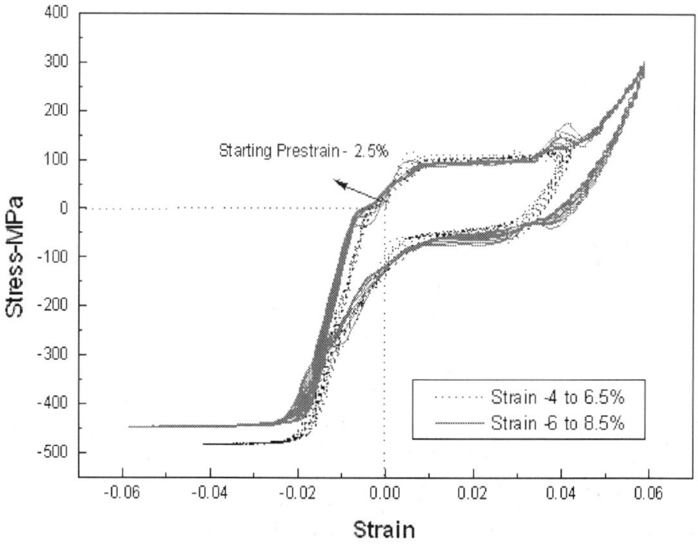

Fig. 11. Effect of pre-strain on the energy dissipation (1.2 mm diameter wires)- frequency of loading 0.5 Hz

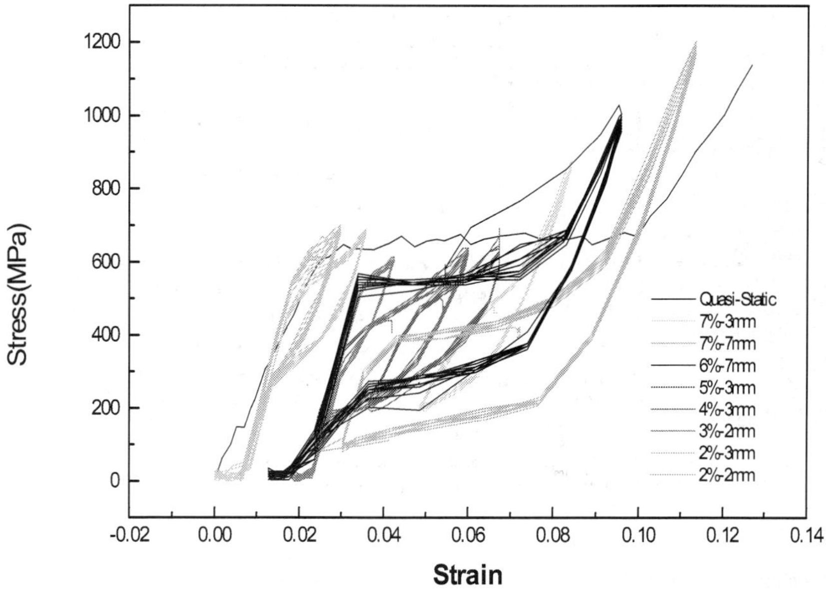

Fig. 12. Cyclic behavior of pre strained wires for varying pre strains. [The legend denotes pre strain %-reverse displacement cycles from that prestrain (mm)]

average equivalent damping ratio of 16% was realized during the test [Sreekala et al, 2008]. It is found to be nearer to half the super-elastic range obtained from quasi-static tests. Results indicate that if the super elastic wires have to play an energy-dissipating role, cycling around the threshold pre-strain value within this range is desirable [Sreekala et al,2010]. In this case the energy dissipation is dependent on the strain value in the super-elastic range from where cycling the entire range fetches maximum, as seen in figure 4.
For wires without pre strain, the equivalent damping ratio observed is in the order of 3%. This is because during cycling, the reverse direction (compressive region) does not contribute for the energy dissipation in non pre-strained wires. Tests on virgin wires shows lesser damping and energy dissipation capacity and these results can be compared with those under MANSIDE project. The design of the SMA based devices can be made by choosing the mechanical properties such as transformation stress levels/ plateau strength as design variables. Fig.14 shows the hysteretic behaviour of the different types of devices that can be developed using the properties of the SMA material. A variety of hysteretic behaviour can be obtained from the material tested and its application can be made suitable for seismic devices like re-centering, supplemental re-centering or in the case of non re-centering devices. These results reveal the suitability of using pre strained SMA wires in vibration control applications, especially for Non Re-Centering Devices-NRCD as seen in Fig.14. Normally pre straining of wires in vibration control devices requires highly skilled labour and maintenance becomes a difficult procedure. As far as the numbers of cycles are concerned, pre strained wires also goes beyond 900 cycles under constant amplitude loading. e.g.:-in 6% pre strained wire, under the maximum pseudo-elastic range goes up to 940 cycles before failure. No reduction in the energy dissipation is observed up to 30 cycles, but towards the failure 30 % reduction is observed as in non-pre strained wire. Usually for seismic applications, the numbers of cycles to be considered are less than 500 cycles as per various codal spectrums for seismic design in different countries. Table 1 summarizes the number of cycles, energy dissipation capacity per cycle, the strain variation and the equivalent damping. These results reveal the suitability of using SMA wires in seismic vibration control applications.
The usable strain range of the order of 10% in the alloys provide them very high energy dissipation per unit mass of material. Re-centering devices gain the best mechanical characteristics of both quasi-elastic devices (e.g., rubber isolators) and elasto-plastic devices (e.g., steel hysteretic dampers). On the one hand, they recover the initial position of the structure, with a good control of the displacements; on the other hand they put a threshold to the force transmitted to the super-structure. The full possibility of designing the mechanical behavior combining the self-centering and the energy dissipation capability, permit to calibrate the desired features and fit the specific needs. The availability of such features opens considerable room for improvement of the structural system design. Fig 7 indicates the various device options for passive vibration control using this material. But the mechanical behavior of SMAs is strongly dependent upon the alloy composition and the thermo-mechanical treatment. Special care should be taken while alloy selection is made.

6. Equivalent viscous damping

Some significant quantities usually employed in earthquake engineering to characterize the dissipation capability of the material under investigation are equivalent viscous damping, secant stiffness and energy loss per cycle. The equivalent viscous damping expresses the effectiveness of the material in vibration damping. It is calculated based on the average energy dissipated per cycle. It is calculated as

$$\xi_{eq.} = W_D / (2\Pi\, K_S \delta^2) \quad (1)$$

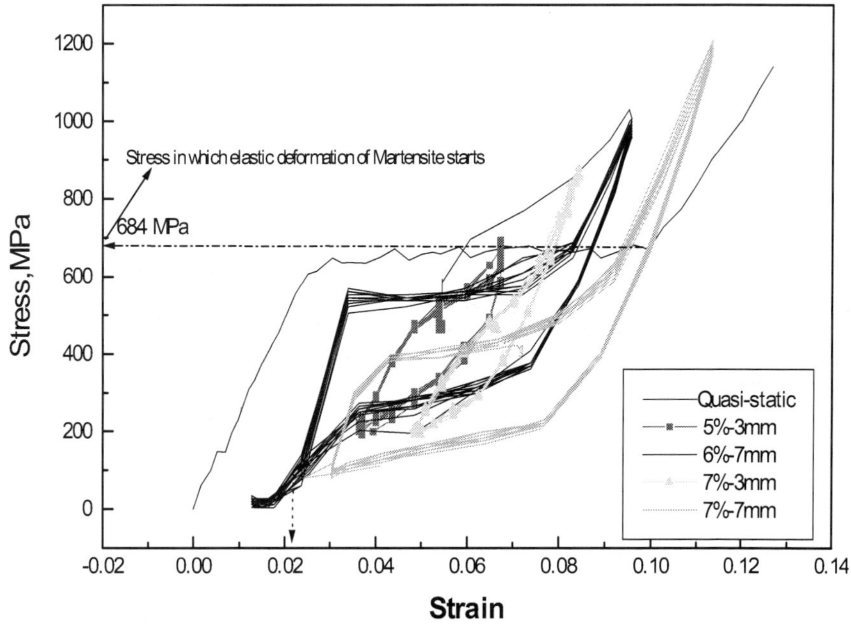

Fig. 13. Cyclic behavior of pre strained wires -maximum energy dissipation when pre-strained almost at the middle of the pseudo-elastic range. [The legend denotes pre strain %-displacement cycles (mm)]

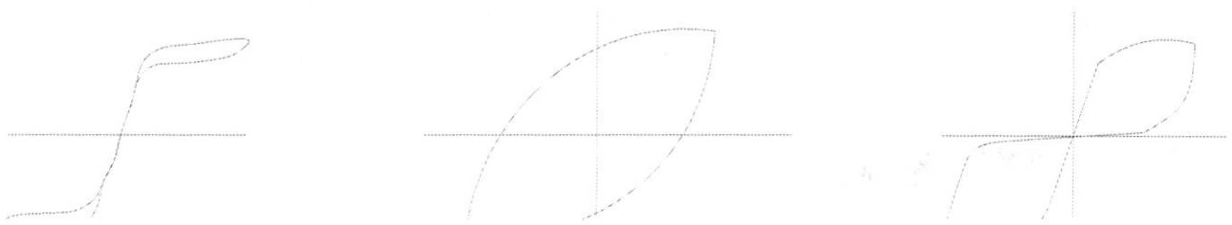

(1) (2) (3)

Fig. 14. Load -deformation behavior (Hysterisis curves) for various types of Devices based on SMA, SRCD-Supplementary Re-Centering Device, NRCD-Non Re-centering Devices, and RCD-Re-centering Devices

where W_D is the energy loss per cycle and δ is the maximum cyclic displacement under consideration. The secant stiffness (K_s) is computed as
$K_s = (F_{max} - F_{min}) / (\delta_{max} - \delta_{min})$ where F_{max} and F_{min} are the forces attained for the maximum cyclic displacements δ_{max} and δ_{min}. The associated energy dissipation and the equivalent viscous damping calculated are included in Table.1.
The effect of pre-strain is found at 2.5% pre strain applied in the form of deformation to the sample, on two cyclic strains namely 4 % and 6 % (Fig.11.). An increase in the energy dissipation can be observed as the cyclic strain amplitude increases. The energy dissipated during cycling in pre-strained wires is considerably high compared to non-pre strained wires as observed from the experiment. It is interested to observe that (Fig.13) leaving the flat plateau; the response follows the elastic curve below 2% strain as obtained from the quasi-static test. Hence the quasi- static behaviour gives the envelope curve for the cyclic actions and a variety of hysteretic behaviour suitable for seismic devices can be obtained with pre straining effect [Sreekala et al, 2010].

7. Modeling the maximum energy dissipation for the material under study

It is observed from the experiment that the wires pre strained to the middle of the strain range gives the maximum energy dissipation(fig. 8). The behaviour can be predicted using the following mathematical expressions.

$$\sigma = E\left[\varepsilon - |\varepsilon|\{1/\left(\frac{\sigma-\beta}{Y}\right)^n\} + \sigma_* \right] \tag{2}$$

$$\varepsilon = \left[\frac{\sigma}{E} + |\varepsilon|\{1/\left(\frac{\sigma-\beta}{Y}\right)^n\} - \frac{\sigma_*}{E} \right] \tag{3}$$

σ_* denotes the stress in the pre strained wire at the beginning of the cycling. Here β can be expressed as

$$\beta = E\alpha\{\ \varepsilon^{in} + f_T|\varepsilon^c|\, erf(\alpha\varepsilon)\left[u\left(-\varepsilon\,\dot{\varepsilon}\right)\right]\ \} \tag{4}$$

β denotes the one dimensional back stress, Y is the "yield" stress ,that is the beginning of the stress- induced transition from austenite to martensite, n is the overstress power. In equation (4) the unit step function activates the added term only during unloading processes. In the descending curve it contributes to the back stress in a way that allows for SMA stress-strain description. The term with (.) shows the ordinary time derivative.

$\alpha = \frac{E_y}{(E - E_y)}$ is a constant controlling the slope of σ - ε, where E is the elastic modulus of austenite and E_y is the slope after yielding.
Inelastic strain, ε^{in} ,is given by

$$\varepsilon^{in} = \varepsilon - \frac{\sigma}{E} \tag{5}$$

The error function, erf(x) ,and the unit step function, u(x) are defined as follows.

$$erf(x) = \frac{2}{\sqrt{\Pi}} \int e^{-t^2} dt \tag{6}$$

$$u(x) = 1 \quad , x \geq 0 \tag{7}$$

$$u(x) = 0 \quad , x < 0 \tag{8}$$

The basic expression obtained here for pre-strained wires is a modified form of the model suggested by wilde et al, [12] which is used for predicting the tensile behaviour of SMA materials. This mathematical expression describes the mechanical response of materials showing hysteresis.

$$\dot{\sigma} = E\left[\dot{\varepsilon} - \left|\dot{\varepsilon}\right|\left(\frac{\sigma - \beta}{Y}\right)^n\right] \tag{9}$$

where σ is the one dimensional stress and ε is the one-dimensional strain, β is the one dimensional back stress , E elastic modulus , Y the yield stress and n the constant controlling the sharpness of the transition from elastic to plastic states.

The modified Cozzarelli model suggested by Wilde et al represents the hardening of the material after the transition from austenite to martensite is completed. Here in this case for pre strained wires for maximum energy dissipation, the hardening branch has not been considered which requires additional terms containing further unit step functions. The model is rate and temperature independent. The requirement of zero residual strain at the end of the loading process motivated the selection of the particular form of back stress through the unit step function. The coefficients f_t, and c are material constants controlling the recovery of the elastic strain during unloading. Since the stress and strain were found to be independent of the rate, the time differential can be eliminated. Hence equation (2) and (3) can be used for predicting the behaviour after many cycles by appropriately changing the values of 'n' and the starting stress value which follows the trend as in fig. 15. The material constants for the wires tested were obtained as

f_t = 0.115, C =0.001, n = 0.25, α = 0.055. Using Equations (2) and (3) the curves are fitted. Fig. 15 gives the fitted curve for the maximum energy dissipation for the material tested.

8. Various seismic response mitigation strategies

Earthquake engineering has witnessed significant development during the course of the last two decades. Seismic isolation and energy dissipation are proved to be the most efficient tools in the hands of design engineer in seismic areas to limit both relative displacements as well as transmitted forces between adjacent structural elements to desired values. Parallel development of new design strategies (the seismic software) and the perfection of suitable mechanical devices to implement the strategies (the seismic hardware) made it possible to achieve efficient seismic response control. An optimal combination of isolator and the energy dissipater ensures complete protection of the structure during earthquake. Energy dissipation and re-centering capability are the two important functions to cater this need.

Most of the devices now in practice have poor re-centering capabilities. Instead of using a single device a combination of devices can provide significant advantages
The chart shown below Fig.18 illustrates the various seismic response mitigation strategies.

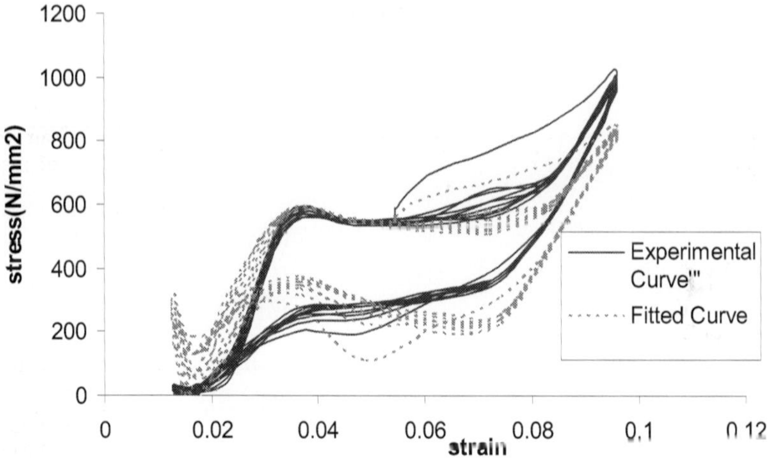

Fig. 15. Cyclic behaviour of prestrained wires -maximum energy dissipation obtained and the fitted curve.

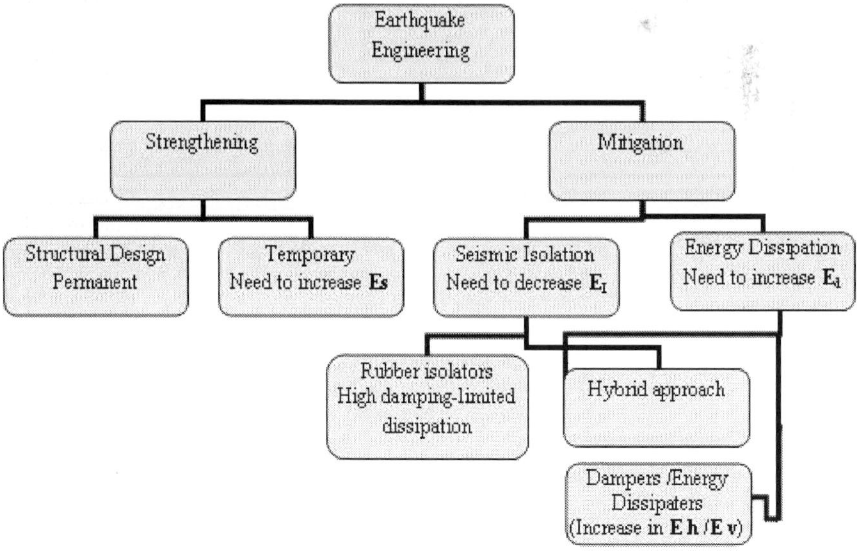

Fig. 16. Various seismic response mitigation strategies

Sl.no:	Nature of Test	Pre- strain in the wire	Cycling Strain range (%)	Average energy dissipated per cycle X10^{-2} J	No: of cycles	Equivalent viscous damping
1	Dynamic Tests Freq. of operation 0.5 Hz. (Sinusoidal cyclic)	7%	6.5-7.5 6.0-8.0 5.0-9.0 4.0-10.0 3.0-11.0 1.3-12.0	00.30 00.60 06.70 18.60 33.00 -	10 10 [a] 10 [a] 10 [a] 10 [a] -	0.02 0.02 0.06 0.07 0.09 -
2		6%	2-10	33.80 27.00 21.00	10 50 [a] 500 [a]	0.16 0.11 0.09
3		5%	4.5-5.5 4.0-6.0 3.0-7.0	00.50 04.60 07.70	10 10 [a] 10 [a]	0.04 0.13 0.12
4		4%	3.5-4.5 3.0-5.0 2.0-6.0	00.33 05.40 10.35	10 10 [a] 10 [a]	0.02 0.16 0.16
5		3%	2.5-3.5 2.0-4.0	01.01 01.91	10 10 [a]	0.05 0.04
6		2%	1.5-2.5 1.0-3.0 0.0-4.0	01.50 04.40 05.10	10 10 [a] 10 [a]	0.10 0.10 0.07
7		Nil	-3 - +3 -3 - +3 -4 - +4	07.30 06.40 05.10	25. 328 [a] 560 [a]	0.03 0.03 0.04
8		Nil	-3 - +3 -7 - +7	05.60 10.90	1500 140 [a]	0.03 0.045

Table 1. Evaluated parameters of the Nitinol wire.
[a] - denotes the number of cycles which are followed by the cumulative previous cycles.

From this diagram it is clear that for seismic mitigation, a combination of Seismic isolation and Energy Dissipation is beneficial. Seismic Isolation can be implemented as explained below:

- Through the reduction of the seismic response subsequent to the shift of the fundamental period of the structure in an area of the spectrum poor in energy content
- Through the limitation of the forces transmitted to the base of the structure. A high level of energy dissipation also characterizes this approach. So it represents a combination of the two strategies of seismic mitigation. Isolation systems must be capable of ensuring the following functions:
  1. transmit vertical loads,
  2. provide lateral flexibility,
  3. provide restoring force,
  4. provide significant energy dissipation.

In each device, the constituent elements assume one or more of the four fundamental functions listed above. Some of the cases hybrid systems prove to be very much beneficial. For example, the strategy need to be adopted in suspension bridges is that of isolation and energy dissipation as the vertical cables did not provide energy dissipation characteristics. Hence dampers need to be provided and the hybrid system provide adequate protection during seismic response. The Table 2 below provide various energy dissipators/dampers along with their principle of operation

Classification	Principles of operation
Hysteritic devices	Yielding of metals Friction
Visco elastic devices	Deformation of visco elastic solids Deformation of visco elastic fluids Fluid orificing
Re centering Devices	Fluid pressurization and orificing Friction spring action **Phase transformation of metals (Shape memory alloys belong to this category)**
Dynamic vibration absorbers	Tuned mass dampers Tuned liquid dampers

Table 2.Various energy dissipation devices/damper

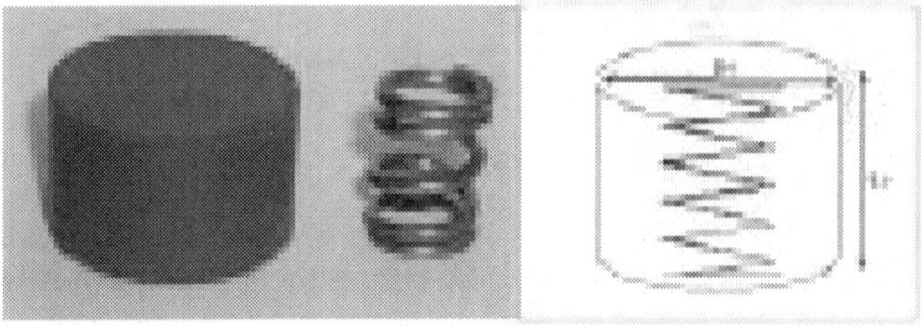

Fig. 17. Example of Composite Rubber/SMA spring damper

The unique constitutive behaviors of Shape Memory Alloys have attracted the attention of researchers in the civil engineering community. The collective results of these studies suggest that they can be used effectively for vibration control of structures through vibration isolation and energy absorption mechanisms. Possible applications of SMA based devices on Various structures for vibration control is shown in Fig.18,

For the analysis of structures, an equivalent Single Degree Of Freedom (SDOF) system can be utilized. The following mechanical model can represent the behavior of the energy dissipating system (Fig.19a). If re centering device is utilized the hysterisis should be adequately represented using mathematical models.

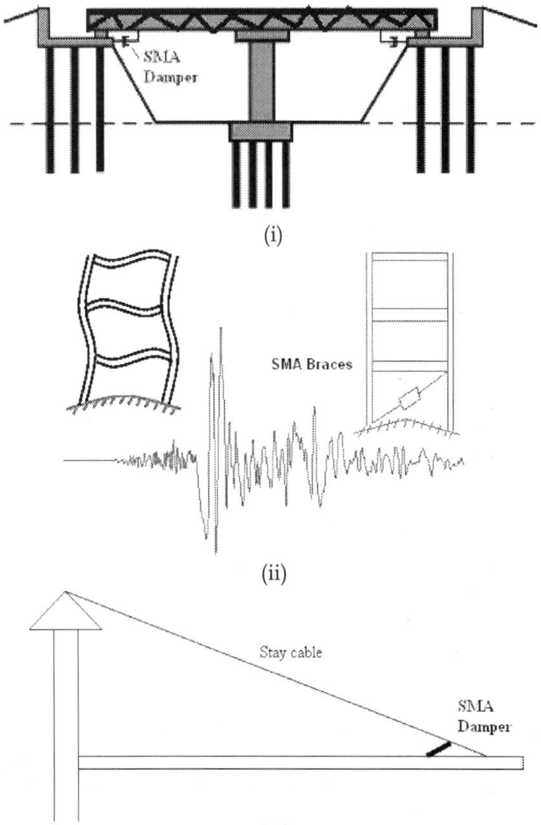

(iii)

Fig. 18. Possible Applications of the Devices (i) Restrainers in Bridges (ii) Diagonal braces in buildings (iii) Cable stays (iv) combination of isolators and dampers in bridges

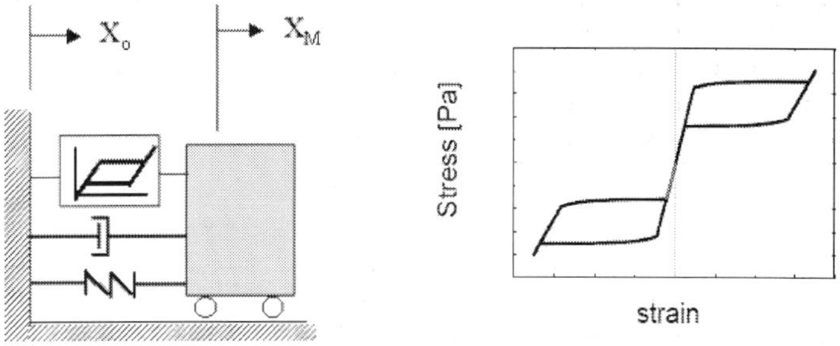

a) b)

Fig. 19. a) Equivalent mechanical model of the SDOF scaled structure
b) The hysterisis /energy dissipation behavior of the re-centering device

9. Conclusion

There have been considerable research efforts in seismic response control for the past several decades. Due to the distinctive macroscopic behaviour like super elasticity, Shape Memory Alloys are the basis for innovative applications such as devices for protecting buildings from structural vibrations. Super elastic properties of Nitinol wires have been established from the experiments conducted and the salient features to be highlighted from the study are

- The material's application can be made suitable for seismic devices like recentering, supplementally recentering or in the case of non-recentering devices, as a variety of hysteretic behaviors were obtained from the tests.
- Cyclic behavior of the non pre strained wires especially energy dissipation capability, equivalent viscous damping and secant stiffness are not very sensitive to the number of cycles in the frequency range of interest (0.5- 3Hz.) as observed from constant amplitude loading.
- Pre-strained super elastic wires shows higher energy dissipation capability and equivalent damping when cycled around the midpoint of the strain range obtained from quasi-static curve. It is found from the experiment that the pre strain value of 6 %, with amplitude cycles which covers 2-10% gives higher energy dissipation.
- For possible application of vibration control devices in structural systems, a judicious selection of the wire under tension mode can be selected between pre strained and non-pre strained wires. However application of pre strained wires in the system provides excellent energy dissipation characteristics but it requires skilled and sophisticated mechanism to maintain/provide the required pre strain.
- The mathematical model predicts the maximum energy dissipation capability of the material namely pre strained nitinol wires under study.
- The test results shows immense promise on SMA based devices which can be used for vibration control of variety of structures(New designs and restoration of structures). SMA structural elements/devices can be located at key locations of the structure to reduce the seismic vibrations.

10. Acknowledgment

The paper has been published with the kind approval of Director, CSIR-Structural Engineering Research Centre, Chennai. The constant encouragement, and support provided by the Director General-CSIR, Dr. Sameer K Brahmachari is gratefully acknowledged. The help and support provided by all colleagues of Advanced Seismic Testing and Research Laboratory in carrying out the experimental work deserve acknowledgement.

11. References

Birman, V (1997) Review of Mechanics of Shape Memory Alloy Structures *Applied Mechanics Review* 50 629-645

Birman, V (1997) Effect of SMA dampers on nonlinear vibrations of elastic structures *Proceedings of SPIE* 3038 ,268-76

Clark, P W; Aiken, I D; Kelly, J M; Higashino, M and Krumme, R C(1996) Experimental and analytical studies of shape memory alloy damper for structural control *Proc. Passive damping (San Diego,CA 1996)*

Cardone D, Dolce M, Bixio A and Nigro D 1999 Experimental tests on SMA elements *MANSIDE Project (Rome, 1999) (Italian Department for National Technical Services)* II85-104 Da GZ, Wang TM, Liu Y, Wamg CM(2001) Surgical treatment of tibial and femoral fractures with TiNi Shape memory alloy interlocking intra medullary nails *The international conference on Shape Memory and Superelastic Technologies and Shape Memory materials, Kunming, China*

Dolce M (1994) Passive Control of Structures Proceedings of the 10th European Conference on Earthquake Engineering, Vienna, 1994.

Duerig, T; Tolomeo, D and Wholey, M. (2000), An overview of superelastic stent design *Minimally Invasive Therapy & Allied Technologies. , 2000:9(3/4) 235–246.*

Eaton, J P. (1999) Feasibility study of using passive SMA absorbers to minimize secondary system structural response , *Master Thesis ,Worcester polytechnic Institute, M A*

Humbeeck, JV (2001) Shape Memory Alloys: a material and a technology *Advanced Engieering Materials* 3 837-850

Miyazaki, S; Imai, T; Igo, Y. And Otsuka, K(1986b), Effect of cyclic deformation on the pseudoelasticity characteristics of Ti–Ni alloys. *Metall. Trans. A.* 17115–120.

Pelton, A; DiCello,J; and Miyazaki,S. , (2000), Optimisation of processing and properties of medical grade nitinol wire. *Minimally Invasive Therapy & Allied Technologies.* , 2000:9(1) 107–118.

Sreekala, R; Avinash, S; Gopalakrishnan, N; and Muthumani, K(2004) Energy Dissipation and Pseudo Elasticity in NiTi Alloy Wires *SERC Research Report* MLP 9641/19, October 2004

Sreekala, R; Avinash, S; Gopalakrishnan, N; Sathishkumar, K and Muthumani, K(2005) Experimental Study on a Passive Energy Dissipation Device using Shape Memory Alloy Wires, *SERC Research Report* MLP 9641/21, April 2005

Sreekala,. R;. Muthumani, K; Lakshmanan, N; Gopalakrishnan, N & Sathishkumar, K (2008),. Orthodontic arch wires for seismic risk reduction, *Current Science,* ISSN:0011-3891, Vol. 95, No:11, pp 1593-1599.

Sreekala,. R,;. Muthumani,. K(2009),. Structural Application of Smart materials,. In:. *Smart Materials,.* Edited by Mel Shwartz, . CRC press, Taylor & Francis Group,. pp. 4-1 to 4-7,. Taylor &Francis Publications,. ISBN-13:978-1-4200-4372-3,. Boca raton, FL,USA

Sreekala,. R;. Muthumani, K; Lakshmanan, N; Gopalakrishnan, N; Sathishkumar, K; Reddy, G,R & Parulekar Y M. (2010),. A Study on the suitability of NiTi wires for Passive Seismic Response Control *Journal of Advanced Materials,* ISSN 1070-9789, Vol. 42, No:2,pp. 65-76.

Stöckel,D. and Melzer, *A. ,, Materials in Clinical Applications.* , ed. P. Vincentini Techna Srl. ,1995, 791–98.

Wilde, K; Gardoni, P. , and Fujino Y. ,(2000) Base isolation system with shape memory alloy device for elevated highway bridges *Engineering Structures* 22 222-229.

Control Design Methodologies for Vibration Mitigation on Wind Turbine Systems

Ragnar Eide and Hamid Reza Karimi
Department of Engineering, Faculty of Engineering and Science University of Agder
Postboks Grimstad
Norway

1. Introduction

The world's energy consumption from the beginning of the industrial revolution in the 18 th. century and until today has increased at a tremendous degree. Since a large part of the energy has come from sources like oil and coal have the negative impacts on the environment increased proportionally. Therefore, more sustainable and climate friendly energy production methods are emphasized among researchers and environmentalists throughout the world. This is the reason why renewable energies, and wind power particularly, have now become an essential part of the energy programs for most of governments all over the world. One example is seen by the outcome of the European Conference for Renewable Energy in Berlin in 2007 where EU countries defined ambitious goals when it comes to the increase in use of renewable energy resources. One of the goals was that by 2020, the EU would seek to get 20% of energy consumption from renewable energies.

Wind power, in conjunction with other renewable power production methods, has been suggested to play a more and more important role in the future power supply (Waltz, 2008) and (Lee & Kim, 2010). One of the reasons for these expectations is the enormous available potential when it comes to wind resources. One of the most comprehensive study on this topic (Archer & Jacobson, 2005) found the potential of wind power on land and near-shore to be 72TW, which alone could have provided over five times the world's current energy use in all forms averaged over a year.

The World Wind Energy Association (WWEA) estimates the wind power investment worldwide to expand from approximately 160 GW installed capacity at the end of 2010 to 1900 GW installed capacity by 2020. One example is from the USA, where the current contribution of electricity from wind power is merely 1,8% in (2009). However, the U.S Department of Energy is now laying a framework to get as much as 20% contribution by the year 2030.

Due to the economical advantages of installing larger wind turbines (WTs), the typical size of utility-scale turbines has grown dramatically over the last three decades. In addition to the increasing turbine-sizes, cost reduction demands imply use of lighter and hence more flexible structures. If the energy-price from WTs in the coming years are to be competitive with other power production methods, an optimal balance must be made between maximum power capture on one side, and load-reduction capability on the other side. To be able to obtain this is a well defined control-design needed to improve energy capture and reduce dynamic loads. This combined with the fact that maintenance and constant supervision of WTs at offshore locations is expensive and very difficult, which has further increased the need of a reliable control system for fatigue and load reduction. New advanced control approaches must be designed such as to achieve to the 20- to 25-year operational life required by todayŠs machines (Wright, 2004).

In this paper an above rated wind speed (Region III) regulation of a Horizontal Axis Wind Turbine (HAWT) is presented. The first method is Disturbance Accommodating Control (DAC) which is compared to the LQG controller. The main focus in this work is to use these control techniques to reduce the torque variations by using speed control with collective blade pitch adjustments. Simulation results show effects of the control methodologies for vibration mitigation on wind turbine systems.

The paper is organized as follows. Section 2 will be devoted to the modeling phase. The aim is to come up with a simplified state space model of the WT appropriate to be used in the control design in the subsequent sections. The control system design is covered in Section 3. After a historical overview and a state-of-the-art presentation of WT control are the elements involved in the practical control designed merged with a theoretical description of each topic. The model will then be implemented into a simulation model in the in the MATLAB/Simulink environment in Section 4. Finally, the conclusions and further improvement suggestions are drawn in Section 5.

2. Modeling of the wind turbine

2.1 Introduction

There are different methods available for modeling purposes. Large multi-body dynamics codes, as reported in (Elliot & Wright, 2004), divide the structure into numerous rigid body masses and connect these parts with springs and dampers. This approach leads to dynamic models with hundreds or thousands of DOFs. Hence, the order of these models must be greatly reduced to make them practical for control design (Wright & Fingersh, 2008). Another approach is an assumed modes method. This method discretize the WT structure such that the most important turbine dynamics can be modeled with just a few degrees of freedom. Designing controllers based on these models is much simpler, and captures the most important turbine dynamics, leading to a stable closed-loop system (Wright & Fingersh, 2008). The method is for instance used in FAST, a popular simulation program for design and simulation of control system (Jonkman & Buhl, 2005).

This section presents a simplified control-oriented model. In this approach, a state space representation of the dynamic system is derived from of a quite simple mechanical description of the WT. This state space model is totally non-linear due to the aerodynamics involved, and will thus be linearized around a specific operation point. As reported in (Wright & Fingersh, 2008) with corresponding references, good results are obtained by using linearized time invariant models for the control design.

When modeling a WT one may need to combine different models, each representing interacting subsystems, as Figure 1 below depicts. Here we can see how the WT is simplified to consist of the aerodynamic-, mechanical- (drive train), and electric subsystem, and that the blade pitch angle reference and the power reference in this case are controllable inputs.

The WT is a complicated mechanical system with many interconnecting DOF. However, some of the couplings are rather weak and can be neglected (Ekelund, 1997). For instance, the

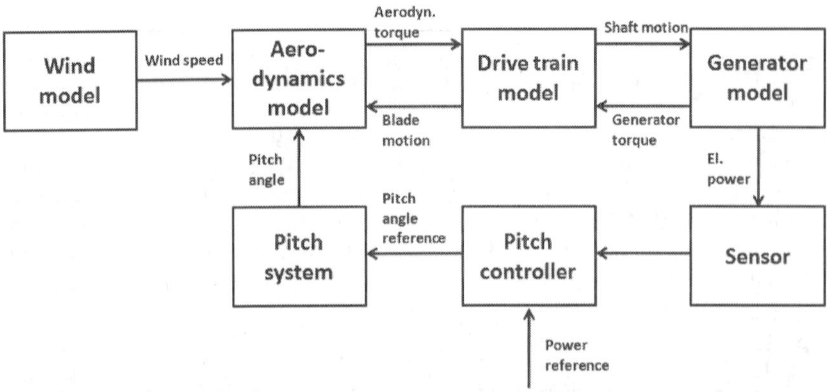

Fig. 1. WT subsystems with corresponding models

connection between the dynamics of the transmission and the tower is neglected in modeling of the mechanical system. The dynamics of the generator and the electrical system are also neglected by regarding the reaction torque from the generator as a fixed value. Also when considering the wind, the approach is to model the wind as simply a scalar input affecting the rotor state.

2.2 State space representation

The nonlinearities of a WT system, for instance due to the aerodynamics, may bring along challenges when it comes to the control design. Since the control input gains of a pitch control usually is the partial derivative of the rotor aerodynamic torque with respect to blade pitch angle variations, these input gains will depend on the operating condition, described by a specific wind and rotor speed. A controller designed for a turbine at one operating point may give poor results at other operating conditions. In fact, a controller which has shown to stabilize the plant for a limited range of operation points, may cause unstable closed-loop behavior in other conditions.

A method which bypasses the challenges of directly involving the nonlinear equations is by using a linear time invariant system (LTI) on state space form. Such a system relates the control input vector u and output of the plant y using first-order vector ordinary differential equation on the form

$$\begin{aligned} \dot{x} &= Ax + Bu + \Gamma u_D \\ y &= Cx + Du \end{aligned} \tag{1}$$

where x is the system states and matrices A, B, Γ, C and D are the state-, input-, disturbance-, output- and feedthrough- matrix, respectively, and u_D is the disturbance input vector.

The representation of the system as given in Eqns. (1) has many advantages. Firstly, it allows the control designer to study more general models (i.e., not just linear or stationary ODEs). Having the ODEs in state variable form gives a compact, standard form for the control design. State space systems also contain a description of the internal states of the system along with the input-output relationship. This helps the control designer to keep track of all the modes

(which is important to do since a system can be internally unstable, although it is input-output stable). As shall be shown in the following is a description of the dynamics on state-space a good starting point for the further controller and observer designs.
The WT drive-train modeled with its high and low speed shaft separated by a gearbox is shown in Figure 2. As it's seen, the drive train is modeled as a simple spring-damper configuration with the constants K_r and C_r denoting the spring stiffness and damping in the rotor shaft, and similarly; K_g and C_g as representing the spring stiffness and damping in the generator shaft. Figure 2 also shows the inertia, torque, rot. speed and displacement of the rotor and generator shafts. The parameters named as T1, ω_1, q1, N1,I1 are the torque, speed, displacement, number of teeth, and inertia of gear 1, and similarly for gear 2.

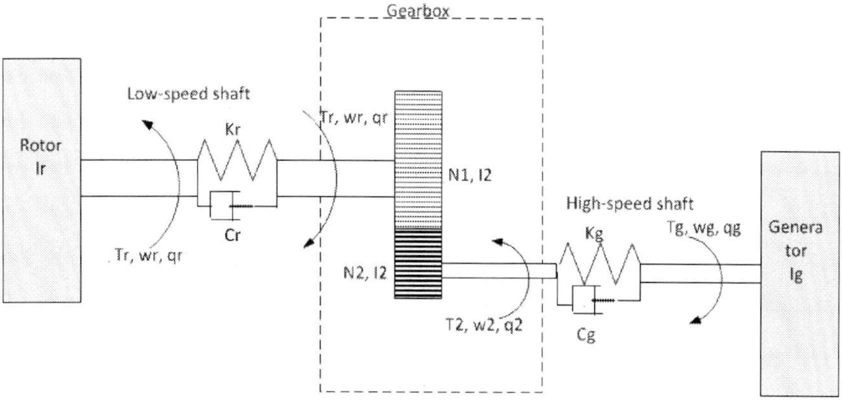

Fig. 2. Model of the drive train with the high and low speed shafts. (For definition of the parameters, see text)

The model in Figure 2 results in the following equation of motion for the rotor torque

$$T_r = I_r\ddot{q}_r + K_r(q_r - q_1) + C_r(\dot{q}_r - \dot{q}_1) + T_1 + I_1\ddot{q}_1 \tag{2}$$

where the factor $K_r(q_r - q_1) + C_r(\dot{q}_r - \dot{q}_1)$ is the reaction torque in the low speed shaft. Equivalently, the equation for generator motion is as follows

$$T_g = I_g\ddot{q}_g + K_g(q_g - q_2) + C_g(\dot{q}_g - \dot{q}_2) + T_2 + I_2\ddot{q}_2 \tag{3}$$

where the factor $K_g(q_g - q_2) + C_g(\dot{q}_g - \dot{q}_2)$ is the reaction torque at the high speed shaft.
The relationship between T_1 and T_2 is derived based on the equation describing a constrained motion between two gears in the following way

$$\frac{T_2}{T_1} = \frac{\omega_1}{\omega_2} \rightarrow T_1 = \frac{\omega_1}{\omega_2}T_2 \tag{4}$$

From Eqn. (3) we find

$$T_2 = T_g - I_g\ddot{q}_g - K_r(q_g - q_2) - C_r(\dot{q}_g - \dot{q}_2) - I_2\ddot{q}_2 \tag{5}$$

then, the following equation for the rotor rotation holds

$$T_r = I_r\ddot{q}_r + K_r(q_r - q_1) + C_r(\dot{q}_r - \dot{q}_1) + \frac{\omega_1}{\omega_2}(T_g - I_g\ddot{q}_g - K_r(q_g - q_2) - C_r(\dot{q}_g - \dot{q}_2) - I_2\ddot{q}_2) + I_1\ddot{q}_1 \quad (6)$$

Since the goal of the modeling is to use it for control design, this equation will be simplified in the following. First, it can be assumed that the the high speed shaft is stiff. This will imply that $T_g = T_2$, $\omega_g = \omega_2$ and so on. Secondly, the gearbox can be assumed lossless, hence the terms involving I_1 and I_2 can be omitted. This reduces Eqn. 6 to the following equation

$$T_r = I_r\ddot{q}_r + K_r(q_r - q_1) + C_r(\dot{q}_r - \dot{q}_1) + \frac{\omega_1}{\omega_2}(T_g - I_g\ddot{q}_g) \quad (7)$$

The model is now of three degrees of freedom (DOF); rotor speed, generator speed, and a DOF describing the torsional spring stiffness of the drive train. These DOFs correspond with the three states shown in Figure 3

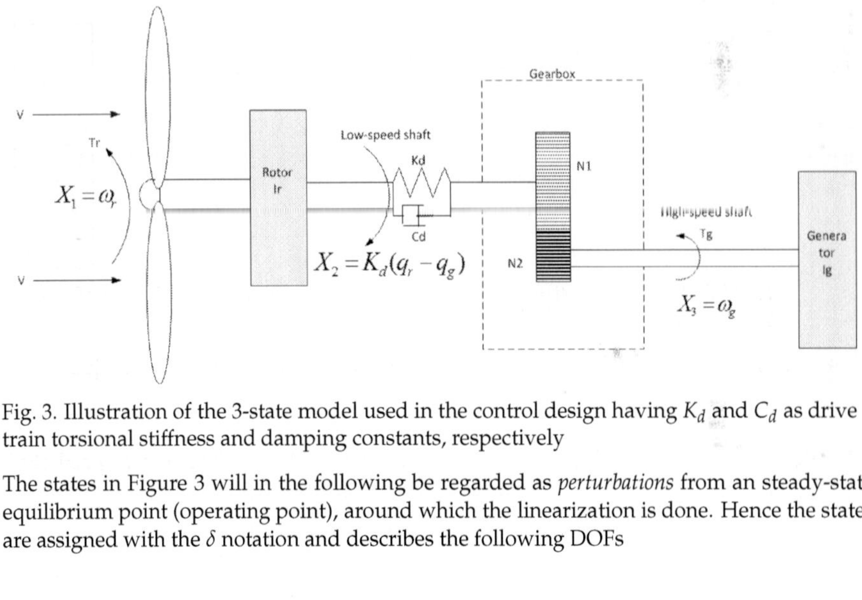

Fig. 3. Illustration of the 3-state model used in the control design having K_d and C_d as drive train torsional stiffness and damping constants, respectively

The states in Figure 3 will in the following be regarded as *perturbations* from an steady-state equilibrium point (operating point), around which the linearization is done. Hence the states are assigned with the δ notation and describes the following DOFs

$$\begin{aligned} X_1 &= \delta\omega_r \text{is the perturbed rotor speed} \\ X_2 &= K_d(\delta q_r - \delta q_g) \text{is the perturbed drive train torsional spring stiffness} \\ X_3 &= \delta\omega_g \text{is the perturbed generator speed} \end{aligned}$$

where $\delta\dot{q}_r = \delta\omega_r$ and $\delta\dot{q}_g = \delta\omega_g$.
Now, from the Newtons second law, the following relation holds

$$I_r\ddot{q}_r = T_r - T_{sh} \quad (8)$$

where the left hand side expresses the difference between the aerodynamic torque on the rotor caused by the wind force, and the reaction torque in the shaft. This reaction torque can be expressed according to Eqn. (7) as

$$T_{sh} = K_d(q_r - q_g) + C_d(\dot{q}_r - \dot{q}_g) = K_d(q_r - q_g) + C_d(X_1 - X_3) + \frac{\omega_1}{\omega_2}(T_g - I_g\ddot{q}_g) \tag{9}$$

This equation can, when expressed in terms of deviations from the steady state operation point, be written as:

$$\delta T_{sh} = K_d(\delta q_r - \delta q_g) + C_d(\delta\dot{q}_r - \delta\dot{q}_g) + \frac{\omega_1}{\omega_2}(T_g - I_g\delta\ddot{q}_g) \tag{10}$$

Hence the BEM theory provides a way to calculate the power coefficient C_P based on the combination of a momentum balance and a empirical study of how the lift and drag coefficients depend on the the collective pitch angle, β, and tip speed ratio, λ . In this way an expression of the aerodynamic torque can be found to be

$$T_r(V, \omega_r, \beta) = \frac{1}{2}\frac{\pi\rho R^2 C_P(\beta, \lambda)}{\lambda}V^2 \tag{11}$$

where ρ is the air density, R is the rotor radius, and V is the wind speed.
Let us now assume an operating point at $(V_0, \omega_{r,0}, \beta_0)$ such that Eqn. (11) can be written as

$$T_r = T_r(V_0, \omega_{r,0}, \beta_0) + \delta T_r \tag{12}$$

where δT_r is deviations in the torque from the equilibrium point ($\delta T_r = T_r - T_{r,0}$) and consists of partial derivatives of the torque with respect to the different variables, i.e Taylor series expansion (Henriksen, 2007) and (Wright, 2004), in the following way

$$\delta T_r = \frac{\partial T_r}{\partial V}\delta V + \frac{\partial T_r}{\partial \omega_r}\delta\omega_r + \frac{\partial T_r}{\partial \beta}\delta\beta \tag{13}$$

where $\delta V = V - V_0$, $\delta\omega = \omega_r - \omega_{r,0}$, and $\delta\beta = \beta - \beta_0$. By assigning α, γ, and ζ to denote the partial derivatives of the torque at the chosen operating point $(V_0, \omega_{rot,0}, \beta_0)$, Eqn. (12) becomes

$$T_r = T_r(V_0, \omega_{r,0}, \beta_0) + \alpha(\delta V) + \gamma(\delta\omega_r) + \zeta(\delta\beta) \tag{14}$$

If the above expression is put into Eqn.(8), it follows that

$$I_r\ddot{q}_r = T_r(V_0, \omega_{r,0}, \beta_0) + \delta T_r - T_{sh,0} - \delta T_{sh} \tag{15}$$

At the operation point is $T_r(V_0, \omega_{r,0}, \beta_0) = T_{sh,0}$ since this is a steady state situation. This reduces Eqn. (15) to

$$I_r\ddot{q}_r = \alpha(\delta V) + \gamma(X_1) + \zeta(\delta\beta) - X_2 - C_d(X_1 - X_3) \tag{16}$$

when substituting with the corresponding state equations. The following expressions for the derivatives of the state variables can now be set up

$$\dot{X}_1 = \frac{(\gamma - C_d)X_1 - X_2 + C_d X_3 + \gamma(\delta\beta) + \alpha(\delta\omega_r)}{I_r} \tag{17}$$

$$\dot{X}_2 = K_d(\delta\dot{q}_r - \delta\dot{q}_g) = K_d X_1 - K_d X_3 \tag{18}$$

$$\dot{X}_3 = \frac{C_d X_1 + X_2 - C_d X_3}{I_g} \tag{19}$$

Note that in the derivative of the generator speed state X_3 is it used that $I_g \ddot{q}_g = I_g X_3 = \delta T_{sh} - \delta T_g = \delta T_{sh}$ when assuming a constant generator torque.
The dynamic system can now be represented in a state space system on the form as described by Eqns (1) yielding

$$\begin{bmatrix} \dot{X}_1 \\ \dot{X}_2 \\ \dot{X}_3 \end{bmatrix} = \begin{bmatrix} \frac{\gamma - C_d}{I_r} & -\frac{1}{I_r} & \frac{C_d}{I_r} \\ K_d & 0 & -K_d \\ \frac{C_d}{I_g} & \frac{1}{I_g} & -\frac{C_d}{I_g} \end{bmatrix} \begin{bmatrix} X_1 \\ X_2 \\ X_3 \end{bmatrix} + \begin{bmatrix} \frac{\zeta}{I_r} \\ 0 \\ 0 \end{bmatrix} \delta\beta + \begin{bmatrix} \frac{\alpha}{I_r} \\ 0 \\ 0 \end{bmatrix} \delta V$$

$$y = \begin{bmatrix} 0 \ 0 \ 1 \end{bmatrix} \begin{bmatrix} X_1 \\ X_2 \\ X_3 \end{bmatrix} \tag{20}$$

where the disturbance input vector u_D from the general form in Eqns. (1) now is given as δV, which is the perturbed wind disturbance (i.e deviations from the operating point, $\delta V = V - V_0$), and the control input vector u from Eqns. (1) now given as $\delta\beta$ (i.e perturbed (collective) pitch angle, $\delta\beta = \beta - \beta_0$).
The parameters will be assigned when coming to Section 4. It is worth no notice that the the measured output signal here is the generator speed. This can be seen from the form of the output vector C. It will be shown later that this will lead to a non-minimum phase plant, i.e a plant with an asymptotically unstable zero in the right complex plane (Lee, 2004). Such plant is in general not suitable for the LTR approach such that a revision of this is required.

3. Wind turbine control

3.1 Introduction

In order to have a power production which ensures that speed, torque and power are within acceptable limits for the different wind speed regions, is it necessary to control the WT. This control system should be complex enough to meet the intended control objectives, but at the same time simple enough to easy interpret the results. A frequently used approach, which will also be applied in this work, is to start with a simple model and a simple controller that can be developed further by adding more degrees of freedom into the model.
Optimally speaking, if the control system shall be able to meet the requirement of reduced energy cost, it must find a good balance between (a) a long working life without failures and (b) an efficient (optimal power output) and stable energy conversion. The first point can be regarded as the main focus in this work. This requirement can be further crystallized to the following properties which the control system should possess

- good closed loop performance in terms of stability, disturbance attenuation, and reference tracking, at an acceptable level of control effort.
- low dynamic order (because of hardware constraints)
- good robustness

To meet the objective of long working life should the control system be designed to be efficiently mitigate loads in order to reduce the fatigue stresses (especially in shaft and blade rotor) due to varying wind disturbances. Before going into the specific control design, an historical overview of WT control with a special view toward the objective of load reduction, will be presented in the following.

3.2 History of WT control

The history of WT control and research within this field have emerged from the simplest form of passive stall control to advanced controllers, like so-called smart rotor control. The latter is well described in for instance (Wilson, Berg, Barone, Berg, Resor & Lobitz, 2009a) and (Lackner & van Kuik, 2010). As these references portrays does this control scheme involve active aerodynamic flow control by implementation of numerous sensors and actuators, bringing along a high level of complexity. Although these advanced control methods have been investigated for ten to fifteen years, most commercial systems are still implemented using multiple single-input-single-output (SISO) loops with classical PID controllers (Bossanyi, 2000). Actually, as reported in (Bossanyi, 2004), PID showed to give competitive results compared to some of the new advanced techniques.
The traditional way of controlling a WT with multiple control objectives, such as speed control for maximum power tracking and load mitigation by pitch control, is to design independent control loops in a way illustrated in Figure 4(a).

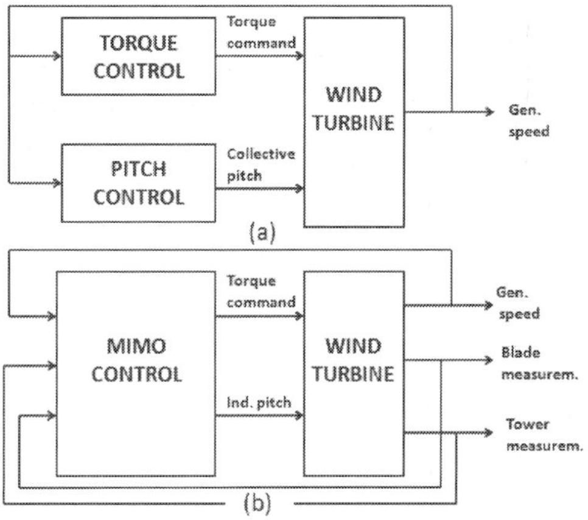

Fig. 4. Illustration of difference between SISO and MIMO controllers

The PID controllers (i.e. the SISO controllers shown in Figure 4 (a)) are traditionally used for the individual torque and pitch control and have shown to have a good effect when carefully tuned and adjusted to its specific application. One disadvantage is, however, that the PID control loops must be designed not to interfere with each other. If this happen to be the case will the result often be a destabilized turbine. This problem can be solved in an efficient manner within the modern and so-called advanced control techniques (Wright & Fingersh, 2008) using MIMO controllers (cf. Figure 4 (b)). In these more advanced control designs, multiple control objectives is seen to be met with fewer control loops leading to stable closed loop behavior (Wright & Stol, 2008).

With increasing turbine sizes, much research is done to find new and better ways of load control compared to the classical methods (see (Wright & Stol, 2008) and (Bossanyi, 2003) with references). Large turbine sizes will give rise to loads that vary along the blade and change quickly due to wind gusts and other varying wind conditions. Rapidly changing loads can cause fatigue damage and reduce the life of the WT, which in turn may decide the lifetime of the other turbine components. Because of the inertia of the system, as well as the limitations of the actuators, active pitch control alone can only control ŞaverageŤ loads on the blade. On the other hand, passive load control strategies cannot respond to local load variations. Active aerodynamic load control (AALC) is therefore suggested to have a good potential as an addition the existing control strategies when it comes to load reduction (Wilson, Berg, Barone, Berg, Resor & Lobitz, 2009b). One approach where AALC has been combined with an individual blade pitch control scheme has shown to reduce the root flap bending moment significantly (Wilson, Berg, Resor, Barone & Berg, 2009).

A well known MIMO controller, the Linear Quadratic Gaussian (LQG) regulator, is suggested in (Selvam, 2007) to be used in load control. This paper presents that this regulation policy can have a good load reduction capability for a large frequency range. Reports as (Bossanyi, 2003) and (Bossanyi, 2004) suggest load mitigation when using individual pitch actuators. A thorough study has been performed by Bossanyi in (Bossanyi, 2003) where the classical PI control is compared with a multi-variable LQG control approach. Although this work has shown to yield good results when applying LQG, the design process is not straightforward and the resulting algorithm is relatively difficult.

One of the disadvantages of LQG design, however, is that it cannot directly take into account robustness margins (like the gain- and the phase margin). Due to this poor robustness properties of LQG control it was been necessary to search after other methods which could handle model uncertainties in a better way. One method which can be traced back to the early 1980s is the H_∞ approach. In this approach does the control designer from the very beginning specify a model of system uncertainties, which can be for instance additive perturbations or output disturbance (D.-W.Gu et al., 2005). Connor, Iyer, Leithead and Grimble were among the first to suggest an application of H_∞ control on a WT model (see Connor et al., 1992). Their main concern in this work were how to reduce the matter of fatigue loading. Although they encountered some overshoot problems did they prove that the method was applicable to a WT control problem. Suggestions of how H_∞ control can be used for load reduction has later been reported in (Bianchi et al., 2004). The method has also been shown to be applicable in advanced power control to enhance a better power capture for a wider range of wind speeds (Liu et al., 2008).

One particular problem the WT control designers must be able to handle is that WT control generally involves a multi-objective optimization problem, each objective having different goals. For instance, the control objective can be alleviate loads due to large scale gusts over the whole disk area, while others can be more slowly varying and low frequency loads over only one blade. To handle this problem does (Bottasso et al., 2010) propose an approach based on a multi-layer architecture. In this method are three control layers designed, each aiming at a specific control target and cooperates with the other layers to obtain various control goals. In this way, the choice of the controller used on each layer can be optimized and tailored to the specific control goal of that layer, thereby improving performance and simplify tuning (Bottasso et al., 2010).

What the control designer in all of the above mentioned controlling methods are actually trying to do is to compensate for the stochastic fluctuations in the wind. To do this in a perfect way is of course more or less impossible. One method called Disturbance Accommodating Control (DAC) take into account the fluctuations in the wind by an estimation of the disturbance for an assumed above-rated wind speed situation (Balas et al., 1998). DAC has shown to have a good effect on load mitigation and is therefore used to obtain special attention when it comes to fatigue load reduction. The theory was thoroughly compared with PI control by Alan Wright in his PhD thesis from 2004 (Wright, 2004). He concluded in his paper that the analytic results obtained using DAC appeared to be "extremely promising". For instance, it was shown how DAC has much better performance compared to the PI control in reducing drive-train shaft torsional moments and blade root flapwise-bending moments (Wright, 2004). A brief description of the theory behind the DAC will be given shortly, while an implementation and testing of DAC for the WT system will be done in Section 4.

In addition to the above mentioned control schemes are methods as Model Predictive Control (MPC) (Henriksen, 2007), Generalized Predictive Controller (GPC), and Fuzzy Logic Control (Karimi-Davijani et al., 2009) proposed for WT control. A comparison of different control techniques will not be done in this work (please refer to (Bottasso et al., 2007) for more on this).

3.3 The Linear Quadratic Regulator (LQR)

The LQR controller was among the first of the so-called advanced control techniques used in control of wind turbines. Liebst presented in 1985 a pitch control system for the KaMeWa wind turbine using LQR design (Wright, 2004). The objective of this controller was to alleviate blade loads due to wind shear, gravity, and tower deflection using individual blade pitch control. Results of this work included a reduction of blade and tower cyclic responses as well as the reduction of a large 2-per-revolution variation in power (Liebst, 1985).

The practical application of this technique is limited by the challenges of obtaining accurate measurements of the states needed in the controller. One way to solve this is to add extra sensors and measurement arrangements. However, since this extra sensors and measurements will add considerable cost and complexity to a WT, is this most often not regarded as a good solution. In addition, errors in the measurement of these states can result in poor controller behavior (Wright, 2004). To avoid these problems, a state estimator is added in any practical LQR implementation (Laks et al., 2009). This approach was described by Mattson in his Ph.D thesis as early as in 1984, where power-regulation for a fixed-speed WT using blade pitch is presented (Mattson, 1984). The proposed solution to the WT control problem was here based on linear models containing drive-train torsion and tower fore-aft bending DOF. The same report also describes the use of state estimation to estimate wind speed.

The principles of a LQR controller is given in Figure 5. Here the state space system is represented with its matrices A, B, and C, together with the LQR controller (shown with the $-K$).

The LQR problem rests upon the following three assumptions (Athans, 1981):

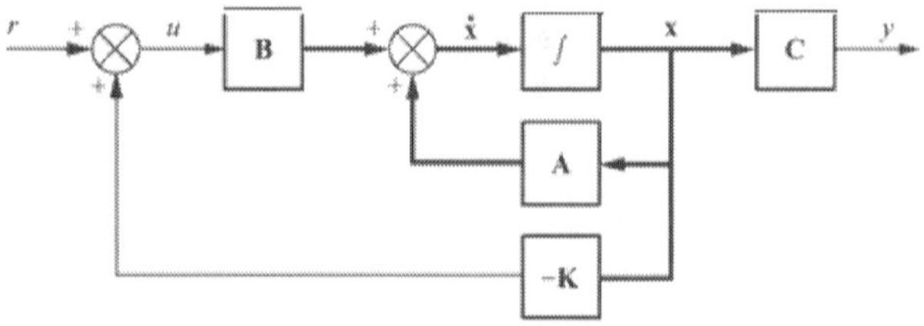

Fig. 5. State space control using a LQR controller where K is the LQR gain matrix

1. All the states are available for feedback, i.e it can be measured by sensors etc.
2. The system are *stabilizable* which means that all of its unstable modes are controllable
3. The system are *detectable* having all its unstable modes observable

LQR design is a part of what in the control area is called *optimal control*. This regulator provides an *optimal control law* for a linear system with quadratic performance index yielding a cost function on the form (Burns, 2001)

$$J = \int_0^{\infty} x^T(t)Qx(t) + u^T(t)Ru(t)\,dt \tag{21}$$

where $Q = Q^T$ and $R = R^T$ are weighting parameters that penalize the states and the control effort, respectively. These matrices are therefore controller tuning parameters.
It is crucial that Q must be chosen in accordance to the emphasize we want to give the response of certain states, or in other words; how we will penalize the states. Likewise, the chosen value(s) of R will penalize the control effort u. As an example, if Q is increased while keeping R at the same value, the settling time will be reduced as the states approach zero at a faster rate. This means that more importance is being placed on keeping the states small at the expense of increased control effort. On the other side, if R is very large relative to Q, the control energy is penalized very heavily. Hence, in an optimal control problem the control system seeks to maximize the return from the system with *minimum cost*.
In a LQR design, because of the quadratic performance index of the cost function, the system has a mathematical solution that yields an *optimal control law* given as

$$u(t) = -Kx(t) \tag{22}$$

where u is the control input and K is the state feedback gain matrix given as $K = R^{-1}B^TS$. It can be shown (see (Burns, 2001))that S can be found by solving the following algebraic Riccati Equation (ARE)

$$SA + A^TS + Q - SBR^{-1}B^TS = 0 \tag{23}$$

The process of minimizing the cost function by solving this equation can easily be done using the built-in MATLAB function `lqr`.
As can be seen from the above description does the LQR algorithm take care of the tedious work of optimizing the controller, which for the case of the PID controller can be a time consuming task. However, the control designer still needs to specify the weighting factors

and compare the results with the specified design goals. This means that controller synthesis will often tend to be an iterative process where the designed "optimal" controllers are tested through simulations and will be adjusted according to the specified design goals.
The optimal control solution from the LQR approach yields some impressive properties. Firstly the controller guarantees at least 60° phase margin in each input channel (Saberi et al., 1993), which means that as much as 60° phase changes can be tolerated in each input channel without violating the stability. In addition, the LQR has infinite gain margin which implies that the LQR is still able to guarantee the stability, even though the gain increases indefinitely. Hence, LQR guarantees a very robust controller, capable of handling various kinds of uncertainties in the modeling. However, when an state estimator is implemented with measurement feedback, these properties just described is often lacking. But, as will be shown later; the robustness properties can be recovered!

3.4 State estimation

As mentioned for the case of the LQR controller, all sensors for measuring the different states are in this case assumed to be available. This is not a valid assumption in practice. A void of sensors means that all states (full-order state observers), or some of the states (reduced order observer), are not immediately available for use in the control task. In such cases, a state observer must be implemented to supply accurate estimations of the states at all wind-rotor positions. The observer state equations are given by

$$\dot{\hat{x}} = A\hat{x} + Bu + L(y - C\hat{x}) \tag{24}$$

$$\hat{y} = C\hat{x} \tag{25}$$

where $\hat{x}$ is the estimate of the actual state x. Furthermore, equations (24) and (25) can be re-written to become

$$\dot{\hat{x}} = (A - LC)\hat{x} + Bu + Ly \tag{26}$$

These are the governing equations for a *full order observer*, having two inputs u and y and one output, $\hat{x}$. Since we already know A, B, and u, observers of this kind is simple in design and provides accurate estimation of all the states around the linearized point. The proportional observer gain matrix, L, can be found for instance by pole placement procedure, or as will be shown in the following; by applying the Kalman filter design.

3.5 Kalman estimator

In the previous design of the state observer, the measurements $y = Cx$ were assumed to be noise free. This is not usually the case in practical life. Other unknown inputs yielding the state equations to be on the general stochastic state space form, which is already defined in Eqn. (1) as

$$\dot{x} = Ax + Bu + \Gamma u_D$$

$$y = Cx + Du + n$$

here the D matrix relating the input u to the output y most commonly is set to zero, the disturbance input vector u_D is stationary, zero mean Gaussian white process noise (i.e wind disturbance), and n is zero mean Gaussian white sensor noise.

The Kalman filter can be applied to yield the estimated state vector $\hat{x}$ and output vector $\hat{y}$ (cf. Eqns (24) and (25), by using the known inputs u and the measurements y. The Kalman filter is optimal in the sense that it minimizes the estimated error covariance when some presumed conditions are met (see Bottasso & Croce 2009). The mean-square estimation error is given by

$$J = E[(x(t) - \hat{x}(t))^T(x(t) - \hat{x}(t))] \tag{27}$$

where

$$E[(x(t) - \hat{x}(t))^T y(t) = 0 \tag{28}$$

The optimal Kalman gain is given by

$$L(t) = S_e(t)C^T R^{-1} \tag{29}$$

where $S_e(t)$ is the same as J given in (3.5). Furthermore, when $t \to \infty$, the algebraic Riccati equation can be written as

$$0 = S_e A^T + AS_e + Q_n - S_e C^T R_n^{-1} C S_e \tag{30}$$

where Q_n and R_n are the process and measurement noise covariances, respectively. Tuning of the Kalman filter are required if these are not known. Finally, the sub-optimal Kalman gain for a steady state Kalman filter can be expressed as $L = S_e C^T R^{-1}$.

3.6 Disturbance accommodating control

This section will explain the principles of DAC and for more details, please refer to e.g. (Johnson, 1976). The DAC method will be utilized together with a LQR regulator and is a straightforward way to model and simulate a system with an assumed-wind disturbance.
Recalling from the state space system described by Eqns. (1), which included the vector Γ. This vector describes the magnitude of the disturbance, while the input signal u_D is the disturbance quantity (which in our case is the wind speed perturbation). The Γ vector will in the DAC procedure be used as a free design parameter.
The basic thought in DAC theory is that disturbances are described by an assumed-waveform model on state space form. Such a wind model can be described according to the following equations (Wright, 2004)

$$\dot{z}_D = F z_D \tag{31}$$

$$u_D = \Theta z_D \tag{32}$$

where z_D denotes the disturbance state, F the state matrix, u_D is disturbance output vector (input to the plant), and Θ relates disturbance model input to disturbance states. Note that it is necessary to know something about the initial condition of the disturbance state, z_D^0, in order to know the amplitude of the disturbance.
Since the state variables x and the disturbance variable z_D cannot be measured directly, observers will be utilized to estimate these. The mathematical description of the observer

will in this case according to Eqns (24) and (25) be given as

$$\dot{\hat{x}} = A\hat{x} + Bu + \Gamma\hat{u}_D + K_x(y - \hat{y}) \tag{33}$$

$$\hat{y} = C\hat{x} \tag{34}$$

where the estimator gains K_x can be chosen according to for instance the pole placement method using the `place` command in MATLAB.
Furthermore, the disturbance observer is described as

$$\dot{\hat{z}}_D = Fz_D + K_D(y - \hat{y}) \tag{35}$$

$$\hat{u}_D = \Theta\hat{z}_D \tag{36}$$

where the disturbance state estimator gain K_D is a scalar.
The estimated variables can then be fed into the controller to minimize the effect of the disturbances. The result of this is that the state feedback now also include the feedback of disturbance observer yielding a "new" feedback control law on the form

$$u(t) = G\hat{x}(t) + G_D\hat{z}_D(t) \tag{37}$$

where the LQR gain G is computed in MATLAB, and the full state feedback gain (i.e disturbance gain) G_D must be chosen as to minimize the norm $|BG_D + \Gamma\Theta|$. Calculated numerical values will be given in Section 4.

3.7 The LQG controller

Another well-known method of handling noise inputs is the Linear Quadratic Gaussian (LQG) controller. This is simply a combination of a Kalman filter and a LQR controller. The separation principle guarantees that these can be designed and computed independently (Morimoto, 1991). LQG controllers can be used both in linear time-invariant (LTI) systems as well as in linear time-variant systems. The application to linear time-variant systems enables the design of linear feedback controllers for non-linear uncertain systems, which is the case for the WT system.
The schematics of a LQG is in essence similar to that depicted in Figure **??** in Section 3.4, the only difference is that the observer gain matrix L in this figure now can be defined as the Kalman gain K_f. However, we also assume disturbances in form of noise, such that the system with the LQG regulator in compressed form can be described as in Figure 6

4. Control design and simulations

4.1 Introduction

The purpose in this project is to investigate the application of modern control theories, such as the LQR and LQG, on WT systems and how a speed controller can be made with these theories to reduce torque variations due to a wind disturbance. In addition, a further idea is to investigate if a better load reduction can be obtained by using a dynamic compensator from the LTR algorithm, and whether factors as the LTR gain will affect this result.
The simulations are done in the MATLAB Simulink environment, enabling the LTI state space model described in Section 2 to be incorporated together with the different controllers in the convenient block diagram approach available in Simulink.

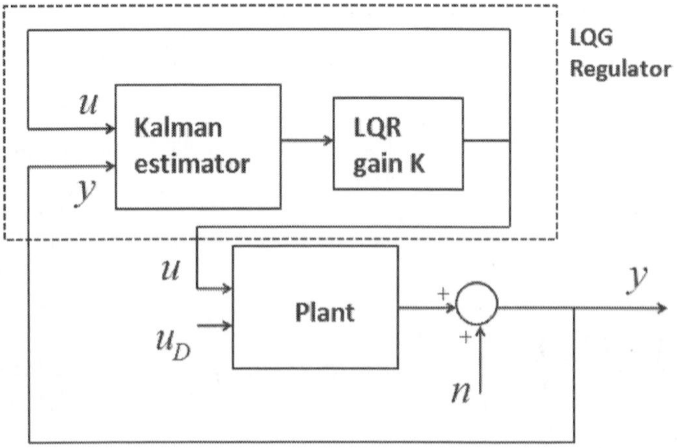

Fig. 6. LQG regulator

This Section presents results from the practical implementation of the WT control design. It is important to note that a test of the controlled system under various operating conditions has not been performed. This is mainly due to the fact that this work is not focused on a real-life application of the controller, but instead on the theoretical aspects of controller design. Also, the linearization of the WT plant at different operation points would have required a computer program, such as FAST[1]. Load mitigation in above rated wind speeds has already been described to be the focus of this work, and the operation point chosen will reflect this goal.

The first control approach that is being tested is the DAC theory, where the wind disturbance modeled in a straightforward way together with the LQR and observer procedures, as described in Section 3. The reason for starting with this method is to investigate if the results are comparable with the results from the LQG method, and if so; what is the difference in the results, and which results can be regarded as the "best" results.

It is important to note that for the simulations in this project are the actuator constraints, such as the dynamics of the pitch-system, not taken into consideration.

4.2 Linearization

As was described in Chaper 2 does the WT compose of a complex non-linear relationship between factors such as the aerodynamic properties of the blades, pitch angle, and wind speeds. The best approach is thus to build the model out from experimental data and a non-linear simulation model of the WT. Although it is possible to build a non-linear WT simulation model also in Simulink and extract linear models by utilizing the built-in `trim` and `linmod` functions, has this not been done in this work. Since the mathematical model presented in this paper is simplified a lot would this approach probably result in a more inaccurate model than with the use of the tailored WT simulation code provided by FAST,

[1] A popular aeroelastic design code for horizontal axis wind turbines developed by National Renewable Energy Laboratory, USA (Jonkman & Buhl, 2005)

where many aspects regarding the aerodynamics and structural dynamics are taken into account. The linearized state space model, which will be used in the control design is based on what is presented in (Wright, 2004) and (Balas et al., 1998).
The operation point, around which the linearization is done, is given as (Wright, 2004),
$V_0 = 18m/s$
$\omega_{rot,0} = 42rpm$
$\beta_0 = 12^\circ$

The corresponding FAST linearization with reference to (Wright, 2004) around this point yields the following state space system

$$\begin{bmatrix} \dot{X}_1 \\ \dot{X}_2 \\ \dot{X}_3 \end{bmatrix} = \begin{bmatrix} -0.145 & -3.108 \cdot 10^{-6} & 0.02445 \\ 2.691 \cdot 10^{7} & 0 & -2.691 \cdot 10^{7} \\ 0.1229 & 1.56 \cdot 10^{-}5 & -0.1229 \end{bmatrix} \begin{bmatrix} X_1 \\ X_2 \\ X_3 \end{bmatrix} + \begin{bmatrix} -3.456 \\ 0 \\ 0 \end{bmatrix} \delta\beta + \begin{bmatrix} 1 \\ 0 \\ 0 \end{bmatrix} (\delta V)$$
$$y = \begin{bmatrix} 0 & 0 & 1 \end{bmatrix} \begin{bmatrix} X_1 \\ X_2 \\ X_3 \end{bmatrix} \tag{38}$$

where the factors C_d/I_{rot} and C_d/I_{gen} are calculated by using the known relation to the other parameters.
Recalling from Section 2 that Γ vector is given as $\frac{\alpha}{I_r}$, where α is describes the relationship between torque variations and wind variations, $\frac{\partial T_r}{\partial V}$. This means that if Γ is increased, it will follow that the torque is more sensitive to wind variations, which is surely a negative effect when the matter of fatigue is concerned. The torque is given as a non-linear function in Eqn (11), which shows the dependency on the wind speed, rotor speed and blade pitch, and will be fixed at the chosen operating point. However, the Γ vector is initially chosen to $[100]^T$.
The transfer function (TF) of the open loop system formed by the linearized state space system in Eqns. (4.2) is

$$G(s) = \frac{1.499 \cdot 10^{-14} s^2 - 0.4247s - 1450}{s^3 + 0.374s^2 + 503.3s + 59.44} \tag{39}$$

extracted by using the MATLAB function `ss2tf`
One of the zeros of this TF is located at $s = 2.83 \cdot 10^{13}$ leading to a non-minimum phase system. Since the LTR procedure applied as an extension to the LQG later on is not feasible for non-minimum phase systems (Saberi et al., 1993) will this vector be formed as to represent rotor speed measurement instead, which changes the output to

$$y = \begin{bmatrix} 1 & 0 & 0 \end{bmatrix} \begin{bmatrix} X_1 \\ X_2 \\ X_3 \end{bmatrix} \tag{40}$$

4.3 Disturbance accommodating control

The DAC procedure has already been described in Section 3.6. The Simulink block digram shown in Figure 7 was set up according to the governing equations described in this section.

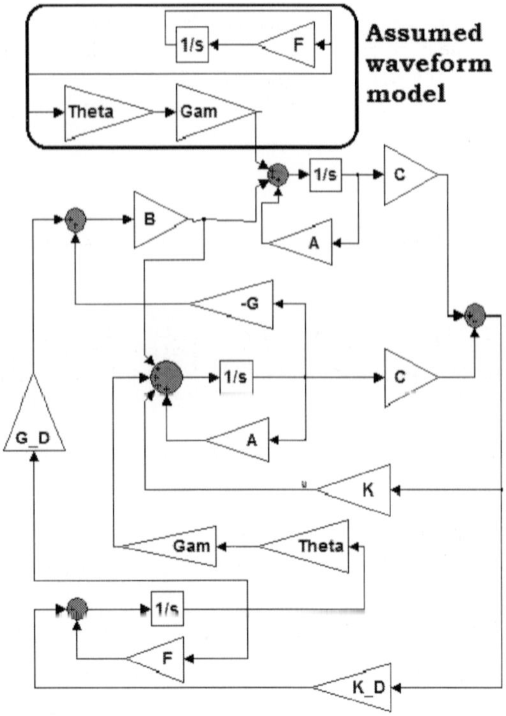

Fig. 7. Simulink block diagram of DAC

The assumed waveform model is shown in Figure 7. It is assumed a step input disturbance from the wind, hence $F = 0$ and $\Theta = 1$ will be chosen to reflect this (Henriksen, 2007). In order to "excite" the controller an initial condition (chosen as 1) were inserted into the integrator of the disturbance model.

4.3.1 Simulation without any control

It is interesting first to have a look at the response without any control at all. Then, after implementing the controller, is it easier to see how good the controller works in the disturbance accommodating task.

Now, by setting both the LQR and the wind disturbance gain equal to zero, perturbations in rotor speed (i.e. deviation from the operation point) shown in Figure 8 when simulating for 50 seconds. According to the simulation result, it is observed that the wind disturbance is not accommodated, and the rotor speed will increase.

4.3.2 Simulation with only wind disturbance gain

The next scenario which is to be simulated is a situation with the LQR control gain set to 0, but with the wind disturbance state gain G_D set to minimize the norm $|BG_D + \Gamma\Theta|$. The response of the rotor speed is shown in Figure 9 below.

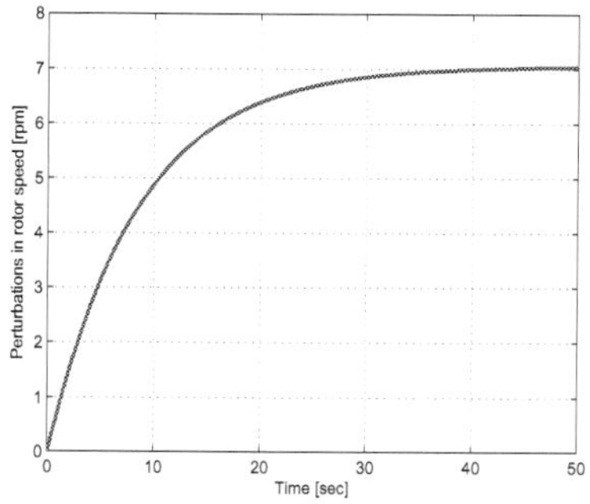

Fig. 8. Perturbations in rotor speed with no control

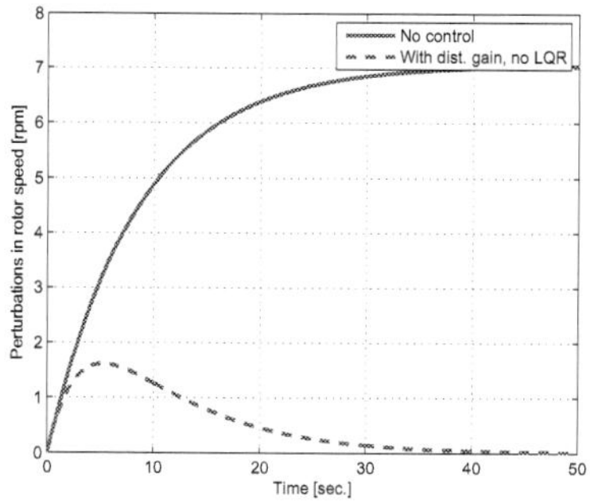

Fig. 9. Perturbations in rotor speed with disturbance gain

As can be seen in this plot does G_D alone manage to regulate the speed, but with a long settling time.

4.3.3 Simulation of DAC with LQR

The final DAC approach will involve a LQR design. Hence, this design will be described in the following, with reference to the theories on LQR presented in Section 3.3. The initial parameters of the wighting functions Q and R are chosen arbitrarily. The next task is then to simulate to check whether the results correspond with the expected performance. After an iterative study while changing Q and R values, the following weighting matrices were chosen:

$$Q = \begin{bmatrix} 100 & 0 & 0 \\ 0 & 0 & 0 \\ 0 & 0 & 100 \end{bmatrix} \tag{41}$$

The control weight of the performance index R was set to 1.
The chosen values in Q will result in a relatively large penalty of the states x_1 and x_3. This means that if x_1 or x_3 is large, the large values in Q will amplify the effect of x_1 and x_3 in the optimization problem. Since the optimization problem are to minimize J, the optimal control u must force the states x_1 and x_3 to be small (which make sense physically since x_1 and x_3

represent the of the perturbations in rotor and generator speed, respectively). On the other hand, the small R relative to the max values in Q involves very low penalty on the control effort u in the minimization of J, and the optimal control u can then be large. For this small R, the gain K can then be large resulting in a faster response.
Since the system has been found to be observable the estimator poles can be arbitrarily placed. Hence, the poles were chosen to be at -10, -30, and -50 yielding the estimator gain matrix

$$K = \begin{bmatrix} 17,235 \\ 1.41 \cdot 10^8 \\ 89.626 \end{bmatrix} \tag{42}$$

The disturbance state estimator gain K_D is a scalar and chosen to 10.
It was found when doing these simulations and checking the results, that an exponential LQR were more appropriate for the current application. The exponential LQR gain were found by applying the code:

```
P=are(A+alpha*eye(3),inv(R)*B*B',Q);
Gdac=inv(R)*B'*P;
```

where alpha is the exponential time-constant chosen to 4. This resulted in the gain matrix:

$$Gdac = \begin{bmatrix} -18.2087 & -1.2265 \cdot 10^{-5} & -0.0425 \end{bmatrix} \tag{42}$$

Simulation of the time response of perturbations in rotor speed for ten seconds is shown in Figure 10.

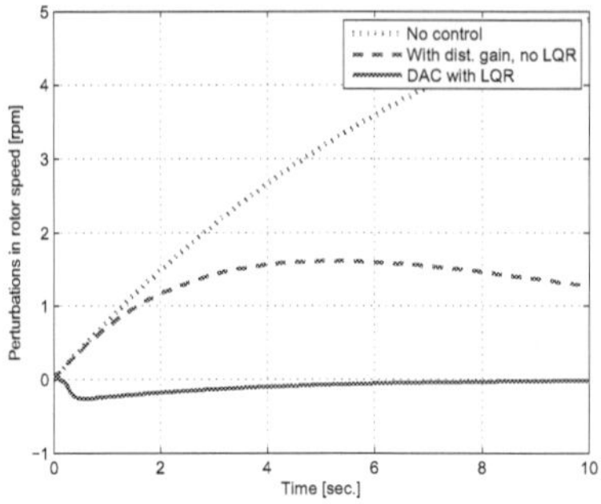

Fig. 10. Comparison of perturbations in rotor speed when applying DAC with LQR

When looking at the plots in Figure 10, it is clear that the disturbance now is much better attenuated while comparing with the first two scenarios. The reason why the speed have this undershoot is due to the right complex plane zeros the LQR controller gives to the closed loop system.
It is also of interest to see how the control signal changes in the three previous situations. This is shown in Figure 11.
As can be seen from Figure 11 does the pitch angle end up having a higher value. This is an expected situation since it is now assumed a constant wind disturbance which must be

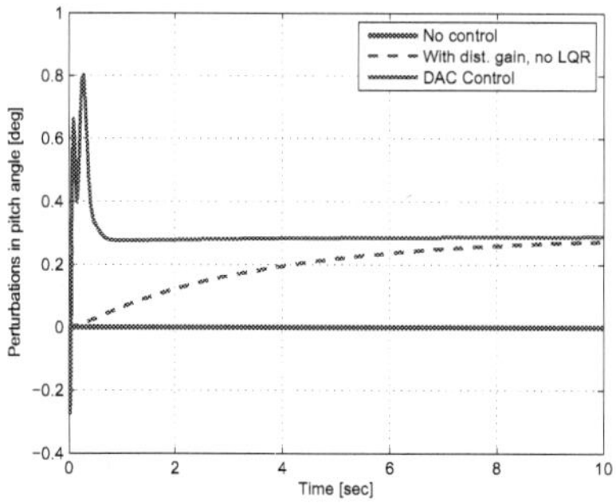

Fig. 11. Comparison of control signal for three previous scenarios

accounted for by an increase in pitch angle. Otherwise, the step in the wind would result in a higher rotor speed, a situation we do not want to occur and the very reason for having a control system in the first place.
Another interesting signal is the error signal, which is the discrepancy between the "measured" output signal y and the estimated signal $\hat{y}$. This is shown in Figure 12 below.

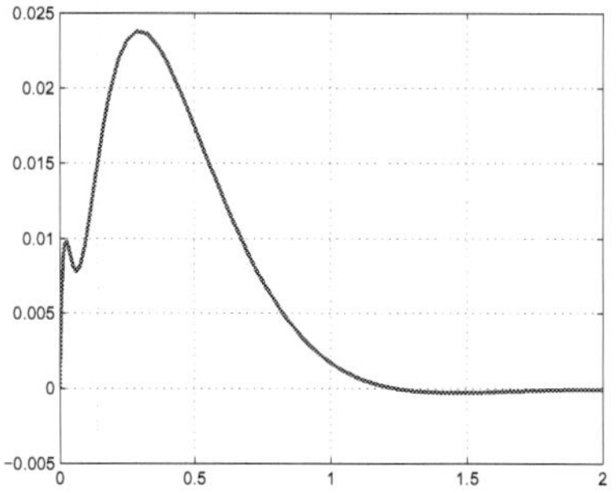

Fig. 12. Error signal for a step input

The DAC simulations in this section have shown that this control method enables to mitigate disturbance due to a wind assigned to affect the rotor state. However, the simulations show that the settling time is relatively long (approximately 10 sec). Whether it is possible to obtained a "better" result using a LQG controller, will be the task in the following section.
The disturbance will now be replaced by two step functions and the disturbance vector Γ in order to make the DAC more comparable with the LQG analysis. This gives the disturbance input signal as shown in Figure 13.
The output response when applying this disturbance model is given in Figure 14

4.4 LQG controller

The control objective is the same as for the DAC; make the perturbations in the rotor speed as small as possible in order to attenuate the wind disturbance. The wind disturbance is made

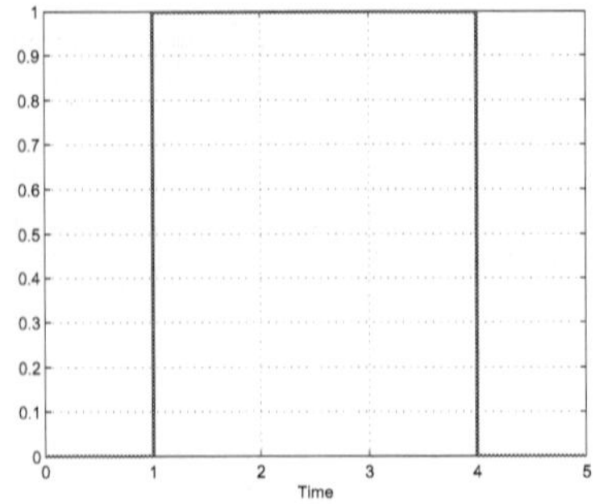

Fig. 13. Step signal for disturbance modeling

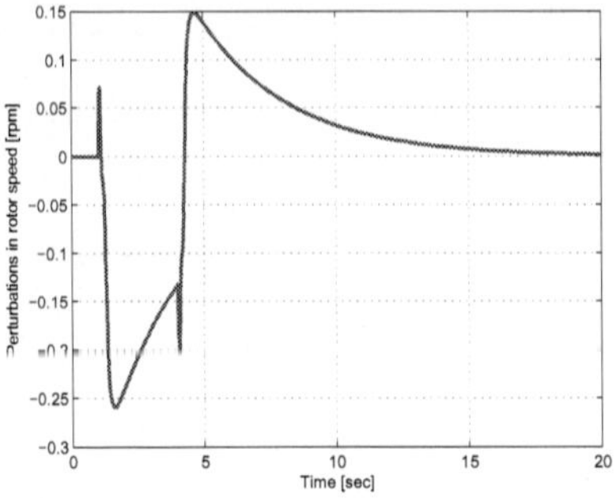

Fig. 14. Output response when applying step input

by using the combination of two step functions as described earlier. This will give a step input affecting the rotor state. Firstly, a situation without control is showed in Figure 15.

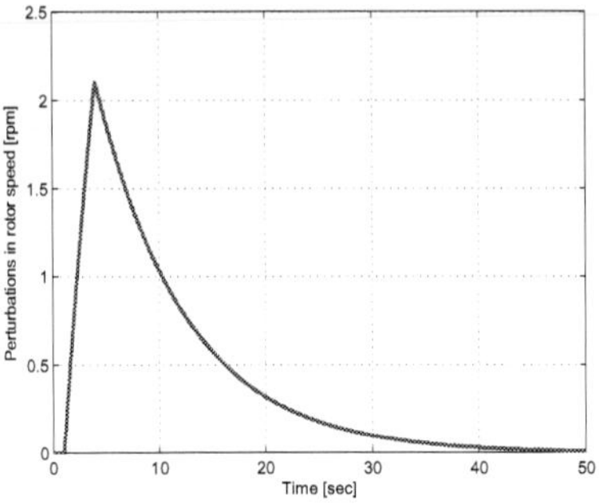

Fig. 15. Response in rotor speed without any control

The sudden change in the rotor speed after one second is due to the step input function. The figure show that without any control will the rotor speed increase to approximately 2 rpm

above the operational speed. After four seconds, when the disturbance is set to zero again, will the speed then decrease and be brought to zero after approximately 50 seconds.
Calculation of the Kalman gain used in the LQG controller is according to the theory discussed in Section 3. There are different methods to do this calculation in MATLAB. One is by applying the `kalman` function, and another is the `lqrc` function. The latter is for instance used in the LTR demo provided with the Robust Control Toolbox in MATLAB. When applying the same weighting as for the LQR in the DAC approach in the `lqrc` function, and the weighting functions

$$Q_n = \begin{bmatrix} 30 & 0 \\ 0 & 30 \end{bmatrix}$$
$$R_n = 1$$

representing the process and measurement noise covariances in the `kalman` function, did the `kalman` function yield much better disturbance attenuation for a given control effort (shown in Figure 16). The corresponding load attenuation for the two cases is shown in Figure 17 below

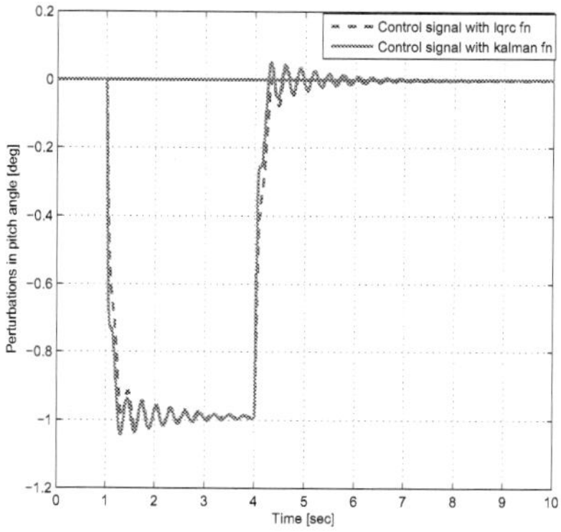

Fig. 16. Comparison of control effort while applying `kalman` and `lqrc` functions in MATLAB

The differences in the output response can be seen through the Kalman gain matrix, which when using the `lqrc` function were calculated to

$$\mathrm{Kf} = \begin{bmatrix} 10.38 \\ -1.76 \cdot 10^{-6} \\ -1.65 \end{bmatrix} \tag{40}$$

Similarly, the Kalman gain matrix when using the `kalman` function is

$$\mathrm{Kf2} = \begin{bmatrix} 57.60 \\ 3.87 \cdot 10^{-6} \\ 1.07 \end{bmatrix} \tag{40}$$

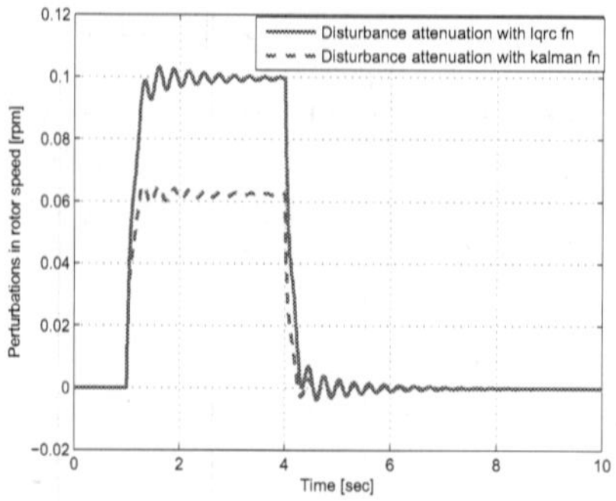

Fig. 17. Comparison of rotor speed for similar control effort while applying `kalman` and `lqrc` functions in MATLAB

The LQR gains were calculated with an exponential LQR with a time-constant α set to 10, and were given as

$$G = \begin{bmatrix} -25.78 & -2.68 \cdot 10^{-5} & -5.70 \end{bmatrix} \tag{40}$$

The perturbation in rotor speed when comparing the LQG with the DAC response is shown in Figure 18.

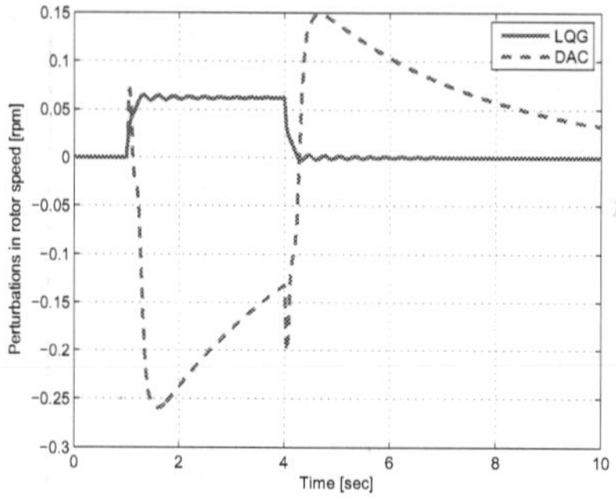

Fig. 18. Comparison of perturbations in rotor speed for LQG and DAC

Figure 18 makes it clear that the effect of the two different control methods is quite different when comparing the two output signals. The LQG controller will not force the disturbance to zero in the same way as the DAC does, which must be regarded as a disadvantage of the LQG. However, the DAC gives a much larger overshoot which is not very positive seen from a controller's perspective.
Similarly, the perturbations in pitch angle for the two methods is shown in Figure 19.
This figure shows that the control effort is much higher for the DAC approach than for LQG. To improve the robustness properties of the LQG will the LTR approach now be introduced as a way to ensure good robustness in spite of uncertainties in the plant model.

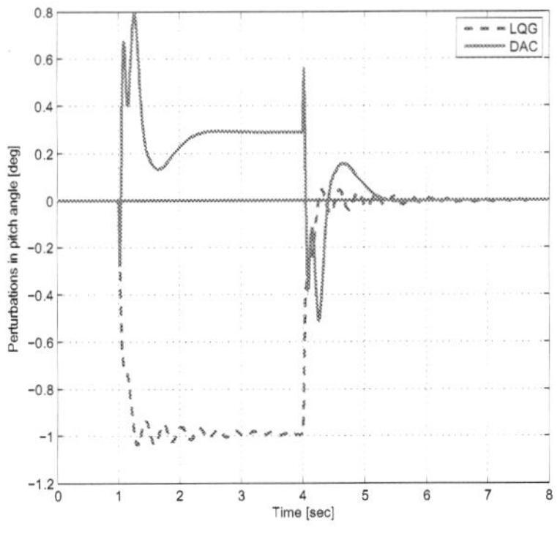

Fig. 19. Comparison of perturbations in pitch angle for the LQG and DAC

5. Conclusions and further works

The purpose of this thesis has been to reduce the loads on a WT for above rated wind speeds (Region III) by speed control when applying different controlling methods. This was first performed by pitch regulation with DAC. This regulation policy showed to have some regulation capacity, but also resulted in some bias in the control signal. It was found when utilizing the DAC controller that although the wind disturbance was well mitigated, the settling time was relatively long(approx. 10 sec). It could have been interesting to extend the work also to include torque regulation in below rated wind speeds. This regulation policy would aim at mitigation of speed and torque variations due to wind disturbances in Region II. The LQG regulator showed to give good speed attenuation, but since DAC and LQG is quite different approaches were a comparison between them not possible. It would have been of major interest to extend the work to also considering pitch actuator constraints to see how this would have affected the results, especially the control input signal.

6. Acknowledgment

This work has been (partially) funded by Norwegian Centre for Offshore Wind Energy (NORCOWE) under grant 193821/S60 from Research Council of Norway (RCN). NORCOWE is a consortium with partners from industry and science, hosted by Christian Michelsen Research.

7. References

Archer, C. L. & Jacobson, M. Z. (2005). Evaluation of global wind power, *Journal of Geophysical Research* Vol. 110: p. 17.

Athans, M. (1981). The Linear Quadratic LQR problem, *Massachusets Institute of Technology* .

Balas, M., Lee, Y. & Kendall, L. (n.d.). Disturbance tracking control theory with application to horizontal axis wind turbines, *Proceeding of the 1998 ASME Wind Energy Symposium, Reno, Nevada, 12-15 January* pp. 95–99.

Bianchi, F., Mantz, R. & Christiansen, C. (2004). Power regulation in pitch-controlled variable-speed WECS above rated wind speed, *Renewable Energy* Vol. 29(No. 11): pp. 1911–1922.

Bossanyi, E. A. (2000). The design of closed loop controllers for wind turbines, *Wind Energy* Vol.3: pp.149–163.

Bossanyi, E. A. (2003). Individual blade pitch control for load reduction, *Wind Energy* Vol. 6: pp.119–128.

Bossanyi, E. A. (2004). Developments in individual blade pitch control, *EWEA conference-The Science of Making Torque from Wind DUWIND Delft University of Technology April 19-21 2004* .

Bottasso, C. & Croce, A. (2009). Cascading kalman observers of structural flexible and wind states for wind turbine control, *Technical report*, Dipartimento di Ingegneria Aerospaziale, Politecnico di Milano, Milano, Italy.

Bottasso, C., Croce, A., Devecchi, D. & Riboldi, C. (2010). Multi-layer control architecture for the reduction of deterministic and non-deterministic loads on wind turbines.

Bottasso, C., Croce, A. & Savini, B. (2007). Performance comparison of control schemes for variable-speed wind turbines, *Journal of Physics: Conference Series 75, The Science of Making Torque from Wind* .

Burns, R. S. (2001). *Advanced Control Engineering*, Butterworth Heinemann.

Connor, B., Iyer, S., Leithead, W. & Grimble, M. (1992). Control of a horizontal axis wind turbine using H-infinity control, *First IEEE Conference on Control Applications*, pp. 117–122.

Gu D.-W., Petkov P.Hr. & Konstantinov, M. (2005). *Robust Control Design With MATLAB*, Springer.

Ekelund, T. (1997). *Modeling and Linear Quadratic Optimal Control of Wind Turbines*, PhD thesis, Chalmers University of Technology, Gothenburg, Sweden.

Elliot, A. S. & Wright, A. D. (2004). Adams/wt: An industry-specific interactive modeling interface for wind turbine analysis, *Wind Energy* .

Henriksen, L. C. (2007). Model Predictive Control of a Wind Turbine, *Informatics and Mathematical Modelling - Technical University of Denmark* .

Johnson, C. D. (1976). Theory of Disturbance- Accommodating Controllers, *Advances in Control and Dynamic Systems* Vol. 12: pp. 387–489.

Karimi-Davijani, H., Sheikholeslami, A., Livani, H. & Karimi-Davijani, M. (2009). Fuzzy logic control of doubly fed induction generator wind turbine, *World Applied Sciences Journal* Vol. 6 (4): pp. 499–508.

Lackner, M. A. & van Kuik, G. A. M. (2010). The performance of wind turbine smart rotor control approaches during extreme loads, *Journal of Solar Energy Engineering* Vol. 132(No.1).

Laks, J. H., Pao, L. Y. & Wright, A. D. (2009). Control of wind turbines: Past, present, and future, *University of Colorado at Boulder, USA* .

Lee, C.-H. (2004). Stabilization of nonlinear nonminimum phase systems: Adaptive parallel approach using recurrent fuzzy neural network, Vol. 34(No. 2): pp. 1075–1088.

Lee, J. & Kim, S. (2010). Wind power generations impact on peak time demand and on future power mix, *Green Energy and Technology* 3: 108–112.

Liebst, B. (1985). A pitch control system for the KaMeWa Wind Turbine, *Journal of Dynamic Systems and Control* Vol. 107(No.1): pp.46–52.

Liu, J.-H., Xu, D.-P. & Yang, X.-Y. (2008). Multi-objective power control of a variable speed wind turbine based in h infinite theory, *International Conference on Machine Learning and Cybernetics*.

Jonkman & Buhl Jr M.L. (2005). Fast's user guide, *National Renewable Energy Laboratory* .

Morimoto, H. (1991). Adaptive LQG regulator via the Separation Principle, *IEEE Transactions on Automatic Control* VOL. 35(No. I): 85–88.

Saberi, A., Chen, B. M. & Sannuti, P. (1993). *Loop Transfer Recovery, Analysis and Design*, Springer.

Mattson S.E. (1984). *Modeling and Control of Large Horizontal Axis Wind Power Plants*, PhD thesis, Lund Institute of Technology, Lund, Sweden.

Selvam, K. (2007). Individual pitch control for large scale wind turbines.

Waltz, E. (2008). Offshore wind may power the future, *Scientific American* .

Wilson, D. G., Berg, D. E., Barone, M. F., Berg, J. C., Resor, B. R. & Lobitz, D. W. (2009a). Active aerodynamic blade control design for load reduction on large wind turbines, *European Wind Energy Conference and Exhibition, 16-19 March*.

Wilson, D. G., Berg, D. E., Barone, M. F., Berg, J. C., Resor, B. R. & Lobitz, D. W. (2009b). Active aerodynamic blade control design for load reduction on large wind turbines, *AWEA Wind Power Conference* .

Wilson, D. G., Berg, D. E., Resor, B. R., Barone, M. F. & Berg, J. C. (2009). Combined individual pitch control and active aerodynamic load controller investigation for the 5mw upwind turbine, *AWEA Wind Power Conference* .

Wright, A. D. (2004). Modern control design for flexible wind turbines, *Technical report*, National Renewable Energy Laboratory.

Wright, A. D. & Fingersh, L. J. (2008). Advanced control design for wind turbines part i: Control design, implementation, and initial tests, *Technical report*, National Renewable Energy Laboratory.

Wright, A. & Stol, K. A. (2008). Designing and testing controls to mitigate dynamic loads in the controls advances research turbine, *Conference Paper 2008 ASME Wind Energy Symposium* .

Vibration Analysis of an Oil Production Platform Submitted to Dynamic Actions Induced by Mechanical Equipment

José Guilherme Santos da Silva, Ana Cristina Castro Fontenla Sieira, Luciano Rodrigues Ornelas de Lima and Bruno Dias Rimola
State University of Rio de Janeiro (UERJ)
Brazil

1. Introduction

Structural engineers have long been trying to develop solutions using the full potential of its composite materials. At this point there is no doubt that the structural solution progress is directly related to an increase in materials science knowledge.
On the other hand, the competitive trends of the world market have long been forcing structural engineers to develop minimum weight and labour cost solutions. A direct consequence of this new trend design is a considerable increase in problems related to unwanted floor vibrations. For this reason, the structural floors systems can become vulnerable to excessive vibrations, for example, produced by impacts such as mechanical equipment (rotating machinery) (Rimola, 2010; Rimola, 2010a; Rimola, 2010b).
This way, the present paper investigated the dynamic behaviour of an oil production platform made of steel and located in Santos basin, São Paulo, Brazil. The structural model consists of two steel decks with a total area of 1915 m^2 (upper deck: 445 m^2 and lower deck: 1470 m^2), supported by vertical sections made of tubular steel members (steel jacket), and piled into the seabed. A variety of mechanical equipment was located on the steel decks of the structural model, related to electrical generators and compressors (Rimola, 2010).
The soil representation was based on the Winkler's Theory (Winkler, 1867). This theory simulates the soil behaviour as a group of independent springs, governed by the linear-elastic model. In the Winkler's model, the soil stiffness was considered as the necessary pressure to produce a unitary displacement (Winkler, 1867).
The proposed computational model, developed for the oil production platform dynamic analysis, adopted the usual mesh refinement techniques present in finite element method simulations implemented in the GTSTRUDL program (GTSTRUDL, 2009). In this finite element model, floor steel girders and columns were represented by three-dimensional beam elements, where flexural and torsion effects were considered. The steel decks were represented by shell finite elements. In this investigation, it was considered that both structural elements (steel beams and steel deck plates) have total interaction with an elastic behaviour.
The structural model dynamic response was determined through an analysis of its natural frequencies and peak accelerations. The results of the dynamic analysis were obtained from

an extensive numerical study, based on the finite element method using the GTSTRUDL program (GTSTRUDL, 2009). In this investigation, dynamic loadings coming from the rotating machinery (electrical generators and compressors) were applied on the steel decks of the structural system (production platform).
A numerical analysis was performed, in order to obtain the dynamic impacts on the deck structure coming from the electrical generators and compressors. Based on the peak acceleration values, obtained on the structure steady-state response, it was possible to evaluate the structural model performance in terms of human comfort, maximum tolerances of the mechanical equipment and vibration serviceability limit states of the structural system, based on the design code recommendations (CEB 209/91, 1991; ISO 1940-1, 2003; ISO 2631-1, 1997; ISO 2631-2, 1989; Murray et al., 2003).

2. Vibration analysis of steel floors

Besides the evaluation of the structural systems behaviour when submitted to dynamic loads, the causes and effects of vibration on people have been subject of many studies and experiments, due to the fact that they affect people in different ways, causing discomfort, health problems, reduced ability concentration and efficiency at work or sickness, in the case of vibrations of very low frequencies.
(Reiher & Meister, 1946) developed a scale used to determine levels of acceptable vibration in floors. This scale was developed based on experiments in which a group of people was submitted to vibration, whose frequency varied from 1 Hz to 100 Hz. According to this scale, the vibration levels can be classified into several levels, depending on the amplitude and frequency.
(Srinivasulu & Vaidyanathan, 1976) presented the principles of analysis, design and construction of machines of different types. The authors investigated several factors to be considered during the design of machine foundations, in order to obtain the best solution, leading to better operation and reduce the undesirable effects of vibrations on the structure.
(Bachmann & Ammann, 1987) have studied the necessary procedures for the analysis of structures under dynamic loads coming from machines, including machines with rotating parts. The authors treat from the load formulation, also dealing with the effects of machinery induced vibrations in structures and measures to avoid such a problem. In their work are also included the acceptance criteria, both from the point of view of the structure and the point of view of human comfort.
(Griffin, 1996) indicates some reasons to measure human exposure to vibration, especially the following: development of standardized documentation on vibrations in the human body; determination of vibration levels and reduction in frequency range that can prejudice the human body and providing data that may be used for comparison between two or more occupational environments.
(Lenzen, 1996) observed during the development of his research, that the scale developed by (Reiher & Meister, 1946) did not take into account the influence of damping on the human perception of vibration. Through his studies in the laboratory, it was modified the scale of (Reiher & Meister, 1946). This modified scale showed satisfactory results in floors with damping ratios of up to 5%.
According to (Vasconcelos, 1998), establishing the concept of human vibration discomfort can be a difficult task. There are several factors that can influence the subjective feeling of discomfort, such as the socio-cultural people, the type of activity performed, the person's

psychological state at the time of the event, environmental factors, noise, etc. It is not easy to simulate these conditions in the laboratory to reduce the variability of individual responses. Thus, the limit of comfort of people subjected to vibration can be regarded as a rather subjective measure, generating some controversy as to the acceptable values of accelerations imposed.

(Zhou & Shi, 2001) considered that the elimination of vibration of rotating machinery is an important engineering problem. In their study, they presented a detailed review of the developed research that deals with the active balancing of rotors in real time and active control of vibration of rotating machinery, as well as dynamic modelling and analysis techniques for rotating systems. The authors report that the major problem found by the scheme of active control of vibration is the limited number of actuators to control an unlimited number of vibration modes.

(Pereira, 2005) presents a study on the vibration related to human comfort and perception, focusing on the suitability of buildings for vibration levels, aiming the generation of curves related to the perception and human comfort and vibration by means of laboratory experiments and comparing the results to the limits of vibration to other investigations and the design codes (ISO 2631-2, 1989).

The experimental tests developed by (Pereira, 2005) considered 30 volunteers, 15 men and 15 women exposed to vertical vibration at a frequency band ranging from 12 to 80 Hz in sitting and standing posture. The author also performed an analysis on the uncertainty of the outcome of the limit of perception, verifying the existence of a range of vibration in which individuals are not sure whether or not they are able to detect the vibration. It also aimed to know the vertical vibration levels that people find uncomfortable at their home environment, to determine the relationship between the perception threshold and comfort. According to these results, it was proved that the reduction in amplitude of the movement to higher frequency vibration becomes more difficult to be detected, reducing the sensitivity of people.

(Milet, 2006) discusses the basic concepts of dynamic analysis of machine foundations, investigating some analytical strategies and numerical methods available for designing. In this work, some design recommendations were presented and discussed.

(Souza et al., 2007) developed a prototype that allows, looking through a simple system, based on an unbalanced rotor, possible structural data caused by the resonance phenomenon, also allowing comparisons to be made with more complex structural systems. Furthermore, the experiment presented as being practical and simple, can serve as an analytical tool in the classroom, thus giving a better understanding of phenomena related to the vibration system.

(Assunção, 2009) addressed the issues and the most important conditions for a dynamic analysis of elevated frame structures, where equipment were allocated for industrial processes. The author developed a study related to the main causes of dynamic actions coming from industrial equipment and examined a framed structure supporting an unbalanced machine. Through this investigation, the author demonstrated that the developed computational model was appropriate to simulate the transmission of efforts and contribution of the vibrating mass on the responses of the framed structures.

3. Loads generated by mechanical equipment (rotating machinery)

Knowledge about the dynamic behaviour of rotors of rotating machinery, even in the design phase, has become an increasingly crucial factor, considering that it is not desirable to take corrective actions after the beginning of activities. Material costs and delivery of services are relatively high when compared with the profits increasingly reduced according to the rules imposed by the market and the fact that such corrective actions still involve a period in which the equipment will need to become inoperative, which means losses, because it will not generate any profit for the period.

According to (Dias Junior, 2009), among several factors that contribute to the transmission of rotational energy to the vibratory movements of the machine, the well-known is undoubtedly the unbalance of the rotor. The rotor is the rotating part of a machine or engine which may be coupled elements as disks, generators, gears, etc. According to (Dias Junior, 2009), due to the unbalance, the force that acts at the rotor gravity centre, pull the shaft out of the line joining the two bearings, forcing the shaft to rotate stressed. This movement is called precession movement.

Rotors are supported on bearings, which are the elements responsible for connecting the movable and fixed structure of a rotating machine. In addition to this point, to absorbing energy, another function of the bearings is to guide or restrict movement during the rotation axis (Silva, 2004).

The process of balancing a rotor is a key factor to minimize the vibrations generated by electric motors. Depending on the vibration level of these engines, the structural system that supports the equipment can be compromised by fatigue or even premature failure. The balancing process is intended to improve the distribution of mass of a body, so that by turning around their bearings, produces no unbalance forces, keeping the vibrations and dynamic loads within suitable limits for the machine operation.

The balance can be achieved up to a certain limit, since after this process the rotor still presents imperfection in the mass distribution, called residual unbalance. It is worth mentioning that there is a direct relationship between the residual unbalance and vibration level of the machine, which depends on many factors (mass housing and the foundation, stiffness of the bearings and foundation, occurrence of resonances etc.). Anyway, there are allowable levels of residual unbalance, consistent with good practice of machine design.

3.1 Excitation forces: Unbalanced mass

Unbalanced mass is defined as a mass located at a distance d measured from the geometric centre of the shaft. The mass remains in a plane perpendicular to the axis y and it is a constant coordinate, as illustrated in Figure 1.

Based on Figure 1, it can be deduced that the force caused by unbalanced mass, acting on the axis, according to directions in the shown coordinate system, can be written as presented in Equations (1) and (2). In Equations (1) and (2), the dynamic forces generated by unbalanced mass have a frequency similar to the rotational frequency of the axis

$$F_u = m_u \Omega^2 d \cdot \sin(\Omega t) \tag{1}$$

$$F_w = m_u \Omega^2 d \cdot \cos(\Omega t) \tag{2}$$

3.2 Unbalance quantification

As previously mentioned, the unbalance is characterized by a mass located at a certain distance from the rotation axis, see Figure 1. Therefore, the unbalance is always measured by a product mass x distance, see Equation (3). The rotor must be subjected to a balancing procedure, in order to achieve a tolerable minimum. This value is called the Permissible Residual Unbalance and designated by the symbol U.

$$U = \text{mass} \times \text{distance} \tag{3}$$

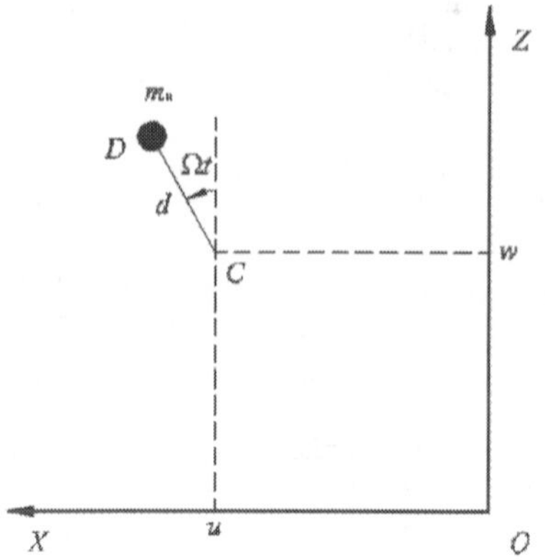

Fig. 1. Unbalanced mass (López, 2002)

It was observed in Equation (3), that the allowable residual unbalance is directly proportional to the mass of the rotor, i.e., as heavier was the rotor, greater it will be the permissible residual unbalance. It is appropriate to relate the allowable residual unbalance, U, and the rotor mass, m, in terms of the Specific Allowable Residual Unbalance, as shown in Equation (4).

$$e = \frac{U}{m} \tag{4}$$

As larger is the rotation speed, smaller should be the residual unbalance, since the centrifugal force, F_{cent}, increases with the square of the speed of it, as shown in Equation (5). In Equation (5) the centrifugal force, F_{cent} is expressed in N.

$$F_{cent} = m \cdot e \cdot \Omega^2 \tag{5}$$

Based on many years of experience, experts decided that the product of the angular velocity of rotor rotation and the specific allowable residual unbalance must be constant, i.e., to increase the speed of rotation it is necessary to reduce the specific residual unbalance.
This product is called Balance Quality Grade and it is designated by the letter G, see Table 1. To find a wide variety of existing rotors it was necessary to assign, depending on the type of rotor and its application, a value for this constant, see Table 1. Table 1 reproduces the G values which deals with quality of balancing rotating rigid bodies (ISO 1940-1, 2003).

3.3 Determination of unbalanced forces

As mentioned before, the unbalance of the rotor produces a dynamic load that depends on the mass, the equipment angular velocity and the eccentricity between the equipment

G	e x ω (mm/s)	Rotor Types - General Examples
G 4000	4000	Crankshaft/drives of rigidly mounted slow marine diesel engines with uneven number of cylinders.
G 1600	1600	Crankshaft/drives of rigidly mounted large two-cycle engines.
G 630	630	Crankshaft/drives of rigidly mounted large four-cycle engines; Crankshaft/drives of elastically mounted marine diesel engines.
G 250	250	Crankshaft/drives of rigidly mounted fast four-cylinder diesel engines.
G 100	100	Crankshaft/drives of fast diesel engines with six or more cylinders; Complete engines (gasoline or diesel) for cars, trucks and locomotives.
G 40	40	Car wheels, wheel rims, wheel sets, drive shafts; Crankshaft/drives of elastically mounted fast four-cycle engines with six or more cylinders; Crankshaft/drives of engines of cars, trucks and locomotives.
G 16	16	Drive shafts (propeller shafts, cardan shafts) with special requirements and parts of crushing machines; Individual components of engines (gasoline or diesel) for cars, trucks and locomotives; Crankshaft/drives of engines with six or more cylinders under special requirements.
G 6.3	6.3	Parts of process plant machines; Marine main turbine gears (merchant service); Centrifuge drums; Paper machinery rolls; print rolls; Fans; Assembled aircraft gas turbine rotors; Flywheels; Pump impellers; Machine-tool and general machinery parts; Medium and large electric armatures (electric motors having at least 80 mm shaft height) without special requirements; Small electric armatures, often mass produced, in vibration insensitive applications and/or with vibration-isolating mountings; Individual components of engines under special requirements.
G 2.5	2.5	Gas and steam turbines, including marine main turbines (merchant service); Rigid turbo-generator rotors; Computer memory drums and discs; Turbo-compressors; Machine-tool drives; Medium and large electric armatures with special requirements; Small electric armatures not qualifying for one or both of the conditions specified for small electric armatures of balance quality grade G 6.3; Turbine-driven pumps.
G 1	1	Tape recorder and phonograph (gramophone) drives; Grinding-machine drives; Small electric armatures with special requirements.
G 0.4	0.1	Spindles, discs and armatures of precision grinders; Gyroscopes.

Table 1. Balance quality values (ISO 1940-1, 2003)

gravity center and the rotation axis. In sequence, Equation (6) determines the dynamic load amplitude generated by the unbalance of a rotor, as follows:

$$P_0 = m \cdot R \cdot \Omega^2 = m \cdot (R\Omega) \cdot \Omega \tag{6}$$

Where:

P_0 : dynamic loading amplitude;

M : total mass in rotation;

Ω : equipment frequency;

R. Ω = G : equipment balance quality grade, see Table 1: (ISO 1940-1, 2003).

For rotors of electric motors, R.Ω was considered equal to 0.0025 m/s. Substituting this value in Equation (6), it is obtained:

$$P_0 = m \cdot (0.0025) \cdot \Omega \tag{7}$$

Considering an unbalanced load spinning around an axis, the procedure for obtaining the global dynamic force acting on a plane is to apply the force in two orthogonal directions. One of these forces is applied in the horizontal direction with the angle phase ϕ equal to zero and the other in vertical direction with the angle phase ϕ equal to 1/4 of the period of vibration of unbalanced force.

Thus, as time increases, there is a variation of the two forces so that the composition (horizontal and vertical directions) results in a harmonic unbalanced force where one component will be multiplied by sin (Ωt) and the other by sin (Ωt + π/2). This way, when one harmonic component presents maximum values the other one is equal to zero and vice versa. The value of the dynamic force is obtained by the vector sum of the components in the vertical and horizontal directions as presented in Equation (8).

$$P(t) = P_0 \sin(\Omega t) + P_0 \sin(\Omega t + \frac{\pi}{2}) \tag{8}$$

3.3 Dynamic loading modelling

To perform the numerical analysis of the oil production platform developed in this investigation, it was used the data in accordance with Table 2. In sequence, Figure 2 shows the design of the equipment.

Equipment data	
Protective cover	1.2 kN
Coupling	5.3 kN
Gear unit	37.5 kN
Motor swing	15 kN
Rotor weight	10.8 kN
Input frequency	30 Hz
Output frequency	0.94 Hz

Table 2. Equipment data

The dynamic load modelling considered two components related to vertical and horizontal directions. Table 3 shows the dynamic loads applied on the structural system steel deck. These actions were properly combined in order to better represent the dynamic excitation induced by equipment on the structure.

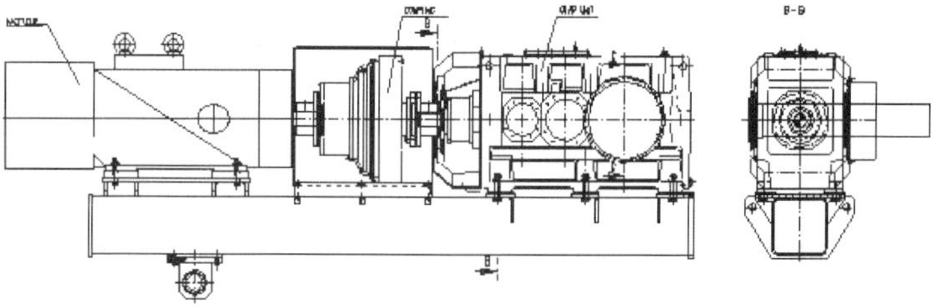

Fig. 2. Driving unit (motor, coupling and gear) supported by a steel beam (Rimola, 2010)

Equipment	Weight (kN)	Frequency (rad/s)	$R\omega$ (m/s)	P_0 (kN)
Rotor	10.80	188.49	0.0025	0.51
Gear	18.75	6.03	0.0025	0.028

Table 3. Dynamic actions related to the equipment

4. Investigated structural model

The structural system investigated in this paper is related to an oil production platform made of steel and located in Santos basin, São Paulo, Brazil. The structural model is supported by vertical sections made of tubular steel members (steel jacket), piled into the seabed by steel piles and consists of two steel decks with a total area of 1915 m^2 (upper deck: 445 m^2 and lower deck: 1470 m^2), see Figures 3 and 4.

Fig. 3. Investigated structural model

The structural model is formed by steel beams and columns and steel deck plates, as presented in Figures 3 and 4. This platform is constituted by a lot of structural elements with very different geometrical characteristics; see (Rimola, 2010). A 2.05x105MPa Young's modulus was adopted for the steel beams, columns and decks.

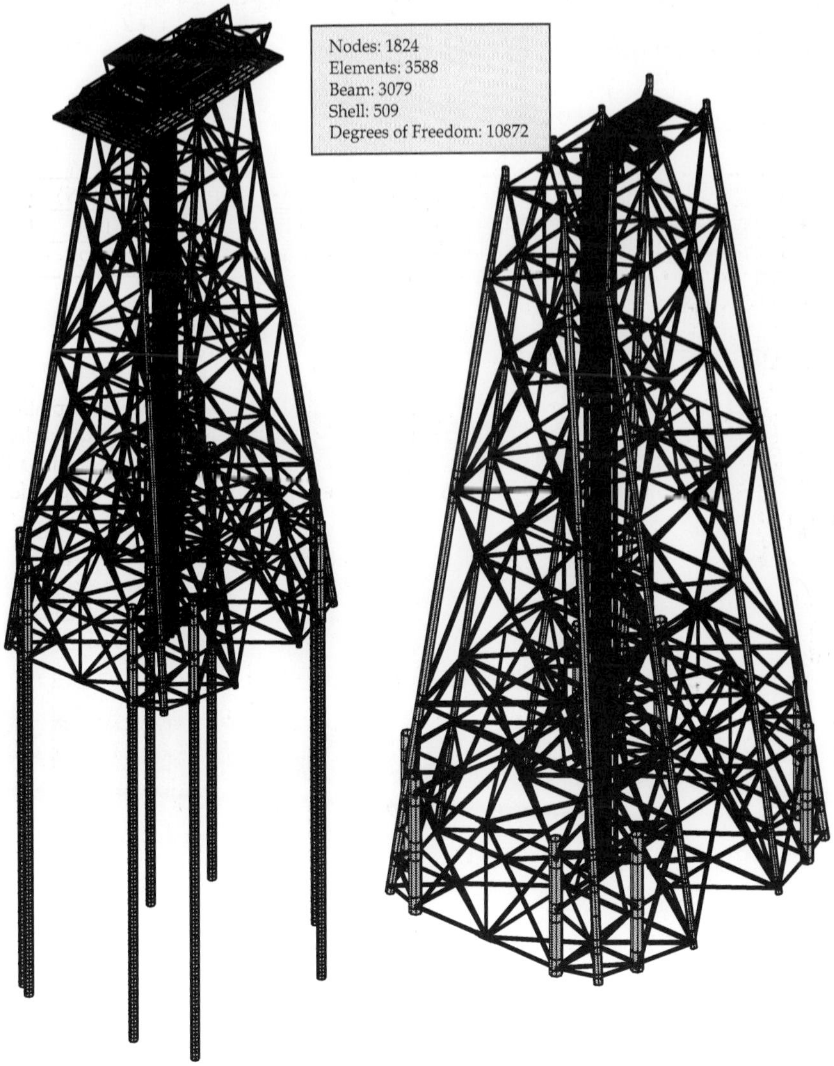

Fig. 4. Finite element model: oil production platform with steel jacket and piles

DEPTH	LAYERS DESCRIPTION	RESISTANCE PARAMETERS
6,090	Medium Sand	$\gamma_{sub} = 9{,}0\ kN/m^3$ $\phi = 30°$
18,280	Stiff Clay	$\gamma_{sub} = 8{,}0\ kN/m^3$ $C = 70{,}75 kN/m^3$
21,940	Fine Sand	$\gamma_{sub} = 8{,}9\ kN/m^3$ $\phi = 30°$
24,990	Medium Clay	$\gamma_{sub} = 8{,}0\ kN/m^3$ $C = 80{,}90 kN/m^2$
34,130	Medium Sand	$\gamma_{sub} = 8{,}9\ kN/m^3$ $\phi = 35°$
42,970	Fine Sand	$\gamma_{sub} = 8{,}9\ kN/m^3$ $\phi = 25°$
54,860	Stiff Clay	$\gamma_{sub} = 8{,}0\ kN/m^3$ $C = 113{,}55 kN/m^2$
67,050	Stiff Clay	$\gamma_{sub} = 8{,}0\ kN/m^3$ $C = 138{,}75 kN/m^2$
79,240	Stiff Clay	$\gamma_{sub} = 8{,}0\ kN/m^3$ $C = 164{,}20 kN/m^2$
103,00	Stiff Clay	$\gamma_{sub} = 8{,}0\ kN/m^3$ $C = 192{,}05 kN/m^2$

Fig. 5. Geotechnical profile

5. Computational modelling

The proposed computational model, developed for the structural system dynamic analysis, adopted the usual mesh refinement techniques present in finite element method simulations implemented in the GTSTRUDL program (GTSTRUDL, 2009).

In this computational model, floor steel girders and columns were represented by a three-dimensional beam element with tension, compression, torsion and bending capabilities. The element has six degrees of freedom at each node: translations in the nodal x, y, and z directions and rotations about x, y, and z axes. On the other hand, the steel deck plates were represented by shell finite elements (GTSTRUDL, 2009).
In this investigation, it was considered that both structural elements (steel beams and steel deck plates) have total interaction with an elastic behaviour. The finite element model has 1824 nodes, 3079 three-dimensional beam elements, 509 shell elements and 10872 degrees of freedom, as presented in Figure 4.

5.1 Geotechnical data

The data related to soil were obtained by means of three boreholes (Standard Penetration Tests: SPT) in a depth range varying from 43.5m to 178.3m. The boreholes allowed the definition of the geotechnical profile (Figueiredo Ferraz, 2004) adopted in the finite element modelling, see Figure 5. Characterisation and resistance tests performed in laboratory provided specific weight, friction angle and cohesion values illustrated in Figure 5. In Figure 5, γ_{sub} is the submerged specific weight (kN/m^3); c represents the soil cohesion in (kN/m^2) and ϕ is soil friction angle (degree).

5.2 Soil-structure interaction

When the study of half-buried columns is considered, the usual methodology to the formulation of the soil-structure interaction problem utilizes the reaction coefficient concept, originally proposed by (Winkler, 1867). In the case of laterally loaded piles, the analysis procedure based on (Winkler, 1867) is analogue to the one used for shallow foundations, see Figure 6.
The soil behaviour is simulated by a group of independent springs, governed by a linear-elastic model. The foundation applies a reaction in the column normal direction and it is proportional to the column deflection. The spring stiffness, designated by reaction coefficient (k_h) is defined as the necessary pressure to produce a unitary displacement (Winkler, 1867), as presented in Equation (9).

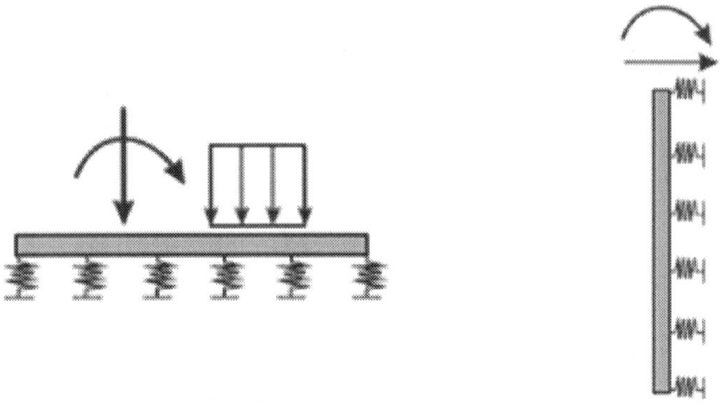

(a) Shallow foundation (b) Laterally loaded pile

Fig. 6. Foundation analysis models

$$k_h = \frac{p}{y} \tag{9}$$

Where:
p : applied pressure, in (N/m^2);
y : soil deflection, in (m).
(Terzaghi, 1955) considered that the reaction coefficient (k_h) for piles in cohesive soils (clays), does not depend on the depth of the pile and may be determined by the following equation.

$$k_h = \frac{0.3048m}{1.5d} \times (k_{s1}) \tag{10}$$

Where:
k_{s1} : modulus to a squared plate with a length of 0.3048m (1 ft), in (MN/m^3);
d : column width (pile), in (m).
Table 4 presents typical values for k_{s1} for consolidated clays.

Clay's consistency	k_{s1} (t/ft^3)	k_{s1} (MN/m^3)
Stiff	75	26.0
Very Stiff	150	52.0
Hard	300	104.0

Table 4. Typical values for k_{s1} (Terzaghi, 1955)

For piles in non-cohesive soils (sand), it is considered that the horizontal reaction coefficient (k_h) varies along the depth, according to Equation (11).

$$k_h = n_h \frac{z}{d} \tag{11}$$

Where:
n_h : stiffness parameter for non-cohesive soils, in (MN/m^3);
d : column width (pile), in (m).
Typical values for n_h obtained by (Terzaghi, 1955), as a function of the sand relative density under submerged and dry condition, are presented in Table 5.

Relative density	n_h (dry) (t/ft^3)	n_h (dry) (MN/m^3)	n_h (submerged) (t/ft^3)	n_h (submerged) (MN/m^3)
Loose	7	2.4	4	1.4
Medium	21	7.3	14	4.9
Dense	56	19.4	34	11.8

Table 5. Typical values for n_h (Terzaghi, 1955)

Based on the horizontal reaction coefficients values (k_h) and the column width (d), the foundation stiffness parameter (k_0) is determined by using Equation (12) (Poulos & Davis, 1980):

$$k_0 = k_h \times d \tag{12}$$

Based on the subsoil geotechnical profile (see Figure 5) and using the analysis procedure based on the Winkler model (Winkler, 1867) the horizontal reaction coefficients on the piles (k_h) were determined as a function of the type of the soil. Applying the horizontal reaction coefficients (k_h) on Equation (12), the foundation stiffness parameters values (k_0) were calculated.

The foundation stiffness parameters values (k_0) were used to determine the spring's stiffness (k) placed in the computational model to simulate the soil behaviour. The spring elements which simulate the piles were discretized based on a range with length equal to 1m (one meter). For each range of 1m it was placed a translational spring in the transversal direction of the pile section axis with a stiffness value equal to the value obtained for the horizontal reaction coefficient evaluated by the Winkler model (Winkler, 1867). Table 6 presents the spring's stiffness coefficients simulating.

Depth (m)	Layer Description	Pile Profile		Spring's stiffness k (kN/m)
		Ø (mm)	Thickness (mm)	
1	Medium sand	2108	55	4850.1163
2				9700.2327
3				14550.3490
4				19400.4653
5				24250.5816
6				29100.6980
7 a 18	Stiff clay	2134	55	5279.6981
19	Fine sand	2134	55	92152.2102
20				97002.3266
21				101852.4429
22				106702.5592
23 a 25	Medium clay	2134	55	281.5839
26	Medium sand	2134	55	126103.0245
27				130953.1408
28				135803.2572
29				140653.3735
30				145503.4898
31				150353.6062
32				155203.7225
33				160053.8388
34				164903.9551
35	Fine sand	2134	55	169754.0715
36				174604.1878
37				179454.3041
38				184304.4204
39				189154.5368
40				194004.6531
41				198854.7694
42				203704.8858
43				208555.0021
44 a 96	Stiff clay	2134	55	5279.6981

Table 6. Stiffness coefficients of soil representative springs

5.3 Structural damping modelling

It is called damping the process which energy due to structural system vibration is dissipated. However, the assessment of structural damping is a complex task that cannot be determined by the structure geometry, structural elements dimensions and material damping (Clough & Penzien, 1995).

According to (Chopra, 2007), it is impossible to determine the damping matrix of a structural system through the damping properties of each element forming the structure with the same way it is determined the structure stiffness matrix, for example. This is because unlike the elastic modulus, which is used in the stiffness evaluation, the damping materials properties are not well established.

Although these properties were known (Chopra, 2007), the resulting damping matrix would not take into account a significant portion of energy dissipated by friction in the structural steel connections, opening and closing micro cracks in the concrete, friction between structure and other elements that are bound to it, such as masonry, partitions, mechanical equipment, fire protection, etc. Some of these energy dissipation sources are extremely difficult to identify.

The physical evaluation of the damping of a structure is not considered properly as if their values are obtained by experimental tests. However, these tests often require time and cost that in most cases is very high. For this reason, damping is usually achieved in terms of contribution rates, or rates of modal damping (Clough & Penzien, 1995).

With this in mind, it is common to use the Rayleigh damping matrix, which considers two main parts, one based on the mass matrix contribution rate (α) and another on the stiffness matrix contribution rate (β), as can be obtained using Equation (13). It is defined **M** as the mass matrix and **K** as the stiffness matrix of the system (Craig Jr., 1981; Clough & Penzien, 1995; Chopra, 2007).

$$\mathbf{C} = \alpha\mathbf{M} + \beta\mathbf{K} \tag{13}$$

Equation (13) may be rewritten, in terms of the damping ratio and the circular natural frequency (rad/s), as presented in Equation (14):

$$\xi_{\mathrm{i}} = \frac{\alpha}{2\omega_{0\mathrm{i}}} + \frac{\beta\omega_{0\mathrm{i}}}{2} \tag{14}$$

Where:

ξ_i : damping ratio related to the ith vibration mode;

ω_{0i} : circular natural frequency related to the ith vibration mode.

Isolating the coefficients α and β from Equation (14) and considering the two most important structural system natural frequencies, Equations (15) and (16) can be written.

$$\alpha = 2\xi_{\mathrm{i}}\omega_{01} - \beta\omega_{01}\omega_{01} \tag{15}$$

$$\beta = \frac{2(\xi_2\omega_{02} - \xi_1\omega_{01})}{\omega_{02}\omega_{02} - \omega_{01}\omega_{01}} \tag{16}$$

Based on two most important structural system natural frequencies it is possible to calculate the values of α and β. In general, the natural frequency ω_{01} is taken as the lowest natural frequency, or the structure fundamental frequency and ω_{02} frequency as the second most important natural frequency.

In the technical literature, there are several values and data about structural damping. In fact, these values appear with great variability, which makes their use in structural design very difficult, especially when some degree of systematization is required. Based on the wide variety of ways to considering the structural damping in finite element analysis, which, if used incorrectly, provide results that do not correspond to a real situation, the design code CEB 209/91 (CEB 209/91, 1991) presents typical rates for viscous damping related to machinery support of industrial buildings, as shown in Table 7.

Construction Type	Minimum	Mean	Maximum
Reinforced Concrete	0.010	0.017	0.025
Prestressed Concrete	0.007	0.013	0.020
Composite Structures	0.004	0.007	0.012
Steel	0.003	0.005	0.008

Table 7. Typical values of damping ratio ξ for industrial buildings (CEB 209/91, 1991)

Based on these data, it was used a damping coefficient of 0.5% (ξ = 0.5% or 0.005) in all modes. This rate takes into account the existence of few elements in the oil production platform that contribute to the structural damping. Table 8 presents the parameters α and β used in the forced vibration analysis to model the structural damping in this investigation.

f_{01} (Hz)	f_{02} (Hz)	ω_{01} (rad/s)	ω_{02} (rad/s)	α	β
0.674	0.716	4.2364	4.4968	0.02172	0.00115

Table 8. Values of the coefficients α and β values used in forced vibration analysis

6. Natural frequencies and vibration modes

The production platform natural frequencies were determined with the aid of the numerical simulations, see Table 9, and the corresponding vibration modes are shown in Figure 7 and 8. Each natural frequency has an associated mode shape and it was observed that the first vibration modes presented predominance of the steel jacket system.

Natural Frequencies f_{0i} (Hz)		Vibration Modes	
f_{01}	0.67	Mode 1	Vibration modes with predominance of the steel jacket system.
f_{02}	0.71	Mode 2	
f_{03}	1.20	Mode 3	
f_{08}	1.99	Mode 8	Vibration modes with predominance of the steel deck displacements.
f_{17}	2.61	Mode 17	
f_{49}	4.14	Mode 49	

Table 9. Production platform natural frequencies

It can be observed in Figure 7, that the three first vibration modes presented predominance of displacements in the steel jacket system. In the 1st vibration mode there is a predominance of translational displacements towards the "y" axis in the finite element model. In the 2nd vibration mode a predominance of translational effects towards the axis "x" of the numerical model was observed. The third vibration mode presented predominance of torsional effects on the steel jacket system with respect to vertical axis "z". Flexural effects were predominant on the steel deck system and can be seen only from higher order vibration modes, see Figure 8.

However, flexural effects were predominant in the steel deck plate (upper and lower), starting from the eighth vibration mode (f_8 = 1.99 Hz - Vibration Mode 8), see Table 9. It is important to emphasize that torsional effects were present starting from higher mode shapes, see Table 9. Figures 7 and 8 illustrated the mode shapes corresponding to six natural frequencies of the investigated structural system.

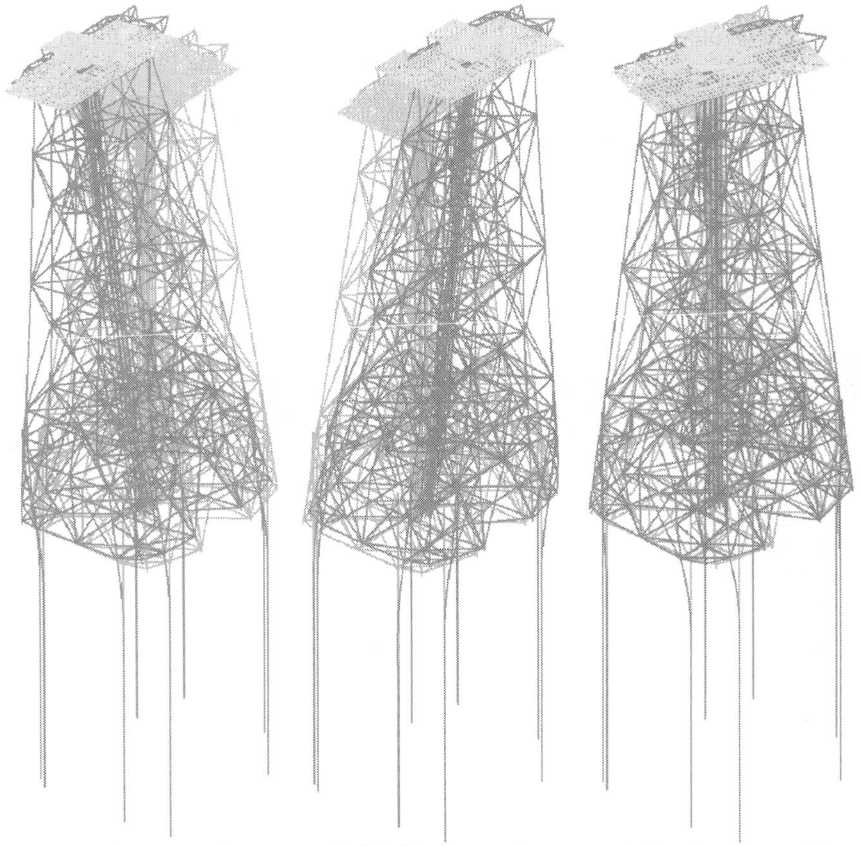

a) 1st vibration mode b) 2nd vibration mode c) 21st vibration mode

Fig. 7. Vibration modes with predominance of the steel jacket system

7. Structural system dynamic response

The present investigation proceeds with the evaluation of the steel platform's performance in terms of vibration serviceability effects, considering the impacts produced by mechanical equipment (rotating machinery). This strategy was considered due to the fact that unbalanced rotors generate vibrations which may damage their components and supports and produce dynamic actions that could induce the steel deck plate system to reach unacceptable vibration levels, leading to a violation of the current human comfort criteria for these specific structures. For this purpose, forced vibration analysis is performed through using the computational program (GTSTRUDL, 2009). The results of forced vibration models are obtained in terms of the structural system displacements, velocities and peak accelerations.

For the structural analysis, it was considered the simultaneous operation of three machines on the steel deck. The nodes of application of dynamic loads in this situation are shown in Figure 9. With respect to human comfort, some nodes of the finite element model were selected near to the equipment in order to evaluate the steel deck dynamic response (displacements, velocities and accelerations). These nodes are also shown in Figure 9.

The dynamic loading related to the rotor was applied on the nodes 9194, 9197 and 9224 and the corresponding dynamic loading associated to the gear was applied on the nodes 9193, 9196 and 9223, as presented in Figure 9. The description of the dynamic loadings (rotor and gear) was previously described on item 3. It should be noted that the positioning of the machines was based on the equipment arrangement drawings of the investigated platform (Figueiredo Ferraz, 2004).

The analysis results were compared with limit values from the point of view of the structure, operation of machinery and human comfort provided by international design recommendations (CEB 209/91, 1991; ISO 1940-1, 2003; ISO 2631-1, 1997; ISO 2631-2, 1989; Murray et al., 2003). It must be emphasized that only the structural system steady-state response was considered in this investigation.

The frequency integration interval used in numerical analysis was equal to 0.01 Hz ($\Delta\omega$ = 0.01 Hz). It was verified that the frequency integration interval simulated conveniently the dynamic characteristics of the structural system and also the representation of the proposed dynamic loading (Rimola, 2010).

In sequence of the study, Tables 10 to 12 present the vertical translational displacements, velocities and accelerations, related to specific locations on the steel deck, near to the mechanical equipment, see Figure 9, calculated when the combined dynamic loadings (rotor and gear) were considered.

These values, obtained numerically with the aid of the proposed computational model, were then compared with the limiting values proposed by design code recommendations (CEB 209/91, 1991; ISO 1940-1, 2003; ISO 2631-1, 1997; ISO 2631-2, 1989; Murray et al., 2003). Once again, it must be emphasized that only the structural system steady state response was considered in this investigation.

The allowable amplitudes are generally specified by manufactures of machinery. When manufacture's data doesn't indicate allowable amplitudes, design guides recommendations (ISO 1940-1, 2003) are used to determine these limiting values for machinery performance, see Table 10. The maximum amplitude value at the base of the driving support (Node 9197: see Figure 9), on the platform steel deck was equal to 446 μm (or 0.446 mm or 0.0446 cm), see Table 10, indicating that the recommended limit value was violated and the machinery performance can be inadequate (0.446 mm > 0.06 mm) (ISO 1940-1, 2003).

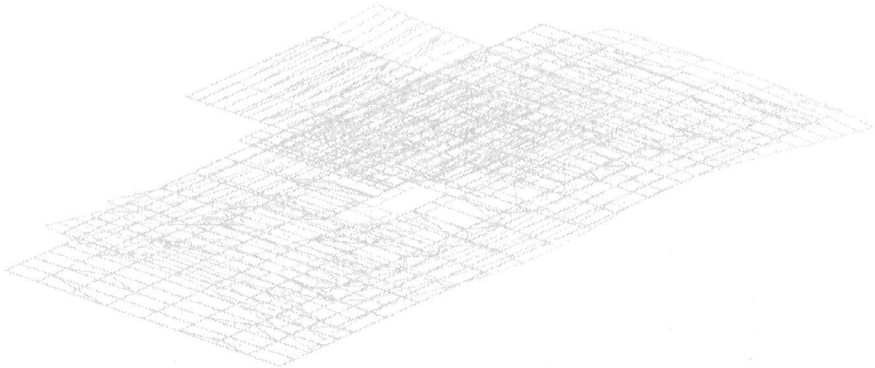

a) 8th vibration mode f_{08}=1.99 Hz

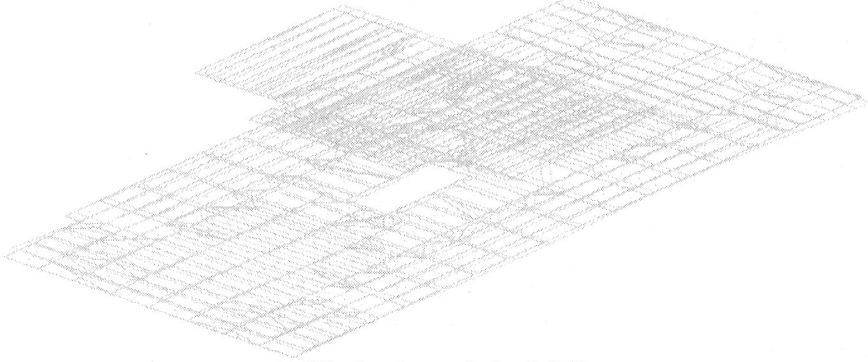

b) 17th vibration mode f_{17}=2.61 Hz

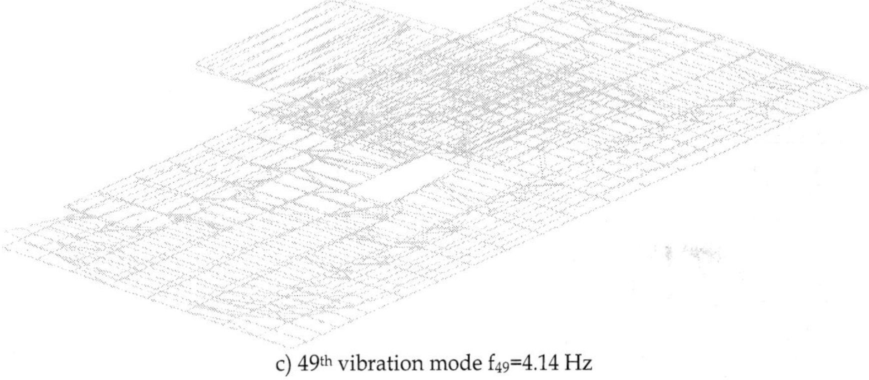

c) 49th vibration mode f_{49}=4.14 Hz

Fig. 8. Vibration modes with predominance of the steel deck system

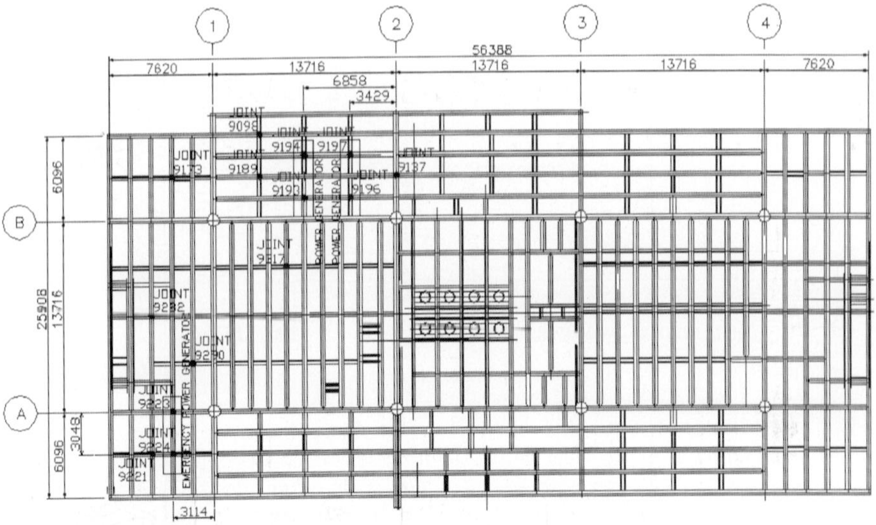

Fig. 9. Selected nodes for application of the dynamic loads

Rotor Support (Node: 9194) (µm)	Rotor Support (Node: 9197) (µm)	Rotor Support (Node: 9224) (µm)	Amplitude Limit Values (µm)
54	**446**	7	40 to 60*
Gear Support (Node: 9193) (µm)	Gear Support (Node: 9196) (µm)	Gear Support (Node: 9223) (µm)	
77	432	38	
* For vertical vibration for high speed machines (>1500 rpm) (ISO 1940-1, 2003).			

Table 10. Vertical displacements related to the combined dynamic loading (driving)

Rotor Support (Node: 9194) (mm/s)	Rotor Support (Node: 9197) (mm/s)	Rotor Support (Node: 9224) (mm/s)	Velocity Limit Values (mm/s)
10.18	**84.12**	11.89	0.70 to 2.80*
Gear Support (Node: 9193) (mm/s)	Gear Support (Node: 9196) (mm/s)	Gear Support (Node: 9223) (mm/s)	
14.49	81.46	7.27	
* Tolerable velocity for electrical motors according to (ISO 1940-1, 2003).			

Table 11. Velocities related to the combined dynamic loading (driving)

The maximum velocity value calculated at the base of the driving support (Node 9197: see Figure 9), on the platform steel deck was equal to 84.12 mm/s, see Table 11. The allowable velocity considering a perfect condition to machinery performance is located in the range of 0.7mm/s to 2.8 mm/s, see (ISO 1940-1, 2003), as presented in Table 11. This velocity value is not in agreement with those proposed by the design codes (84.12 mm/s > 2.8 mm/s) (ISO 1940-1, 2003), violating the recommended limits.

Steel Deck (Node: 9098) (m/s²)	Steel Deck (Node: 9137) (m/s²)	Steel Deck (Node: 9173) (m/s²)	Steel Deck (Node: 9189) (m/s²)	Acceleration Limit Values (m/s²)
5.62	6.16	4.67	**27.33**	0.315 to 1.0*
Steel Deck (Node: 9221) (m/s²)	Steel Deck (Node: 9282) (m/s²)	Steel Deck (Node: 9290) (m/s²)	Steel Deck (Node: 9317) (m/s²)	
11.62	8.19	11.77	0.61	
* Acceptable acceleration values for human comfort in accordance with (ISO 2631-1, 1997).				

Table 12. Accelerations related to the combined dynamic loading (driving)

People working temporarily near to driving could be affected in various degrees (human discomfort). The allowable acceleration value when the human comfort is considered (ISO 2631-1, 1997) is located in the range of 0.315 m/s^2 to 1.0 m/s^2, as illustrated in Table 12. The peak acceleration value calculated on the platform steel deck was equal to 27.33 m/s^2, see Table 12. This maximum acceleration value violated the recommended limits proposed by the design codes (27.33 m/s^2 > 1.0 m/s^2) (ISO 2631-1, 1997), causing human discomfort.

Based on the numerical results obtained with the present investigation, it was clearly demonstrated that the investigated structural model presented problems due to excessive vibration. This way, some changes on the horizontal steel bracing system were proposed by the authors, in order to reduce the vibration effects.

This way, Figure 10 shows the proposed steel bracing members which were added to the original design of the investigated structural model. The central idea was to propose a bracing system that produces an increasing in the steel deck stiffness and, consequently, can cause a reduction of the steel deck system dynamic response, when submitted to the rotor and gear dynamic loads.

In sequence, Figure 11 shows the vibration modes of the production platform steel deck system before and after the addition of the proposed bracing members shown in Figure 10. Comparing the presented vibration modes, it may be concluded that modal amplitudes have been modified, leading to a reduction of the structural system dynamic response, in terms of displacements, velocities and accelerations.

Figure 12 shows the response spectra of vertical translational displacements, for the supports and rotors, and, for particular nodes of the steel deck of the platform. Only three graphs were presented due to the fact that this figure (Figure 12) represents, in general, the dynamic response of the system. Based on the structure dynamic response, the peak of interest for the forced vibration numerical analysis, associated with the excitation frequency of the equipment (f = 30 Hz) is indicated in the Figure 12.

Tables 13 to 15 present the vertical translational displacements, velocities and accelerations, related to specific locations on the steel deck, near of the mechanical equipment, see Figure 9, calculated with the introduction of the proposed steel bracing system, when the combined dynamic loadings were considered.

The values presented in Tables 13 to 15 were obtained numerically with the aid of the developed finite element model. These values were then compared with the limiting values proposed by design code recommendations (CEB 209/91, 1991; ISO 1940-1, 2003; ISO 2631-1,

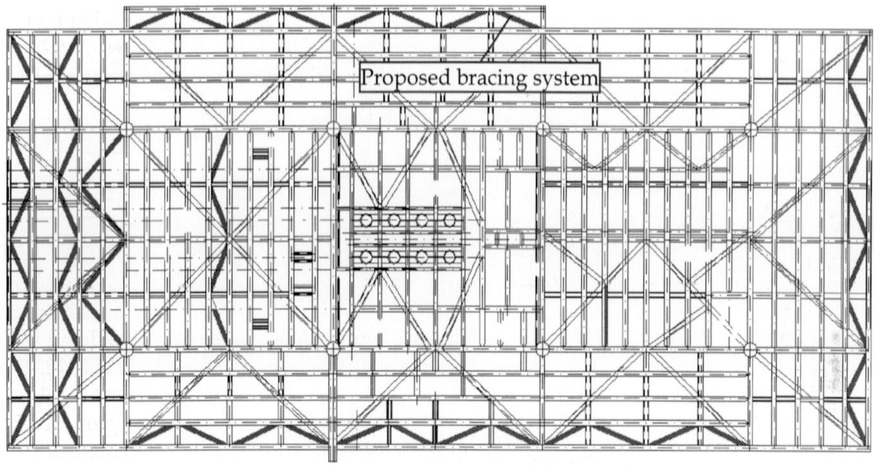

a) Lower deck

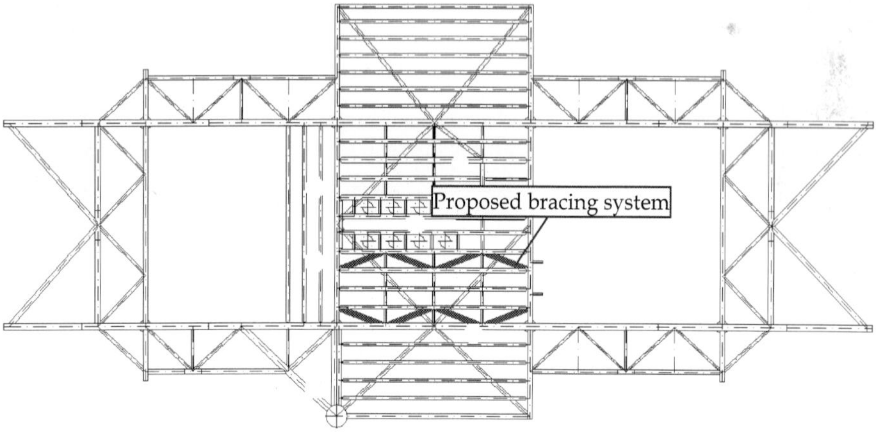

b) Upper deck

Fig. 10. Steel deck with the introduction of the proposed bracing system

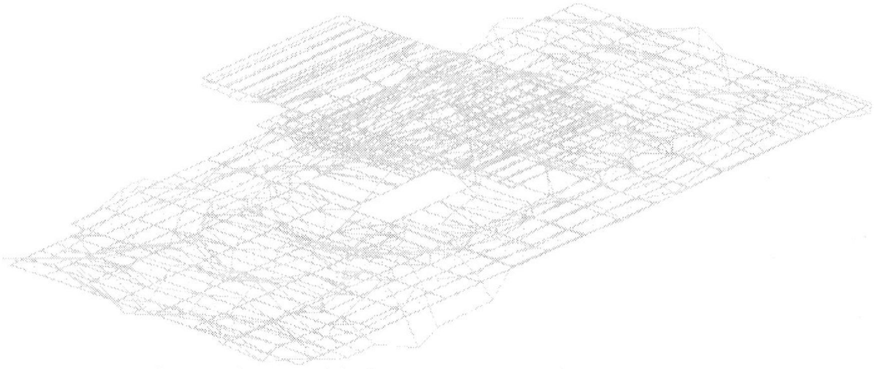

a) Before the introduction of the bracing system. Mode: 1203 - Frequency: 30 Hz

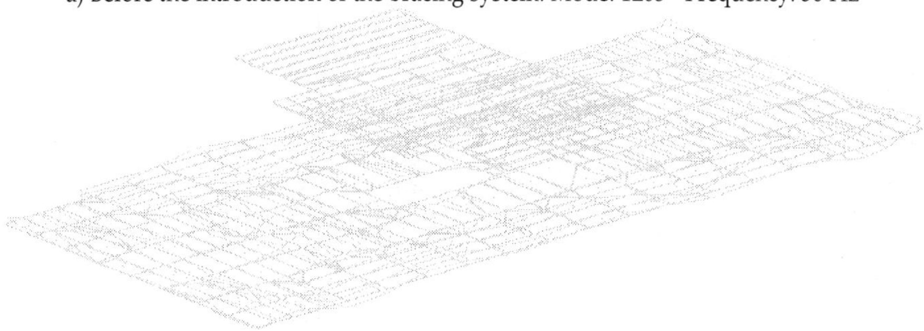

b) After the introduction of the bracing system. Mode: 1195 - Frequency: 30 Hz

Fig. 11. Steel deck vibration modes

1997; ISO 2631-2, 1989; Murray et al., 2003), after the introduction of the proposed bracing system. It must be emphasized, that only the production platform steady state response was considered in this analysis.

Considering the Figure 12 (12a, 12c and 12e), related to the production platform steel deck dynamic response, it can be verified the existence of several energy transfer areas with several energy transfer picks, before the addition of the proposed bracing system.

The peak of interest for the forced vibration numerical analysis, associated with the excitation frequency of the equipment (f = 30 Hz) is clearly indicated in this figure, see Figure 12 (12a, 12c and 12e). However, after the introduction of the proposed bracing system, the energy transfer pick related to the excitation frequency of the equipment was significantly reduced, see Figure 12 (12b, 12d and 12f). This fact demonstrates the efficiency of the proposed bracing system and indicates that the structural system dynamic response can be significantly reduced.

Considering the same strategy previously presented, the design recommendations (see Tables 10 to 12) were utilized, in order to determine the limiting values for machinery performance, related to allowable amplitudes, velocities and accelerations (ISO 1940-1, 2003; ISO 2631-1, 1997).

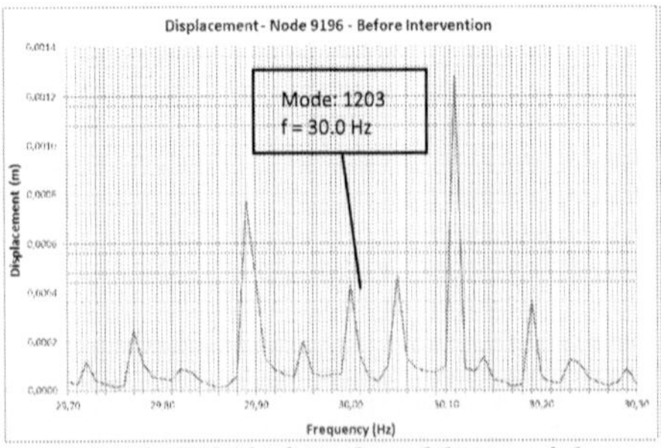

a) Node 9196 before the addition of the bracing system

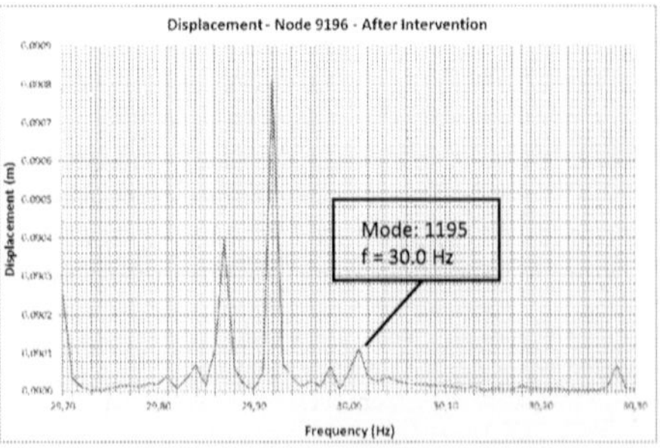

b) Node 9196 after the addition of the bracing system

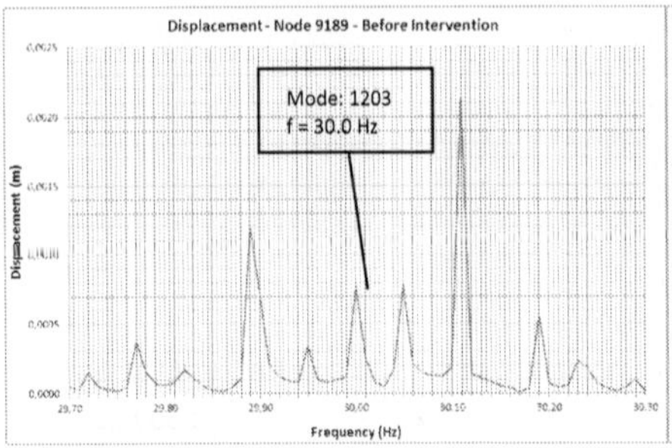

c) Node 9189 before the addition of the bracing system

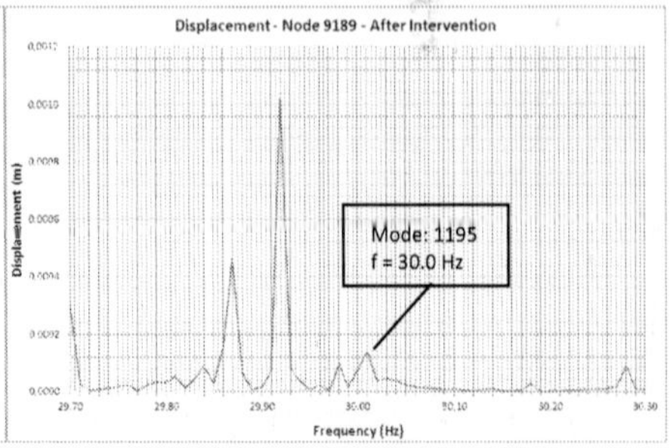

d) Node 9189 after the addition of the bracing system

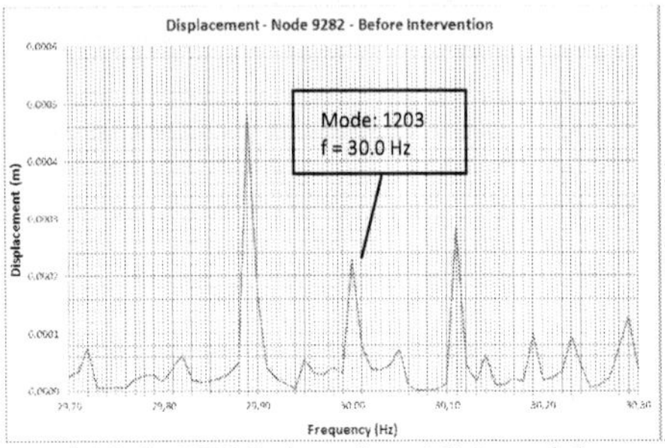

e) Node 9282 before the addition of the bracing system

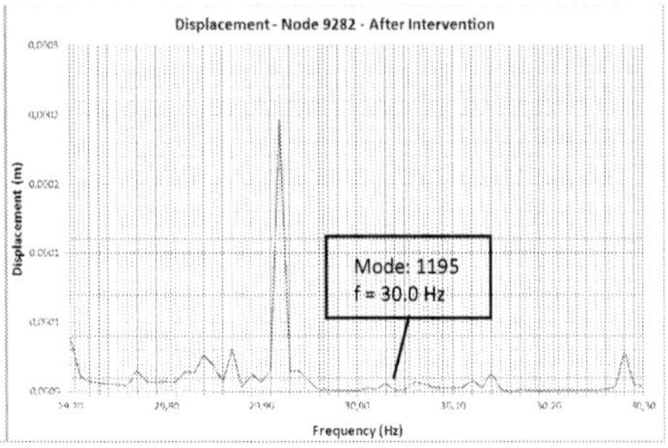

f) Node 9282 after the addition of the bracing system

Fig. 12. Vertical displacements response spectra of the steel deck selected joints before and after the introduction of the proposed bracing system

Rotor Support (Node: 9194) (μm)	Rotor Support (Node: 9197) (μm)	Rotor Support (Node: 9224) (μm)	Amplitudes Limit Value (μm)
35	**96**	54	40 to 60*
Gear Support (Node: 9193) (μm)	Gear Support (Node: 9196) (μm)	Gear Support (Node: 9223) (μm)	
33	52	5	
* For vertical vibration for high speed machines (>1500 rpm) (ISO 1940-1, 2003).			

Table 13. Vertical displacements related to the combined dynamic loading (driving), after the addition of the proposed bracing system

Rotor Support (Node: 9194) (mm/s)	Rotor Support (Node: 9197) (mm/s)	Rotor Support (Node: 9224) (mm/s)	Velocities Limit Value (mm/s)
6.68	6.92	**10.36**	0.70 to 2.80*
Gear Support (Node: 9193) (mm/s)	Gear Support (Node: 9196) (mm/s)	Gear Support (Node: 9223) (mm/s)	
6.23	9.80	2.71	
* Tolerable velocity for electrical motors according to (ISO 1940-1, 2003).			

Table 14. Velocities related to the combined dynamic loading (driving), after the addition of the proposed bracing system

Steel Deck (Node: 9098) (m/s^2)	Steel Deck (Node: 9137) (m/s^2)	Steel Deck (Node: 9173) (m/s^2)	Steel Deck (Node: 9189) (m/s^2)	Amplitudes Limit Value (m/s^2)
1.39	0.50	2.63	**4.96**	0.315 to 1.0*
Steel Deck (Node: 9221) (m/s^2)	Steel Deck (Node: 9282) (m/s^2)	Steel Deck (Node: 9290) (m/s^2)	Steel Deck (Node: 9317) (m/s^2)	
3.44	0.20	1.23	2.75	
* Acceptable acceleration values for human comfort in accordance with (ISO 2631-1, 1997).				

Table 15. Accelerations related to the combined dynamic loading (driving), after the addition of the proposed bracing system

This way, after the addition of the bracing system, see Figure 10, the structural system dynamic response, submitted to the dynamic loadings (driving) coming from the machinery presented the following results: the maximum amplitude value at the base of the driving support (Node 9197: see Figure 9) on the platform steel deck was reduced of 446 μm (or 0.446 mm or 0.0446 cm) to 96 μm (or 0.096 mm or 0.0096 cm), see Tables 10 and 13; the maximum velocity value calculated at the base of the driving support (Node 9197: see Figure 9), on the steel deck was reduced of 84.12 mm/s to 6.92 mm/s, see Tables 11 and 14; the peak acceleration value calculated on the structure steel deck was reduced of 27.33 m/s^2 to 4.96 m/s^2, see Tables 12 and 15.

It must be emphasized that despite the fact that the limiting values for machinery performance, related to allowable amplitudes, velocities and accelerations (ISO 1940-1, 2003; ISO 2631-1, 1997) were violated, even after the addition of the proposed bracing system, the steady-state structural system dynamic response was considerably reduced.

8. Final remarks

This paper investigated the dynamic behaviour of an oil production platform made of steel and located in Santos basin, São Paulo, Brazil. The structural model consists of two steel decks with a total area of 1915 m^2 (upper deck: 445 m^2 and lower deck: 1470 m^2), supported by vertical sections made of tubular steel members (steel jacket), and piled into the seabed, when subjected to impacts produced by mechanical equipment (rotating machinery). The main objective of the paper was to assess the dynamic impacts on the steel deck structure coming from the electrical generators and compressors.

Based on the peak acceleration values and maximum displacements and velocities, obtained on the structure steady-state response, it was possible to evaluate the structural model performance in terms of human comfort, maximum tolerances of the mechanical equipment and vibration serviceability limit states of the structural system, based on the design code recommendations.

It was concluded that the oil production platform displacements response spectra presented several energy transfer areas with several energy transfer picks. However, after the introduction of the horizontal steel bracing system proposed by the authors, the energy transfer pick related to the excitation frequency of the equipment was significantly reduced, demonstrating the efficiency of the idealized system.

After the addition of the horizontal bracing system the structural system dynamic response presented the following results: the maximum amplitude value at the base of the driving support on the steel deck was reduced of 0.0446 cm to 0.0096 cm; the maximum velocity value calculated at the base of the driving support on the deck was reduced of 84.12 mm/s to 6.92 mm/s and the maximum acceleration value calculated on the structure steel deck was reduced of 27.33 m/s^2 to 4.96 m/s^2.

However, even considering the improvement on the structural model bracing system and a considerably reduction on the structure dynamic response, the results obtained throughout this investigation indicated that the analysed platform steel deck violated the human comfort criteria, as well as its vibration serviceability limit states, inducing that individuals working temporarily near the machinery could be affected by human discomfort.

On the other hand, considering the machinery performance, it was also concluded that the platform steel deck design, should be revaluated, due to the fact that the displacements and velocities values related to the machinery supports were very high and violated the recommended limits proposed by design codes.

9. Acknowledgements

The authors gratefully acknowledge the support for this work provided by the Brazilian Science Foundations: CAPES, CNPq and FAPERJ.

10. References

Assunção, T. M. R. C. (2009). *"Considerações sobre efeitos dinâmicos e carregamentos induzidos por fontes de excitação em estruturas"*. MSc Dissertation (In Portuguese), Civil Engineering Post-graduate Programme, Federal University of Minas Gerais, UFMG, Minas Gerais, MG, Brazil.

Bachmann, H.; Ammann, W. (1987). *Vibrations in Structures: Induced by Man and Machines*, Structural Engineering Document 3rd Ed., International Association for Bridges and Structural Engineering.

Chopra, A. K. (2007). *Dynamics of Structures - Theory and Applications to Earthquake Engineering*. 3rd Ed., Pearson Education, Inc., New Jersey.

Clough, R. W.; Penzien, J. (1995). *Dynamics of Structures*, 3rd Ed., Computers and Structures, Inc., Berkeley.

Comité Euro-International Du Beton. (1991). *Vibration Problems in Structures - Practical Guidelines*. Bullitin dínformation n° 209.

Craig JR., R. R. (1981). *Structural Dynamics - An Introduction to Computer Methods*. John Wiley& Sons, Inc., New York.

Dias Junior, M. "Dinâmica de Rotores". Proceedings of IV Congresso da Academia Trinacional de Ciências, Paraná, PR, Brazil, October 2009, Fozdo Iguaçu.

Figueiredo Ferraz Consultoria e Engenharia de Projeto LTDA. (2004). *"Análise em Serviço da Plataforma de Merluza (PMLZ-1)"*, Technical Report (In Portuguese).

Griffin, M. J. (1996). *Handbook of Human Vibration*. Ed. Academic Press, London.

GTSTRUDL. (2009). Structural Design & Analysis Software, Release 29.1.

ISO 1940-1. *Mechanical Vibration. (2003). Balance Quality Requirements for Rotors in a Constant (Rigid) State. Part 1: Specification and Verification of Balance Tolerances.*

ISO 2631-1. *Mechanical Vibration and Shock. (1997). Evaluation of Human Exposure to Whole-body Vibration. Part 1: General Requirements.*

ISO 2631-2. *Mechanical Vibration and Shock. (1989). Evaluation of human exposure to whole-body vibration. Part 2: human exposure to continuous and shock-induced vibrations in buildings (1 to 80Hz).*

Lenzen, K. H. (1996). *Vibration of Steel Joist Concrete Slab Floors*. In: Engineering journal, v.3(3), p. 133-136.

López, E. J. (2002). *Dinámica de Rotores*. Graduation Monography. Universidad Nacional del Comahue.

Milet, R. R., (2006). *"Análise Comparativa de Métodos de Cálculo para Fundações de Máquinas"*, MSc Dissertation (In Portuguese), Civil Engineering Post-graduate Programme, Federal University of Pernambuco, UFPE, Pernambuco, PE, Brazil.

Murray, T. M., Allen, D. E. and Ungar, E. E. (2003). *Floor vibration due to human activity*, Steel design guide series, AISC, Chicago, USA.

Pereira, C. C. G. (2005). *"Curvas de Percepção e Conforto Humano para Vibrações Verticais"*. MSc Dissertation (In Portuguese), Civil Engineering Post-graduate Programme, Federal University of Ouro Preto, UFOP, Minas Gerais, MG, Brazil.

Poulos, H. G.; Davis, E. H. (1980). *Pile Foundation Analysis and Design*. John Wiley & Sons Inc., New York.

Reiher, H. E, Meister, F. J. (1946). *The Effect of Vibration on people*. Translated from Forsch Geb, p 381-386, Ohio.

Rimola, B. D. (2010). *"Análise Dinâmica de Plataformas de Aço a partir da Consideração do Efeito da Interação Solo-Estrutura"*, MSc Dissertation (In Portuguese), Civil Engineering Post-graduate Programme, PGECIV, State University of Rio de Janeiro, UERJ, Rio de Janeiro, RJ, Brazil.

Rimola, B. D. ; Silva, J. G. S. da ; Sieira, A. C. C. F. ; Lima, L. R. O. de ; Neves, L. F. da C. (2010a). Vibration Analysis of a Production Platform Induced by Mechanical Equipments. Proceedings of The Tenth International Conference on Computational Structures Technology, CST 2010, Valência, Spain, Civil-Comp Press, Edinburg. v. 1. p. 1-12.

Rimola, B. D ; Silva, J. G. S. da ; Sieira, A. C. C. F. ; Lima, L. R. O. de. (2010b). Vibration Analysis of an Oil Production Platform. Proceedings of the CILAMCE 2010: XXXI Iberian Latin American on Computational Methods in Engineering and MECOM 2010: IX Argentinean Congress on Computational Mechanics, CILAMCE 2010, Buenos Aires, Argentina. AMCA (Argentine Association for Computational Mechanics). v. XXIX. p. 7529-7540.

Silva, E. L. da. (2004). *"Dinâmica de Rotores: Modelo Matemático de Mancais Hidrodinâmicos"*. MSc Dissertation (In Portuguese), Federal University Paraná. Paraná,PR, Brasil.

Souza, M. G. de; Cicogna, T. R.; Chiquito, A. J. (2007). *"Excitação dos modos normais de um sistema usando um motor desbalanceado"*. Revista Brasileira de Ensaio de Física, v. 29, n. 1, p.5-10, Brazil.

Srinivasulu, P.; Vaidyanathan, C. V. (1976). *Handbook of Machine Foundations*. McGraw-Hill, New Delhi.

Terzaghi, K. (1955). *Evaluation of Coefficients of Subgrade Reaction*. Géotechnique, London, v. 4, n° 4, p. 297-326.

Vasconcelos, R. P. (1998). *"Atenuações de Vibrações em Lajes Nervuradas de Piso em Concreto Armado"*. MSc Dissertation (In Portuguese), Civil Engineering Post-graduate Programme, Federal University of Rio de Janeiro, UFRJ, Rio de Janeiro, RJ, Brazil.

Winkler, E. (1867). *"Die Lehre von Elastizitat und Festigkeit"* (On Elasticity and Fixity), Dominicus, Prague.

Zhou, S.; Shi, J. (2001). *Active Balancing and Vibration Control of Rotating Machinery: A Survey.* The Shock and Vibration Digest - Sage Public ations, Vol. 33, No. 4, p. 361-371.

MIMO Vibration Control for a Flexible Rail Car Body: Design and Experimental Validation

Alexander Schirrer[1], Martin Kozek[1] and Jürgen Schöftner[2]
[1]*Institute of Mechanics and Mechatronics, Vienna University of Technology*
[2]*Institute of Technical Mechanics (ACCM), Johannes Kepler University of Linz*
Austria

1. Introduction

Car bodies of modern rail vehicles are designed as lightweight structures with the aim to minimize mass and thus operational energy demand. The central structural design requirements are given by the main static and dynamic loads. However, ride comfort becomes an increasingly important issue because the softer, more compliant structure exhibits low eigenfrequencies that significantly affect perceived passenger ride comfort.
Various approaches have been taken to reduce comfort-relevant vibrations of the car body that can be grouped into vibration isolation and vibration damping approaches. The isolation approaches include passive, semi-active, and active concepts to decouple the car body from the bogeys and wheel sets. The active approaches are more complex and can affect the safety against derailment, but potentially lead to improved isolation performance over their passive counterparts (see Foo & Goodall (2000) and Stribersky et al. (1998)).
The complementary vibration damping approaches intend to increase the elastic eigenmodes' damping ratios. A passive approach is taken in Hansson et al. (2004), active control schemes have been proposed in Kamada et al. (2005) as well as Schandl et al. (2007) and Benatzky (2006). The latter two references treat the same metro configuration and actuation concept as the present work.
This chapter presents LQG and weighted $\mathcal{H}_2$ MIMO control design methods for the vibration control of lightweight rail car body structures. These designs are studied and compared to achieve vibration reduction and passenger ride comfort improvement in a highly flexible metro rail car body. The metro car body structure is directly actuated via locally mounted Piezo stack actuators. Utilizing strain measurement signals, the control law actuates the structure with the aim of minimizing ride comfort-relevant acceleration signals across the car body interior. This system is subject to variations in damping and frequency of the flexible modes which pose a challenge for control design; the two control methods are implemented in a simulation as well as in a scaled experimental setup and their capabilities are investigated.
The work is embedded in a rich series of research and publications treating various aspects of control design for flexible rail car bodies. A robust $\mathcal{H}_\infty$-optimal control approach is surveyed in Kozek et al. (2011), including an overview on modeling, control design, simulation, identification, and experimental results. These research topics are focused on and detailed in Benatzky & Kozek (2005; 2007a;b); Benatzky et al. (2006; 2007); Bilik et al. (2006);

Popprath et al. (2006; 2007) and Schirrer et al. (2008). Further studies focus on Piezo stack actuators, their modeling and control (Schirrer, Kozek & Benatzky, 2008). Schirrer & Kozek (2008) and Schirrer (2010) report co-simulation studies of the multi-body system dynamics, the control law, and a non-linear actuator model. The underlying research project and its results are documented in two doctorate theses (Benatzky, 2006; Schandl, 2005), as well as in a series of master / diploma theses (Bilik, 2006; Schöftner, 2006; Schirrer, 2010). This work is based on the results of Schöftner (2006), Benatzky (2006), and Schirrer & Kozek (2008).
This text should provide the readers with first-hand experience of robust control design and implementation. It is intended to relate control theory results, simulation, and experimental results. This is done with the aim of improving one's understanding of relevant design parameters, caveats, and ways to successfully establish a working control law for challenging control problems. At the example of the application of modern control design methods to a complex, three-dimensional vibration control problem, a reliable methodology for tackling structurally new problems in vibration control is demonstrated. Additionally, the novel application of large-scale Piezo stack actuators for the vibration control of large and heavy flexible structures is studied.
This chapter is structured as follows: In the first section, a system description of the flexible metro car is given. The equations of motions are expressed by a first-order state-space representation obtained by a finite element analysis within the frequency range of interest. Details about the actuation, measurement, and performance variables are introduced which play an essential role for the MIMO controller design. Since the system order of the optimal state-space controller is equal to the order of the open-loop system, system order reduction is often necessary. The control laws designed for low-order plants then have to be validated on the original system for closed-loop stability and performance. Two types of optimal controllers are designed for a reduced-order plant model: an LQG and a weighted $\mathcal{H}_2$ controller. Simulation results are analysed and interpreted.
In an additional section, real-time control laws are designed with both design methodologies for a laboratory scale model of a rail car body. The redesigned controllers are validated in experimental test runs, providing an extended understanding on robustness, design parameter tuning, and real-time implementation issues. Finally, conclusions are drawn from both theoretical and experimental experience, and specific properties of the design methods are discussed.

2. System description - modeling of a flexible rail car body

2.1 Flexible rail car body model

For the different controller design procedures which will be explained in the next chapters a mathematical model of the flexible rail car body is necessary. The equations of motion of a linear time-invariant system can be stated in state-space form as:

$$\begin{aligned} \dot{x} &= Ax + B_1 w + B_2 u \\ z &= C_1 x + D_{11} w + D_{12} u \\ y &= C_2 x + D_{21} w + D_{22} u, \end{aligned} \tag{1}$$

where the first line describes the differential equation of the physical system. The second and third lines are the output equations. The system matrix of the dynamic system and the input matrices for the disturbance w and the actuation u are denoted by A, B_1, and B_2, respectively. The output matrices C_1 and C_2 as well as the feedthrough matrices D_{11}, D_{12}, D_{21}, and D_{22} depend on the locations and on the types of the measurements and the performance outputs. In the literature on $\mathcal{H}_2$ and $\mathcal{H}_\infty$ controller design, z is called the output performance vector to be optimized and y is the measurement vector.

The state-space model description (1) can be established through analytical modeling, typically via the Finite Element (FE) Method, or through data-based identification. These steps are detailed in Kozek et al. (2011) for the rail car body model. In this work, the simulation part is based on an analytic model and the experimental part utilizes a model identified from measured signal data.

Fig. 1 shows the rail car body containing the positions for the performance outputs ($P_1 - P_6$), the disturbances ($E_1 - E_4$), the feedback sensors ($S_1 - S_4$), and the actuators ($A_1 - A_4$). For a detailed discussion on optimal placement of sensors and actuators see Schirrer (2010).

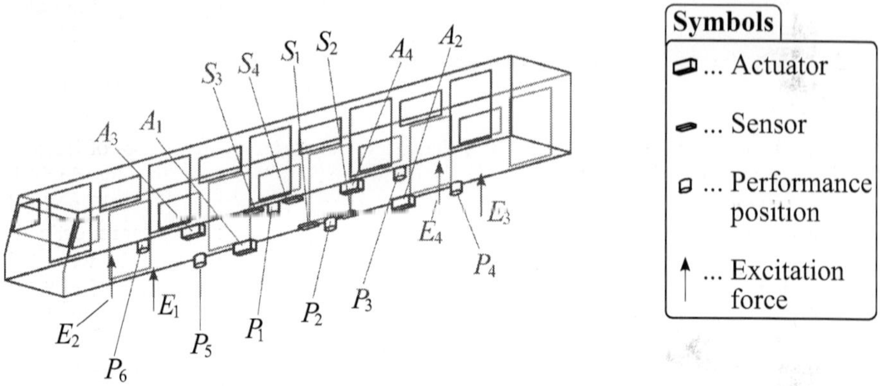

Fig. 1. Flexible rail car body: positions of input and output variables

The mathematical model of the rail car body, obtained by an FE model and subsequent order reduction steps, considers 17 elastic modes and 12 frequency-response-modes (FRM), see Schandl (2005). It is thus of order 58.

Fig. 2 shows the minimum and maximum singular values of the open-loop transfer function matrix T_{dz} which may be interpreted as the best- and the worst-case functional relationship from the white-noise disturbance vector d (related to w as shown below in Sec. 2.2) to the performance vector z. From the singular values of the frequency response (which is identical to the Bode magnitude response in the SISO case), one can observe the elastic modes in the frequency range $\omega = 40 - 200\,\mathrm{rad/s}$ (region 2, blue circles) as well as the FRM in the higher frequency domain (region 3, red squares). Note that no suspension (rigid-body) modes are modeled that typically lie at low frequencies (region 1), because only the flexible car body structure is considered. Full multi-body validation is studied via extensive co-simulations in Schirrer (2010). The main goal in the next sections is to design controllers which suppress elastic modes, but which are not active at other frequencies because the accuracy of the simulation model decreases with increasing frequency. The transfer function is likely to be dominated by other physical effects and nonlinearities for $\omega > 200\,\mathrm{rad/s}$ which have not been considered in obtaining the mathematical model. The FRM are special modes, which are necessary to accurately describe highly localized deformations caused by the structure actuation at the actuator interfaces (see Schandl (2005)). They lie at high frequencies and can be treated here as testing modes, since the actuator action in the high-frequency domain is undesired. High-frequency modes that are always present in continuous structures, but which are not considered in the model used for the controller design may be excited and destabilize the closed-loop system.

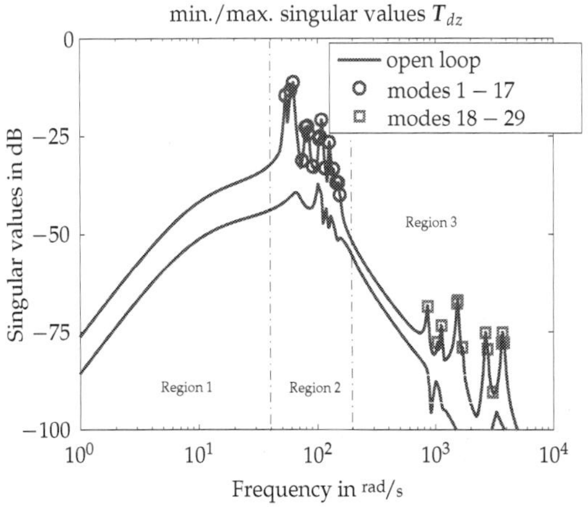

Fig. 2. Minimum and maximum singular values of the open-loop system showing the elastic modes. Modes denoted by squares are the frequency response modes (see Sec. 2.1).

2.2 Excitation

The rail vehicle runs on an imperfect track with misalignments and rail position variations. This track excitation is often characterized by its spatial frequency spectrum and associated filters, see Frederich (1984). For the studied rail car body dynamics, the disturbance $\boldsymbol{w}$ models the excitation of the car body by secondary suspension forces. Its spectrum is known from simulation and measurements and can be modeled by a colored noise sequence resulting from filtering a white noise disturbance input signal $\boldsymbol{d}$ by a second-order lag function with cut-off frequency $\omega_{\text{c.o.}} = 10\,\text{rad/s}$:

$$F(s) = 3000\frac{100}{s^2 + 20s + 100}. \tag{2}$$

The colored noise $\boldsymbol{w}(t)$ is related to the white noise signal $\boldsymbol{d}(t)$ by

$$\boldsymbol{W}(s) = F(s)\boldsymbol{D}(s) \quad \Leftrightarrow \quad \boldsymbol{w}(t) = \mathcal{L}^{-1}\{F(s)\boldsymbol{D}(s)\} \quad \text{with} \quad \boldsymbol{d}(t) = \mathcal{L}^{-1}\{\boldsymbol{D}(s)\}. \tag{3}$$

2.3 Actuators

A novel concept of structural actuation is applied here: Piezo stack actuators, mounted in consoles at the floor support trusses of the car body structure locally introduce bending moments into the structure. This allows direct actuation of the structural vibrations and is thus considered efficient. In this study, four stack actuators are considered in the full-size car body simulations (see Section 3 and Figure 1), whereas two stack actuators are utilized in the experimental setup in Section 3.4, see Figure 25. For more detailed studies on the actuator concept refer to Kozek et al. (2011); the deformation characteristics of a simple console interface structure with a large-scale Piezo stack actuator have been investigated in Schirrer et al. (2008).

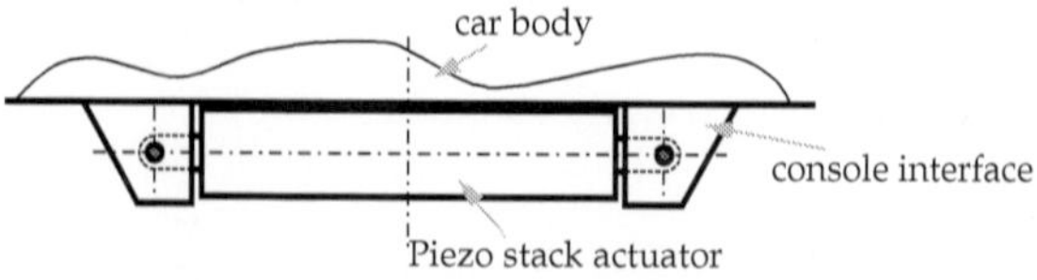

Fig. 3. Actuator mounted in a console to introduce torques into the attached flexible structure (rail car body)

2.4 Measurements

For the controller design, a signal-to-noise ratio (SNR) of about 10 is assumed, see Fig. 4. Two types of measurements are available: either using the acceleration or using strain sensor outputs proportional to the local structure curvature. The measurement positions are either collocated with the actuator or located in the middle of the car body underneath the doors (see Fig. 1). The latter case, when the actuator and sensor positions do not coincide, is called non-collocation. In general, non-collocated systems suffer from a lack of robustness and should not be used if the uncertainty of the system is large, but the controller performance may be better than for collocated systems if a sufficiently accurate mathematical model is available, see Preumont (2006).

In the present work, four strain sensors respectively six acceleration sensors are utilized in the full-size car body simulations. In the experimental setup, two strain sensors respectively one acceleration sensor are utilized for feedback (however, ride comfort performance is evaluated always at six acceleration performance positions as indicated in Figure 1).

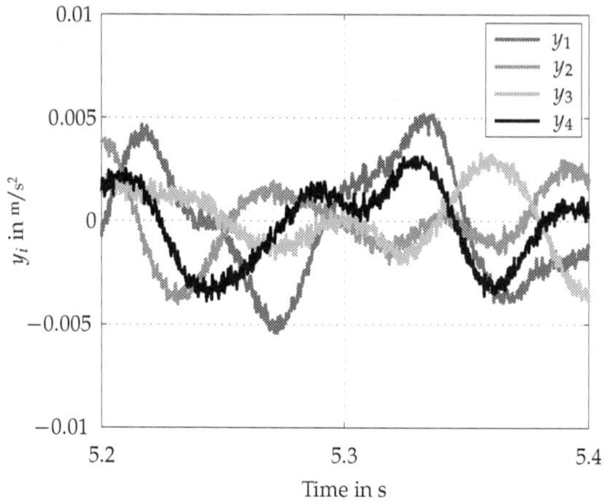

Fig. 4. Measured acceleration signals of the sensors showing a signal-to-noise ratio SNR = 10

2.5 Evaluation of passenger ride comfort according to ISO 2631

Whole-body vibrations are transmitted to the human body of the passengers in a bus, train or when driving a car. The ISO 2631 standard provides an average, empirically verified objective quantification of the level of perceived discomfort due to vibrations for human passengers (ISO, 1997). The accelerations in vertical and horizontal directions are filtered and these signals' root mean square (RMS) are combined into a scalar comfort quantity. Fig. 5 shows the ISO 2631 filter magnitude for vertical accelerations which are considered the only relevant component in the present study. For the heavy metro car, the highest sensitivity of a human occurs in the frequency range of $f \approx 4 - 10\,\mathrm{Hz}$. For the scaled laboratory model, all relevant eigenfrequencies are shifted by a factor of 8 compared to the full-size FEM model. For this reason, the ISO 2631 comfort filters and the excitation spectra are also shifted by this factor. Moreover, only unidirectional vertical acceleration signals are utilized as they represent the main contributions for the considered application.

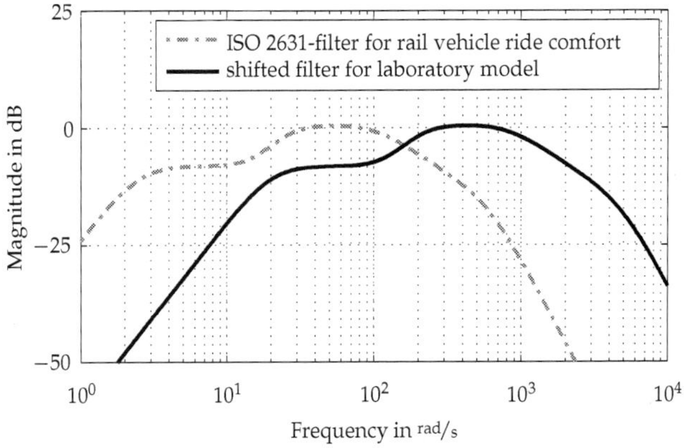

Fig. 5. Filter function according to ISO 2631 (yaw axis)

3. Optimal controller design for the metro car body

Two different methods for controller design are investigated in the following: an LQG and a frequency-weighted $\mathcal{H}_2$ controller are computed for a reduced-order plant model containing only the first 6 eigenmodes. The goal of this study is to obtain a deeper understanding on robustness and controller parameter tuning, since the LQG and the frequency-weighted $\mathcal{H}_2$ control methods are applied to design real-time state-space controllers for the laboratory setup in the next chapter.

3.1 LQG controller for a reduced-order system

3.1.1 Theory

The continuous-time linear-quadratic-gaussian (LQG) controller is a combination of an optimal linear-quadratic state feedback regulator (LQR) and a Kalman-Bucy state observer, see Skogestad & Postlethwaite (1996). Let a continuous-time linear-dynamic plant subject to process and measurement noises be given in state space ($D = 0$ for compactness):

$$\begin{aligned} \dot{x} &= Ax + Bu + Ew \\ y &= Cx + v, \end{aligned} \tag{4}$$

where w and v are assumed to be uncorrelated zero-mean Gaussian stochastic (white-noise) processes with constant power spectral density matrices W and V.
The LQG control law that minimizes the scalar integral-quadratic cost function

$$J = \mathrm{E}\left[\lim_{T\to\infty} \frac{1}{T}\int_0^T l(x,u)\mathrm{d}t\right] \tag{5}$$

with

$$l(x,u) = x^{\mathrm{T}}Qx + u^{\mathrm{T}}Ru \tag{6}$$

turns out to be of the form

$$\dot{\hat{x}} = A\hat{x} + Bu + H(y - C\hat{x}) \tag{7}$$

$$u = \quad -K_{\mathrm{LQR}}\hat{x}. \tag{8}$$

Thereby, $\mathrm{E}[\cdot]$ is the expected value operator, $Q = Q^{\mathrm{T}} \succeq 0$ and $R = R^{\mathrm{T}} \succ 0$ are constant, positive (semi-)definite weighting matrices (design parameters) which affect the closed-loop properties, (7) is the Kalman observer equation, and (8) is the LQR state feedback control law utilizing the state estimate.
The optimal LQR state feedback control law (Skogestad & Postlethwaite, 1996)

$$u = -K_{\mathrm{LQR}}x \tag{9}$$

minimizes the deterministic cost function

$$J = \int_0^\infty l(x,u)\mathrm{d}t \tag{10}$$

and is obtained by

$$K_{\mathrm{LQR}} = R^{-1}B^{\mathrm{T}}X, \tag{11}$$

where X is the unique positive-semidefinite solution of the algebraic Riccati equation

$$A^{\mathrm{T}}X + XA - XBR^{-1}B^{\mathrm{T}}X + Q = 0. \tag{12}$$

The unknown system states x can be estimated by a general state-space observer (Luenberger, 1964). The estimated states are denoted by $\hat{x}$, and the state estimation error ε is defined by

$$\varepsilon := x - \hat{x}. \tag{13}$$

Choosing the linear relation

$$\dot{\hat{x}} = F\hat{x} + Gu + Hy, \tag{14}$$

for state estimation, the following error dynamics is obtained:

$$\dot{\varepsilon} = F\varepsilon + (A - HC - F)x + (B - G)u. \tag{15}$$

If $F = A - HC$ and $G = B$ hold, and if the real parts of the eigenvalues of F are negative, the error dynamics is stable, $\hat{x}$ converges to the plant state vector x, and the observer equation (7) is obtained.

With the given noise properties, the optimal observer is a Kalman-Bucy estimator that minimizes $\mathrm{E}\left[\varepsilon^{\mathrm{T}}\varepsilon\right]$ (see Mohinder & Angus (2001); Skogestad & Postlethwaite (1996)). The observer gain H in (7) is given by

$$H = YC^{\mathrm{T}}V^{-1}, \tag{16}$$

where Y is the solution of the (filter) algebraic Riccati equation

$$AY + YA^{\mathrm{T}} - YC^{\mathrm{T}}V^{-1}CY + EWE^{\mathrm{T}} = 0. \tag{17}$$

Taking into account the separation principle (Skogestad & Postlethwaite, 1996), which states that the closed-loop system eigenvalues are given by the state-feedback regulator dynamics $A - BK$ together with those of the state-estimator dynamics $A - HC$, one finds the stabilized regulator-observer transfer function matrix

$$G_{yu}(s) = -K[s\mathbf{I} - A + HC + BK]^{-1}H\,. \tag{18}$$

Remark: The solutions to the algebraic Riccati equations (12) and (17) and thus the LQG controller exist if the state-space systems $\left(A, B, Q^{\frac{1}{2}}\right)$ and $\left(A, W^{\frac{1}{2}}, C\right)$ are stabilizable and detectable (see Skogestad & Postlethwaite (1996)).

3.1.2 LQG controller design and results for strain sensors / non-collocation

The controller designs are based on a reduced-order plant model which considers only the lowest 6 eigenmodes. The smallest and largest singular values of the system are shown in Fig. 6 and Fig. 7 (compare Fig. 2 for the complete system). The eigenvalues are marked by blue circles. The red lines depict the singular values of the order-reduced $T_{dz,\mathrm{red}}$ (including the shaping filter (2) for the colored noise of the disturbance signal w).

Since a reduced-order system is considered for the controller design, the separation principle is not valid any longer for the full closed-loop system. Neither the regulator gain K_{LQR} nor the estimator gain H is allowed to become too large, otherwise spillover phenomena may occur that potentially destabilize the high-frequency modes. Therefore, the design procedure is an (iterative) trial-and-error loop as follows: in a first step, the weighting matrices for the regulator are prescribed and the resulting regulator gain is used for the full-order system where it is assumed that the state vector can be completely measured. If spillover occurs, the controller action must be reduced by decreasing the state weighting Q. In a second step, the design parameters for the Kalman-Bucy-filter are chosen, considering the fact that the process noise w is no white noise sequence any longer, see (2). Since the process noise covariance is approximately known as $(84.54\,\mathrm{N})^2$ for each channel, the weighting for the output noise V is utilized as a design parameter.

For the optimal regulator the weighting matrices for the states and the input variables are chosen as

$$Q = 9 \cdot 10^8 \cdot \mathbf{I}_{12\times12}, \qquad R = \mathbf{I}_{4\times4}, \tag{19}$$

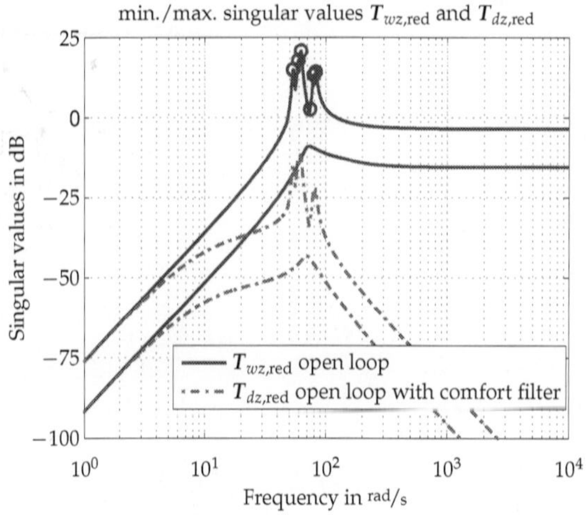

Fig. 6. Smallest and largest singular values of the reduced-order open-loop system (6 modes)

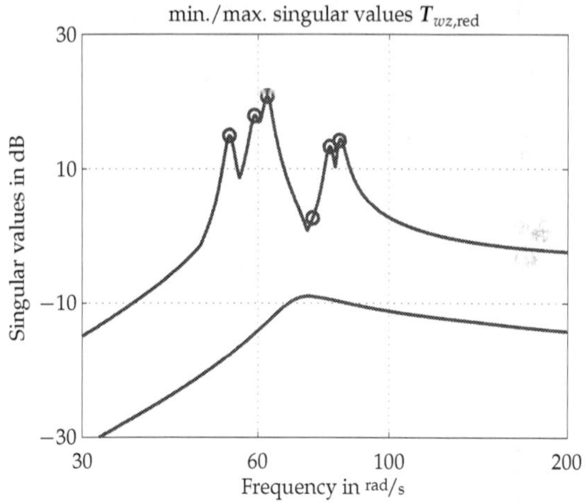

Fig. 7. Smallest and largest singular values of the reduced-order open-loop system (6 modes, zoomed)

where $\mathbf{I}_{n\times n}$ is the identity matrix (n rows, n columns). The observer weightings are chosen to be

$$W = 84.54^2 \cdot \mathbf{I}_{4\times 4}, \qquad V = (1.54 \cdot 10^{-6})^2 \cdot \mathbf{I}_{4\times 4}. \tag{20}$$

Table 1 lists the reduction of the ISO-filtered (see Fig. 5) RMS of each performance variable $z_{1,\mathrm{ISO}}$–$z_{6,\mathrm{ISO}}$ compared to open-loop results. Figures 8–11 contain the maximum/minimum singular values from the white noise input $\boldsymbol{d}$ (which is related to the colored noise input $\boldsymbol{w}$ by (3)) to the performance vector $\boldsymbol{z}$, the time-domain response of two selected performance variables z_1 and z_6, and two pole location plots (overview and zoomed) for the open- and the closed-loop results.

Performance position index i	1	2	3	4	5	6	avg.
RMS reduction $z_{i,\text{ISO}}$ in %	8.44	11.22	29.64	26.53	30.05	31.80	22.94

Table 1. RMS reduction of the performance vector $\boldsymbol{z}$ by LQG control (strain sensors / non-collocation), system order 12

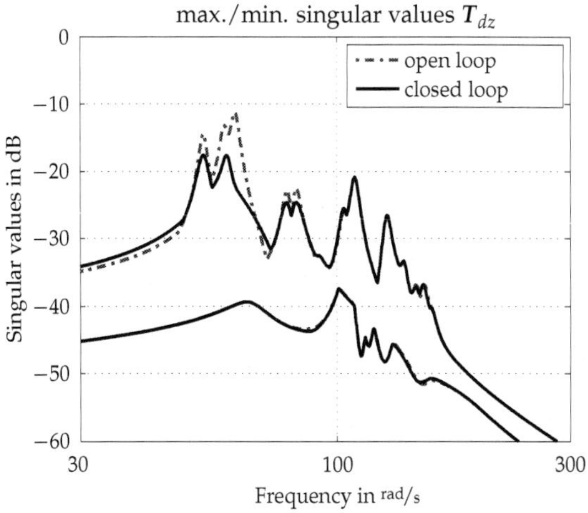

Fig. 8. Reduction of rail car disturbance transfer singular values with non-collocated LQG control

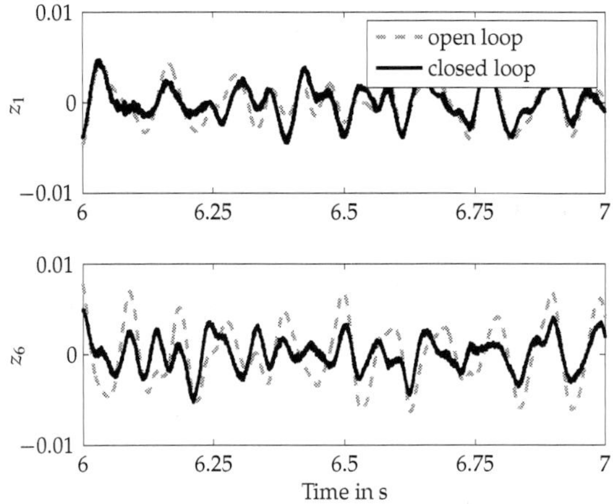

Fig. 9. Acceleration signals z_1 and z_6 without/with non-collocated LQG control

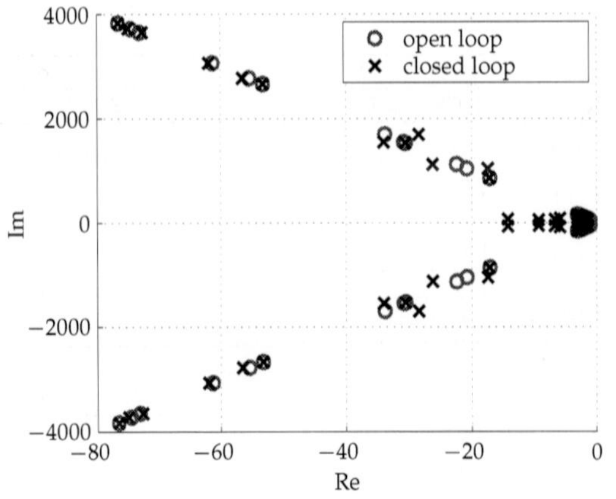

Fig. 10. Rail car model open-loop and non-collocated LQG closed-loop pole locations

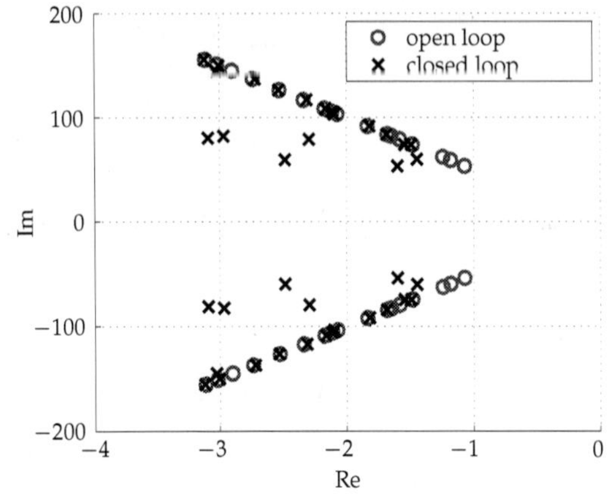

Fig. 11. Rail car model open-loop and non-collocated LQG closed-loop pole locations (zoomed)

3.1.3 Controller design and results for acceleration sensors / collocation

The optimal regulator is designed with the same weighting matrices for the states and the control variables as for the case strain sensors / non-collocation, see (19). The observer weightings are chosen to be

$$W = 84.54^2 \cdot \mathbf{I}_{4\times4}, \qquad V = 0.154^2 \cdot \mathbf{I}_{4\times4}. \tag{21}$$

Table 2 lists the reduction of the ISO-filtered (see Fig. 5) RMS of each performance variable $z_{1,ISO}$–$z_{6,ISO}$ compared to open-loop results. Figures 12–15 contain the maximum/minimum singular values from the white noise input $\boldsymbol{d}$ (which is related to the colored noise input w

by (3)) to the performance vector z, the time-domain response of two selected performance variables z_1 and z_6, and two pole location plots (overview and zoomed) for the open- and the closed-loop results.

Performance position index i	1	2	3	4	5	6	avg.
RMS reduction $z_{i,\mathrm{ISO}}$ in %	7.83	8.36	8.04	7.02	8.79	10.23	8.38

Table 2. RMS reduction of the performance vector z by LQG control (acceleration sensors / collocation), system order 12

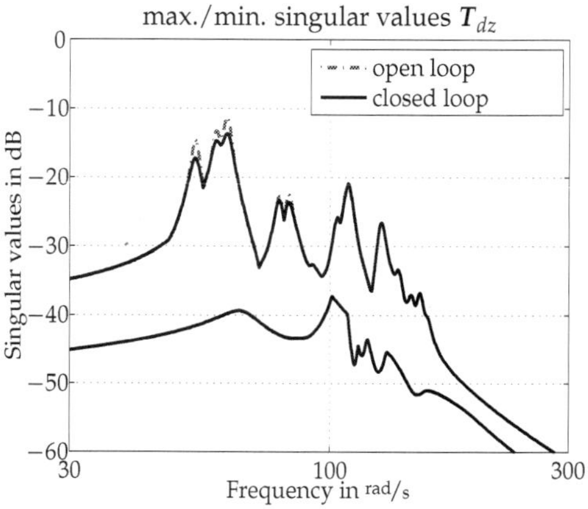

Fig. 12. Reduction of rail car disturbance transfer singular values with collocated LQG control

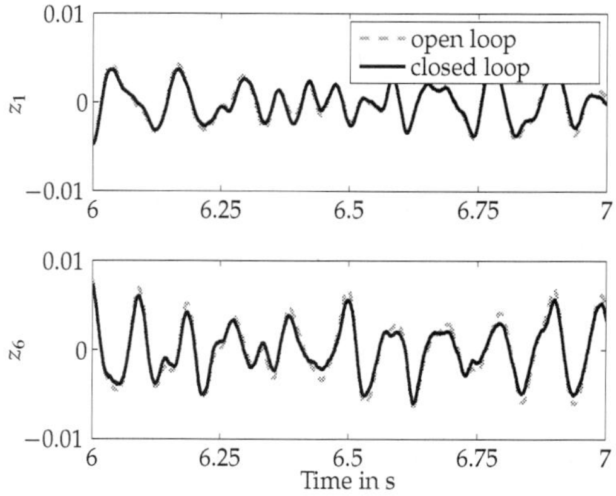

Fig. 13. Acceleration signals z_1 and z_6 without/with collocated LQG control

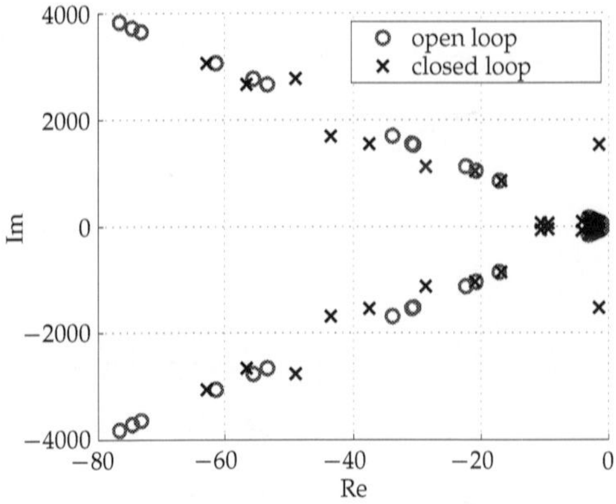

Fig. 14. Rail car model open-loop and collocated LQG closed-loop pole locations

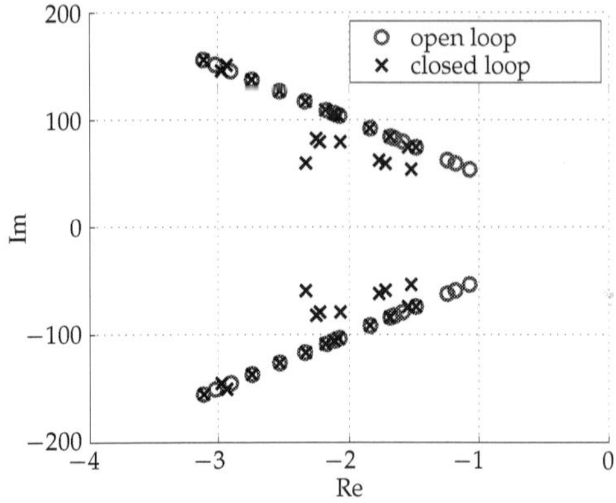

Fig. 15. Rail car model open-loop and collocated LQG closed-loop pole locations (zoomed)

3.2 Frequency-weighted $\mathcal{H}_2$ controller for a reduced-order system

The LQG controllers designed in the previous section do not take into account the performance vector z. The design of the regulator and the estimator gains are a trade-off between highly-damped modes, expressed by the negative real part of the closed-loop poles, and robustness considerations. The generalization of the LQG controller is the $\mathcal{H}_2$ controller, which explicitly considers the performance vector (e.g. one can minimize the deflection 2-norm at a certain point of a flexible system). Another advantage of this type of optimal controller is the possibility to utilize frequency-domain weighting functions. In doing so, the controller action can be shaped for specific target frequency ranges. In turn, the controller can be designed not to influence the dynamic behaviour where the mathematical model is uncertain or sensitive to parameter variations.

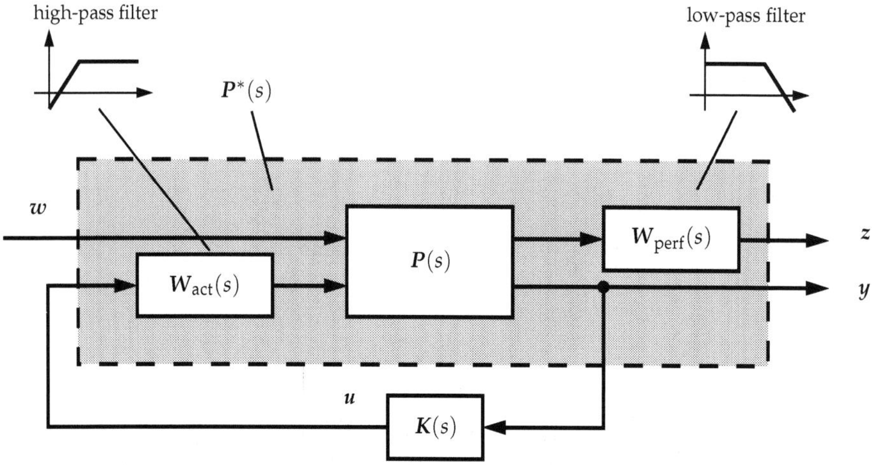

Fig. 16. Closed-loop system $\boldsymbol{P}(s)$ with controller $\boldsymbol{K}(s)$ and actuator and performance weighting functions $\boldsymbol{W}_{\text{act}}(s)$ and $\boldsymbol{W}_{\text{perf}}(s)$

Fig. 16 shows the closed-loop system, where the system dynamics, the controller, and the frequency-weighted transfer functions are denoted by $\boldsymbol{P}(s)$, $\boldsymbol{K}(s)$, $\boldsymbol{W}_{\text{act}}(s)$, and $\boldsymbol{W}_{\text{perf}}(s)$. Taking into account the frequency-weights in the system dynamics, the weighted system description of $\boldsymbol{P}^*$ can be formulated:

$$\begin{bmatrix} z \\ y \end{bmatrix} = \begin{bmatrix} \boldsymbol{P}_{11}^*(s) & \boldsymbol{P}_{12}^*(s) \\ \boldsymbol{P}_{21}^*(s) & \boldsymbol{P}_{22}^*(s) \end{bmatrix} \begin{bmatrix} w \\ u \end{bmatrix}, \tag{22}$$

where $\boldsymbol{P}_{11}^*(s)$, $\boldsymbol{P}_{12}^*(s)$, $\boldsymbol{P}_{21}^*(s)$, and $\boldsymbol{P}_{22}^*(s)$ are the Laplace domain transfer functions from the input variables $\boldsymbol{u}$ and $\boldsymbol{w}$ to the output variables $\boldsymbol{y}$ and $\boldsymbol{z}$.

3.2.1 $\mathcal{H}_2$ control theory

Let the system dynamics be given in the state-space form (1), fulfilling the following prerequisites (see Skogestad & Postlethwaite (1996)):

- $(\boldsymbol{A}, \boldsymbol{B}_2)$ is stabilizable
- $(\boldsymbol{C}_2, \boldsymbol{A})$ is detectable
- $\boldsymbol{D}_{11} = \boldsymbol{0}, \boldsymbol{D}_{22} = \boldsymbol{0}$
- $\boldsymbol{D}_{12}$ has full rank
- $\boldsymbol{D}_{21}$ has full rank
- $\begin{bmatrix} \boldsymbol{A} - \mathrm{j}\omega\mathbf{I} & \boldsymbol{B}_2 \\ \boldsymbol{C}_1 & \boldsymbol{D}_{12} \end{bmatrix}$ has full column rank for all ω

- $\begin{bmatrix} \boldsymbol{A}-\mathrm{j}\omega\mathbf{I} & \boldsymbol{B}_1 \\ \boldsymbol{C}_2 & \boldsymbol{D}_{21} \end{bmatrix}$ has full row rank for all ω

For compactness the following abbreviations are introduced:

$$\begin{aligned}
\overline{\boldsymbol{R}} &= \boldsymbol{D}_{12}{}^{\mathrm{T}}\boldsymbol{D}_{12} \\
\overline{\boldsymbol{S}} &= \boldsymbol{B}_2\overline{\boldsymbol{R}}^{-1}\boldsymbol{B}_2{}^{\mathrm{T}} \\
\overline{\boldsymbol{A}} &= \boldsymbol{A}-\boldsymbol{B}_2\overline{\boldsymbol{R}}^{-1}\boldsymbol{D}_{12}{}^{\mathrm{T}}\boldsymbol{C}_1 \\
\overline{\boldsymbol{Q}} &= \boldsymbol{C}_1{}^{\mathrm{T}}\boldsymbol{C}_1-\boldsymbol{C}_1{}^{\mathrm{T}}\boldsymbol{D}_{12}\overline{\boldsymbol{R}}^{-1}\boldsymbol{D}_{12}{}^{\mathrm{T}}\boldsymbol{C}_1 \succeq \mathbf{0} \\
\overline{\overline{\boldsymbol{R}}} &= \boldsymbol{D}_{21}\boldsymbol{D}_{21}{}^{\mathrm{T}} \\
\overline{\overline{\boldsymbol{S}}} &= \boldsymbol{C}_2{}^{\mathrm{T}}\overline{\overline{\boldsymbol{R}}}^{-1}\boldsymbol{C}_2 \\
\overline{\overline{\boldsymbol{A}}} &= \boldsymbol{A}-\boldsymbol{B}_1\boldsymbol{D}_{21}{}^{\mathrm{T}}\overline{\overline{\boldsymbol{R}}}^{-1}\boldsymbol{C}_2 \\
\overline{\overline{\boldsymbol{Q}}} &= \boldsymbol{B}_1\boldsymbol{B}_1{}^{\mathrm{T}}-\boldsymbol{B}_1\boldsymbol{D}_{21}{}^{\mathrm{T}}\overline{\overline{\boldsymbol{R}}}^{-1}\boldsymbol{D}_{21}\boldsymbol{B}_1{}^{\mathrm{T}} \succeq \mathbf{0},
\end{aligned}$$

where $\succeq \mathbf{0}$ denotes positive-semidefiniteness of the left-hand side. The $\mathcal{H}_2$ control design generates the controller transfer function $\boldsymbol{K}(s)$ which minimizes the $\mathcal{H}_2$ norm of the transfer function $\boldsymbol{T}_{wz}$, or equivalently

$$\|\boldsymbol{T}_{wz}\|_2 = \sqrt{\frac{1}{2\pi}\int_{-\infty}^{\infty}\boldsymbol{T}_{wz}{}^{\mathrm{T}}(\mathrm{j}\omega)\boldsymbol{T}_{wz}(\mathrm{j}\omega)\mathrm{d}\omega} \rightarrow \min. \tag{23}$$

The controller gain $\boldsymbol{K}_c$ and the estimator gain $\boldsymbol{K}_f$ are determined by

$$\boldsymbol{K}_c = \overline{\boldsymbol{R}}^{-1}(\boldsymbol{B}_2{}^{\mathrm{T}}\boldsymbol{X}_2+\boldsymbol{D}_{12}{}^{\mathrm{T}}\boldsymbol{C}_1) \tag{24}$$

and

$$\boldsymbol{K}_f = (\boldsymbol{Y}_2\boldsymbol{C}_2{}^{\mathrm{T}}+\boldsymbol{B}_1\boldsymbol{D}_{21}{}^{\mathrm{T}})\overline{\overline{\boldsymbol{R}}}^{-1}, \tag{25}$$

where $\boldsymbol{X}_2 \succeq \mathbf{0}$ and $\boldsymbol{Y}_2 \succeq \mathbf{0}$ are the solutions of the two algebraic Riccati equations

$$\boldsymbol{X}_2\overline{\boldsymbol{A}}+\overline{\boldsymbol{A}}^{\mathrm{T}}\boldsymbol{X}_2-\boldsymbol{X}_2\overline{\boldsymbol{S}}\boldsymbol{X}_2+\overline{\boldsymbol{Q}} = \mathbf{0}, \tag{26}$$

$$\overline{\overline{\boldsymbol{A}}}\boldsymbol{Y}_2+\boldsymbol{Y}_2\overline{\overline{\boldsymbol{A}}}^{\mathrm{T}}-\boldsymbol{Y}_2\overline{\overline{\boldsymbol{S}}}\boldsymbol{Y}_2+\overline{\overline{\boldsymbol{Q}}} = \mathbf{0}. \tag{27}$$

The state-space representation of the controller dynamics is given by

$$\begin{aligned}
\dot{\widehat{\boldsymbol{x}}} &= (\boldsymbol{A}-\boldsymbol{B}_2\boldsymbol{K}_c-\boldsymbol{K}_f(\boldsymbol{C}_2-\boldsymbol{D}_{22}\boldsymbol{K}_c))\widehat{\boldsymbol{x}}+\boldsymbol{K}_f\boldsymbol{y} \\
\boldsymbol{u} &= -\boldsymbol{K}_c\widehat{\boldsymbol{x}},
\end{aligned} \quad \Rightarrow \boldsymbol{u} = -\boldsymbol{K}(s)\boldsymbol{y}. \tag{28}$$

3.2.2 $\mathcal{H}_2$ controller design and results for strain sensors / non-collocation

The frequency-weighting functions have been specified as

$$\boldsymbol{W}_{\mathrm{act}} = G_{\mathrm{act}}\cdot\mathbf{I}_{4\times4} = 4967\cdot\frac{(s+45)^4\cdot(s^2+6s+3034)}{(s+620)^4\cdot(s+2000)^2}\cdot\mathbf{I}_{4\times4} \tag{29}$$

$$W_{perf} = G_{perf} \cdot \mathbf{I}_{6\times6} = 20 \cdot \mathbf{I}_{6\times6} \tag{30}$$

As in the previous section, the $\mathcal{H}_2$ controller is designed for the reduced-order model (12 states). Considering the shaping filter (2) for the disturbance ($8 = 4 \cdot 2$ states) and the weighting functions (29) and (30) ($24 = 4 \cdot 6$ states), one finds a controller of order 44.
Table 3 lists the reduction of the ISO-filtered (see Fig. 5) RMS of each performance variable $z_{1,ISO}$–$z_{6,ISO}$ compared to open-loop results. Figures 17–20 contain the maximum/minimum singular values from the white noise input $\boldsymbol{d}$ (which is related to the colored noise input $\boldsymbol{w}$ by (3)) to the performance vector $\boldsymbol{z}$, the time-domain response of two selected performance variables z_1 and z_6, and two pole location plots (overview and zoomed) for the open- and the closed-loop results.

Performance position index i	1	2	3	4	5	6	avg.
RMS reduction $z_{i,ISO}$ in %	26.27	27.95	28.71	27.84	30.99	34.31	29.35

Table 3. RMS reduction of the performance vector $\boldsymbol{z}$ by $\mathcal{H}_2$ control (strain sensors / non-collocation), system order 44

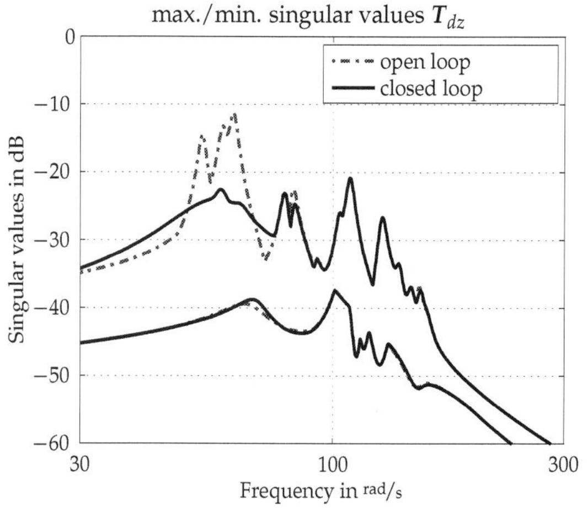

Fig. 17. Reduction of rail car disturbance transfer singular values with non-collocated $\mathcal{H}_2$ control

3.2.3 $\mathcal{H}_2$ controller design and results for acceleration sensors / collocation

The frequency-weighting functions have been specified as

$$W_{act} = G_{act} \cdot \mathbf{I}_{4\times4} = 4967 \cdot \frac{(s+45)^4 \cdot (s^2+6s+3034)}{(s+620)^4 \cdot (s+2000)^2} \cdot \mathbf{I}_{4\times4}, \tag{31}$$

$$W_{perf} = G_{perf} \cdot \mathbf{I}_{6\times6} = 20 \cdot \mathbf{I}_{6\times6}. \tag{32}$$

Table 4 lists the reduction of the ISO-filtered (see Fig. 5) RMS of each performance variable $z_{1,ISO}$–$z_{6,ISO}$ compared to open-loop results. Figures 21–24 contain the maximum/minimum singular values from the white noise input $\boldsymbol{d}$ (which is related to the colored noise input $\boldsymbol{w}$

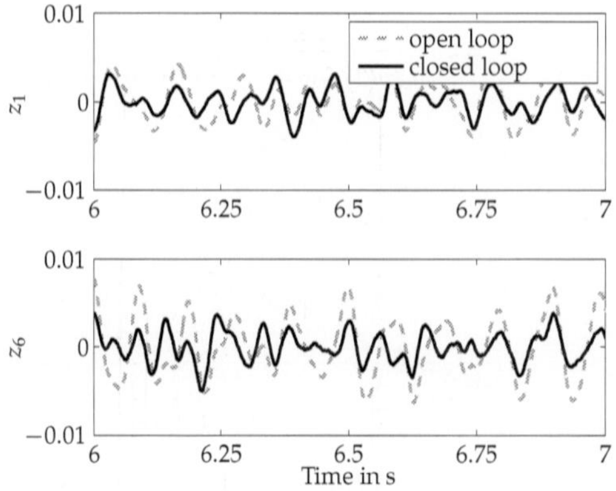

Fig. 18. Acceleration signals z_1 and z_6 without/with non-collocated $\mathcal{H}_2$ control

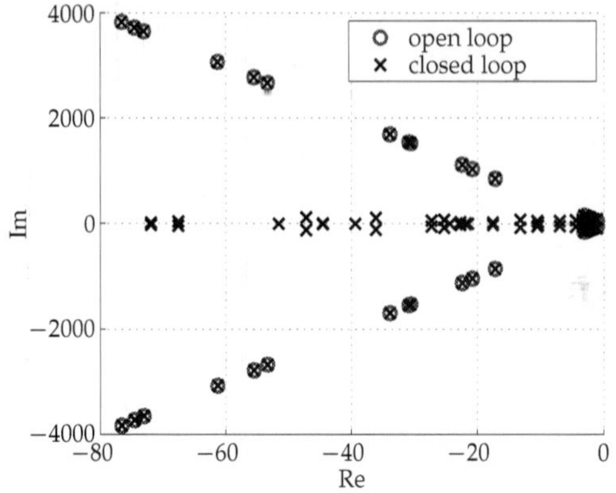

Fig. 19. Rail car model open-loop and non-collocated $\mathcal{H}_2$ closed-loop pole locations

by (3)) to the performance vector $\boldsymbol{z}$, the time-domain response of two selected performance variables z_1 and z_6, and two pole location plots (overview and zoomed) for the open- and the closed-loop results.

Performance position index i	1	2	3	4	5	6	avg.
RMS reduction $z_{i,\mathrm{ISO}}$ in %	23.89	28.12	27.23	24.67	28.85	31.27	27.34

Table 4. RMS reduction of the performance vector $\boldsymbol{z}$ by $\mathcal{H}_2$ control (acceleration sensors / collocation), system order 44

3.3 Interpretation

The main goal for both the LQG and the $\mathcal{H}_2$ controller designs was to increase the damping of the first three eigenmodes. In the present design task, the LQG controller designed for

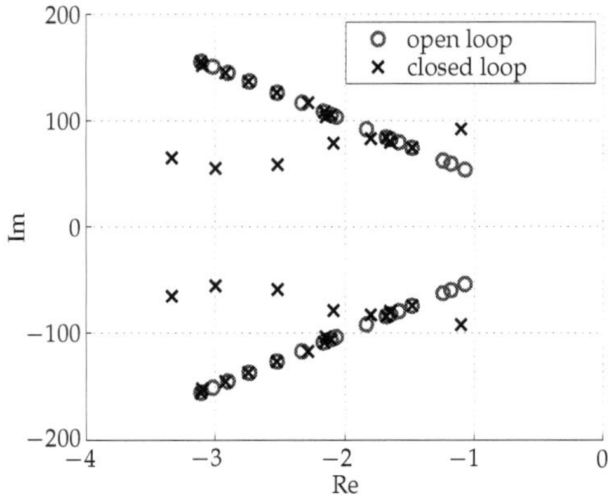

Fig. 20. Rail car model open-loop and non-collocated $\mathcal{H}_2$ closed-loop pole locations (zoomed)

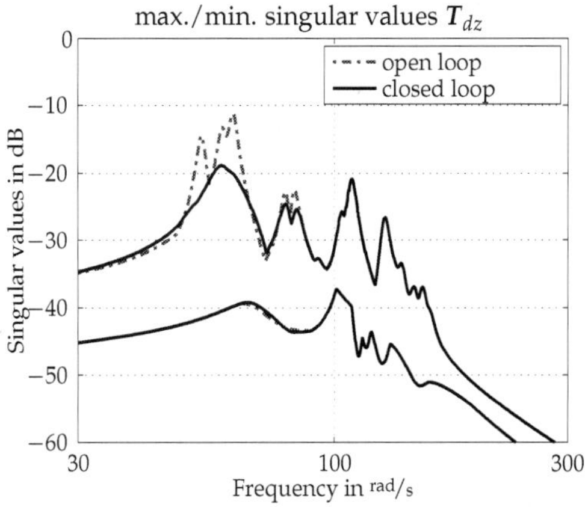

Fig. 21. Reduction of rail car disturbance transfer singular values with collocated $\mathcal{H}_2$ control

collocated acceleration sensors (see Section 3.1.3) did not yield satisfactory performance. The singular value plot shows only marginal magnitude reduction (Figure 12), and also a time-domain analysis of the performance signals z_1 and z_6 (see Figure 13) shows no significant improvement. According to Table 2, the reduction of the filtered performance vector is approximately 8%. However, at $\omega \approx 1500$ rad/s one of the frequency response modes approaches the imaginary axis (Fig. 14 and Fig. 15). Even though the simulated closed loop remains stable, this spillover is critical for operation at an uncertain real plant which possesses unknown high-frequency dynamics.

Considering the LQG design for non-collocated strain sensors in Section 3.1.2, the controller significantly improves the vibrational behaviour. The performance vector is reduced by 23% (Table 1) and a significant reduction is apparent for the time-domain evaluation in Fig. 9. The maximum singular values of the first three eigenmodes are reduced (e.g. third eigenmode

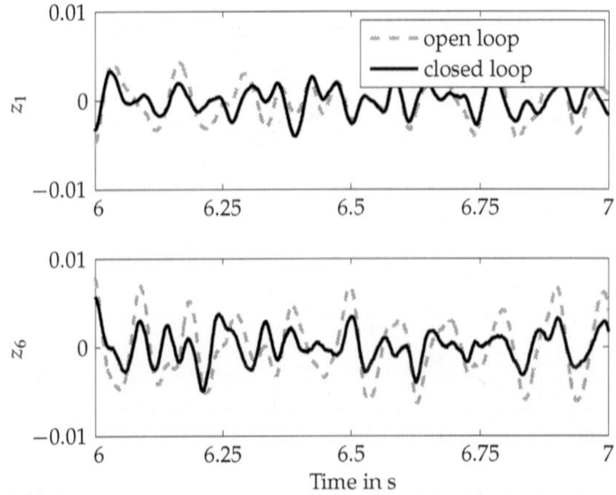

Fig. 22. Acceleration signals z_1 and z_6 without/with collocated $\mathcal{H}_2$ control

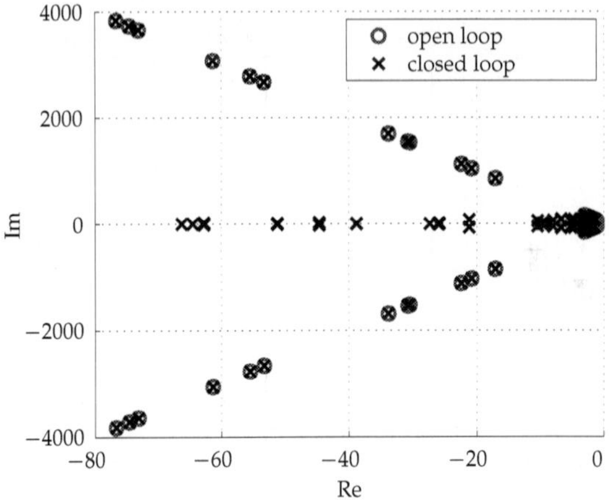

Fig. 23. Rail car model open-loop and collocated $\mathcal{H}_2$ closed-loop pole locations

$-11\,\text{dB}$, see Fig. 8). From the pole location plot one concludes that in the higher frequency domain the frequency response modes remain unchanged (Fig. 10 and Fig. 11).
Both variants of the $\mathcal{H}_2$-optimal controllers (Section 3.2.2 and Section 3.2.3) show significantly higher performance in simulation than the controllers obtained by the LQG design procedure. The main advantage of the $\mathcal{H}_2$ design approach is the possibility to directly incorporate frequency weights to shape the design, see (29) and (31). Specifically, the frequency content of the actuator command signals can be modified. The control law actuates mainly within the frequency range $\omega \approx 50 - 70\,\mathrm{rad/s}$ due to the transmission zeros in the weighting functions W_{act}. In the high-frequency domain, W_{act} is large for both $\mathcal{H}_2$ designs, so only small actuator signal magnitudes result at these frequencies which is especially

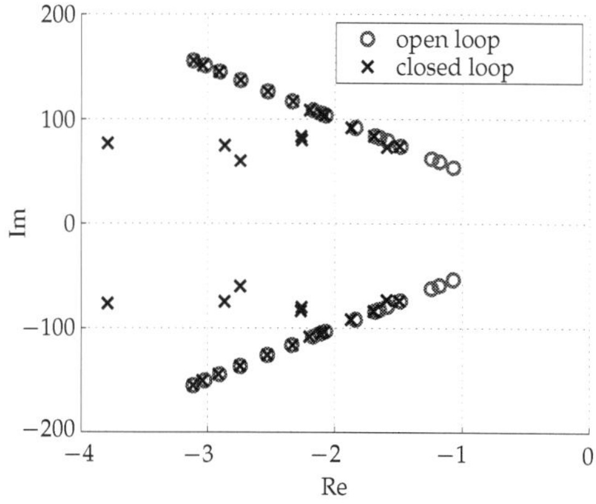

Fig. 24. Rail car model open-loop and collocated $\mathcal{H}_2$ closed-loop pole locations (zoomed)

favorable if the plant dynamics is unknown there. The results for strain/non-collocation and acceleration/collocation control designs are shown in Figs. 17–20 and Table 3 as well as in Figs. 21–24 and Table 4. In the first case the vibrations of the time-domain performance signals $z_{i,\text{ISO}}$ are reduced by 30% (Fig. 18), which is also indicated by the singular values plot (Fig. 17): the lowest three modes are reduced on average by 11 dB. Virtually no spillover occurs at high frequencies ($\omega \approx 150 - 4000\,\text{rad/s}$): The singular values are unchanged (not shown) and also the pole locations remain unchanged for $\omega > 150\,\text{rad/s}$ (seen in Fig. 19 and Fig. 20 where the open-loop poles (blue circles) and closed-loop poles (black crosses) coincide).
The acceleration sensor / collocation simulation results show similar improvement: Only the first three modes are strongly damped (Fig. 21 and Fig. 24), the other ones are hardly affected by the controller action due to the specific choice of the weighting function (31), see Fig. 23. The average reduction of the ISO-filtered performance variables is 27% (Table 4 and Fig. 22).
As a concluding remark, note that the combination of the $\mathcal{H}_2$ method with frequency-weighted transfer functions for the input and the performance signals ($\boldsymbol{W}_{\text{act}}$, $\boldsymbol{W}_{\text{perf}}$) provide satisfactory results, which are characterized by their high robustness and insensitivity to parameter uncertainties. It is shown that the frequency content of the controller action can be tuned by the input weight $\boldsymbol{W}_{\text{act}}$, which affects only the first modes of interest. Higher modes, which are much more difficult to model, are hardly affected due to the roll-off of the $\mathcal{H}_2$ controller. Nevertheless, the LQG controller shows very promising results for the case of non-collocated strain sensors, although the controller is designed for a strongly reduced model containing only 6 modes (note that the full order model has 29 modes). If the acceleration signals are measured and sensor and actuators are collocated, the full-order plant is destabilized by the LQG controller (designed on the reduced-order plant).
Finally, it is noted that so-called reduced-order LQG controllers (see Gawronski (2004)) also have been designed to control the metro vehicle, see Schöftner (2006). By this method an LQG controller has been directly designed for the full-order plant model with 29 modes. Then, the controller transfer functions are evaluated (dynamic systems of order 58) and transformed to the Gramian-based input/output-balanced form. Hardly observable or controllable states, indicated by small Hankel singular values, are truncated, yielding a

low-order controller. While this procedure works well for academic problems (for example, a simply-supported beam), for the metro car body no low-order controller with good vibration reduction performance could be found, see Schöftner (2006).

3.4 Experimental setup of scaled metro car body

3.4.1 General remarks

Fig. 25 and Fig. 26 show the laboratory testbed in which the metro car body scale model is operated. The aluminum structure is excited via an electrodynamic shaker, two Piezo patches measure local structure strain, and two Piezo stack actuators, mounted in consoles on the structure, provide an efficient means of structural actuation. Fig. 26 also shows the actuation and measurement setup symbolically with actuator amplifier (AA), shaker amplifier (SA), antialiasing filters (AF), measurement amplifier (MA) and the laboratory computer (Lab.-PC) on which the real-time control algorithms are implemented.

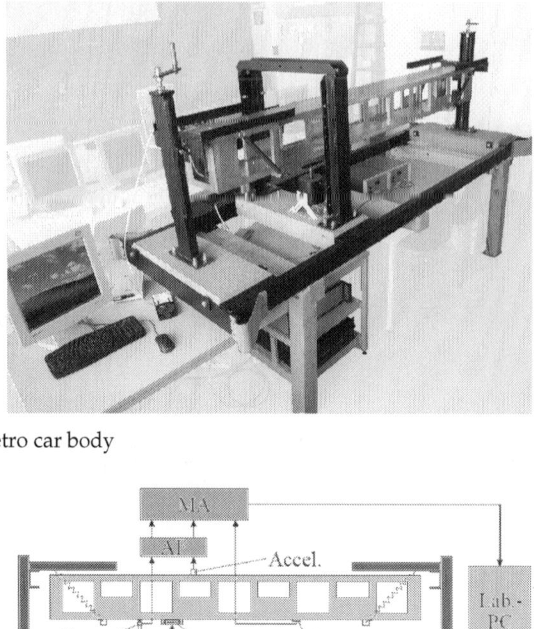

Fig. 25. Scaled metro car body

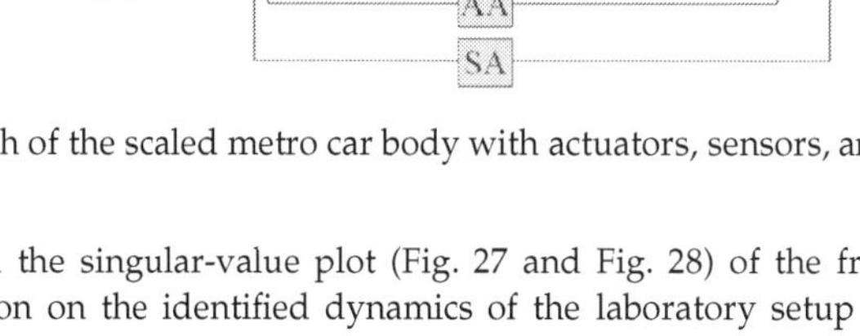

Fig. 26. Basic sketch of the scaled metro car body with actuators, sensors, and performance variables

The pole plot and the singular-value plot (Fig. 27 and Fig. 28) of the frequency response provide information on the identified dynamics of the laboratory setup (200 modes): the modes relevant for the control problem are the bending mode at $f \approx 65\,\text{Hz}$ and the torsional mode at $f \approx 75\,\mathrm{Hz}$. The majority of the poles are either negligible high-frequency modes or other local oscillatory modes.

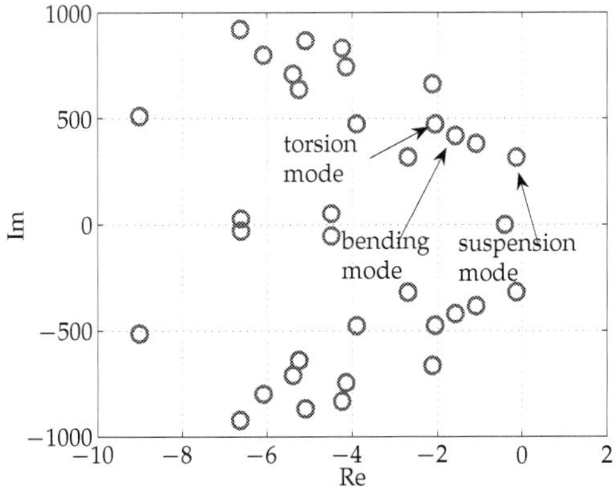

Fig. 27. Pole plot of the identified scaled metro car body (zoomed)

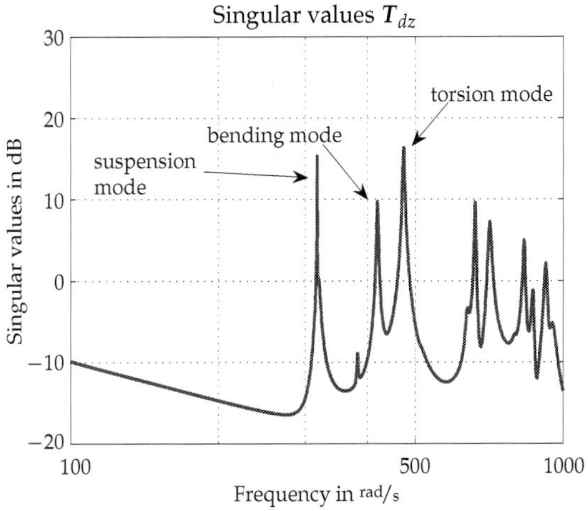

Fig. 28. Singular values of the identified scaled metro car body

The goal is to significantly dampen the torsional and the bending modes without destabilizing other oscillatory modes. For an objective evaluation of the active vibration control problem, the RMS of the frequency-filtered performance variables $z_{i,\mathrm{ISO}}$ ($i = 1, \ldots, 6$) are compared in the open-loop and closed-loop responses. These six performance quantities represent a quantification of passenger ride comfort. A more detailed analysis can be found in Kozek & Benatzky (2008) and Schirrer (2010).

3.4.2 LQG controller design

An LQG controller is designed for a reduced-order plant model with 28 states (the system is identified with 200 states, see Kozek et al. (2011)). Hence the system is transformed into a modal state space representation and all eigenmodes with an eigenfrequency higher than $f \approx 75\,\mathrm{Hz}$ are truncated and not considered for the controller design. The weighting functions for the regulator and the estimator design are

$$Q = \begin{bmatrix} \mathbf{I}_{24\times 24} & \mathbf{0} \\ \mathbf{0} & 6 \cdot 10^{6} \cdot \mathbf{I}_{4\times 4} \end{bmatrix}, \qquad R = \mathbf{I}_{2\times 2} \tag{33}$$

$$W = 0.0054, \qquad V = 2 \cdot 10^{-4} \mathbf{I}_{2\times 2}. \tag{34}$$

Note that Q in (33) is chosen such that only the bending and the torsional vibrations should be significantly damped. A discussion of the results for the closed-loop system is given in Section 3.4.4.

3.4.3 $\mathcal{H}_2$ controller design

The $\mathcal{H}_2$ controller is designed for a plant model which only considers 3 eigenmodes (two of them describe the bending and torsional behaviour). For an efficient control design the frequency-dependent actuator and performance functions are specified as

$$W_{\mathrm{act}} = G_{\mathrm{act}} \cdot \mathbf{I}_{2\times 2} = 188.5 \cdot 10^{9} \cdot \frac{(s^2 + 77.91s + 151800)^4}{(s + 10000)^8} \mathbf{I}_{2\times 2}. \tag{35}$$

$$W_{\mathrm{perf}} = G_{\mathrm{perf}} \cdot \mathbf{I}_{6\times 6} = \mathbf{I}_{6\times 6} \tag{36}$$

Note that the transmission zeros of W_{act} are near the two target modes to be damped, causing the actuator action to be a maximum for these frequencies. A discussion of the results for the closed-loop system are given in Section 3.4.4.

3.4.4 Results

Table 5 and Figures 29, 30, 31, and 32 show the damping ability of both types of controllers. In both cases, the vibrations of the actively controlled system are significantly reduced compared to the open-loop response of the system. The accelerations at both ends of the structures, expressed by the performance variables z_1, z_2, z_5 and z_6 can be significantly reduced, whereas z_3 and z_4 are close to the open-loop response. This is explained due to the fact that the first torsional mode dominates the bending vibrations for the scale laboratory setup. For the LQG controller the singular values only differ for the torsional and the bending vibrations ($-14\,\mathrm{dB}$ and $-10\,\mathrm{dB}$). It is evident that only the eigenvalues of the two targeted flexible modes are affected by the controller. The unchanged mode at lower frequency is an almost uncontrollable suspension mode, while the higher flexible modes are not adversely affected by the control action.

Analogous results could be obtained using the frequency-weighed $\mathcal{H}_2$-optimal control design methodology: the achieved RMS reductions of the performance variables are approximately the same as for the LQG control method. Note that the bandwidth of the frequency-weighted controller is narrow around $f = 60 - 75\,\mathrm{Hz}$ where the actuator weightings are small. Contrary to the LQG approach, the target modes as well as other modes with a higher negative real part are positively influenced. This indicates that the model quality is sufficiently high and that the control laws are insensitive to the occurring differences between design plant and actual system.

Performance position index i	1	2	3	4	5	6	avg.
RMS reduction $z_{i,\mathrm{ISO}}$ (LQG) in %	41.53	34.69	8.83	6.02	37.28	36.94	27.55
RMS reduction $z_{i,\mathrm{ISO}}$ ($\mathcal{H}_2$) in %	41.31	35.69	-4.00	4.59	34.89	36.16	24.77

Table 5. Laboratory testbed results: RMS reduction of the performance vector $\boldsymbol{z}$ by an LQG (system order 28) and an $\mathcal{H}_2$ controller (system order 23) utilizing strain feedback sensors

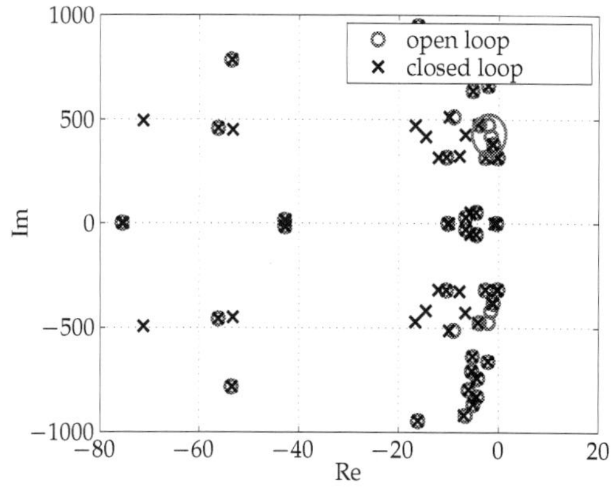

Fig. 29. Pole plot with/without LQG controller

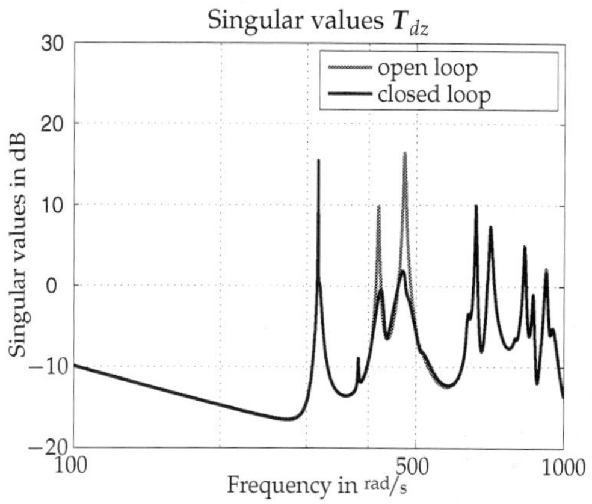

Fig. 30. Singular values of the frequency response plot with/without LQG controller

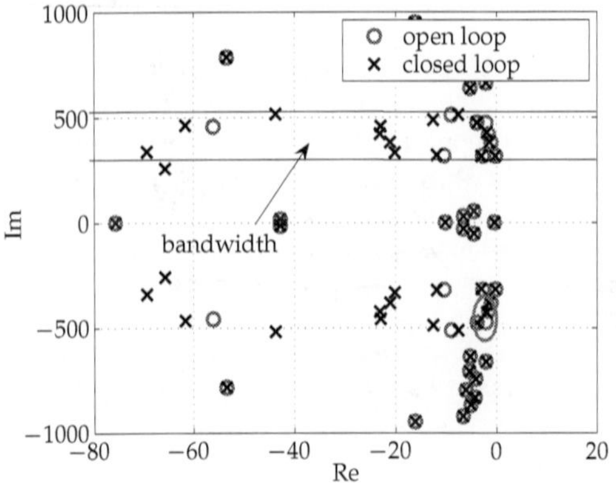

Fig. 31. Pole plot with/without $\mathcal{H}_2$ controller

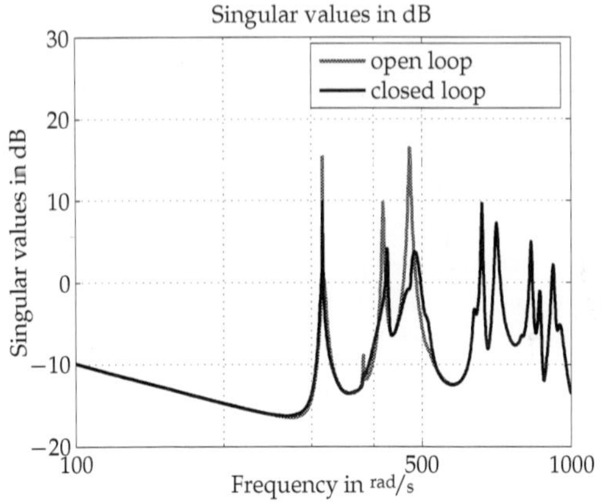

Fig. 32. Singular values of the frequency response plot with/without $\mathcal{H}_2$ controller

4. Conclusions

This chapter presents a case study on the design of MIMO control laws to reduce vibrations in a flexible metro rail car body and thus to improve passenger ride comfort. Direct structural actuation by Piezo actuators is considered and two sensor concepts – strain sensors (in a non-collocated setting) and acceleration sensors (collocated) – are evaluated. One part of the outlined studies focused on a simulation model of a full-size lightweight metro rail car body; the other part tests the control concepts on a laboratory testbed with a scale model of the car body. The control laws have been designed by LQG and by frequency-weighted $\mathcal{H}_2$-optimal control design methodologies. Both design methods are first studied in the simulation and compared. It is found that the weighted $\mathcal{H}_2$ designs yield controllers that perform satisfactorily in the presence of model uncertainty and independent of the sensor concept (strain sensors / non-collocation or acceleration sensors / collocation): the first three modes of interest are significantly attenuated and the unknown modes in the high-frequency domain are hardly affected by the controller action, thus increasing the ride comfort for the passengers. The LQG controller minimizes the vibrations only for strain sensors in the non-collocated setup. Finally, both design methods, which have been studied for the lightweight rail car body simulations, are successfully implemented in a scaled laboratory setup: it is demonstrated that the target modes (torsion and bending) have been significantly damped by both controller types. A further advantage of the weighted $\mathcal{H}_2$ controller is that the controller action can be tuned for a specific bandwidth in the frequency domain, which is essential if the dynamics of the structure under consideration is uncertain or a control input is not desired for certain frequencies. The studies' results show the applicability of weighted $\mathcal{H}_2$ control for partially uncertain flexible-structure systems. The control goal of improving ride comfort is directly formulated as a weighted $\mathcal{H}_2$ minimization problem which justifies the presented study. However, a range of related publications show the design and application of robust $\mathcal{H}_\infty$-optimal controllers for this application, which can give robustness guarantees based on the structured singular value.

5. References

Benatzky, C. (2006). *Theoretical and experimental investigation of an active vibration damping concept for metro vehicles*, PhD thesis, Institute for Mechanics and Mechatronics, Division of Control and Process Automation, Vienna University of Technology, Austria.

Benatzky, C. & Kozek, M. (2005). Effects of local actuator action on the control of large flexible structures, *Proceedings of the 16th IFAC World Congress*, Prague, Chech Republic.

Benatzky, C. & Kozek, M. (2007a). An actuator fault detection concept for active vibration control of a heavy metro vehicle, *Proceedings of the 14th International Congress on Sound and Vibration (ICSV14), Cairns, Australia.*

Benatzky, C. & Kozek, M. (2007b). An identification procedure for a scaled metro vehicle - flexible structure experiment, *Proceedings of the European Control Conference ECC 2007, Kos, Greece*, Kos, Greece.

Benatzky, C., Kozek, M. & Bilik, C. (2006). Experimental control of a flexible beam using a stack-bending-actuator principle, *Proceedings of the 20th Scientific Conference*, Hanoi, Vietnam.

Benatzky, C., Kozek, M. & Jörgl, H. (2007). Comparison of controller design methods for a scaled metro vehicle - flexible structure experiment, *Proceedings of the 26th American Control Conference, New York, USA.*

Bilik, C. (2006). *Aufbau und Inbetriebnahme des Prüfstandmodelles eines Schienenfahrzeug-Wagenkastens zum Nachweis von aktiver Schwingungsdämpfung*, Diploma thesis, Vienna University of Technology, Vienna.

Bilik, C., Benatzky, C. & Kozek, M. (2006). A PC-based multipurpose test bed environment for structural testing and control, *Proceedings of the 3rd International Symposium on Remote Engineering and Virtual Instrumentation*, Maribor, Slovenia.

Foo, E. & Goodall, R. M. (2000). Active suspension control of flexible-bodied railway vehicles using electro-hydraulic and electro-magnetic actuators, *Control Engineering Practice* 8(5): 507–518.

Frederich, F. (1984). *Die Gleislage - aus fahrzeugtechnischer Sicht*, Vol. 108 (12) of *Gleislauftechnik*, Siemens Verlagsbuchhandlung, pp. 355 – 361.

Gawronski, W. (2004). *Advanced structural dynamics and active control of structures*, Springer, New York.

Hansson, J., Takano, M., Takigami, T., Tomioka, T. & Suzuki, Y. (2004). Vibration Suppression of Railway Car Body with Piezoelectric Elements, *JSME International Journal Series C* 47(2): 451–456.

ISO (1997). ISO2631-1: Mechanical vibration and shock - evaluation of human exposure to whole-body vibration. Part 1: General requirements, International Organization for Standardization. Corrected and reprinted July 15th, 2007.

Kamada, T., Tohtake, T., Aiba, T. & Nagai, M. (2005). Active vibration control of the railway vehicle by smart structure concept, *in* S. Bruni & G. Mastinu (eds), *19th IAVSD Symposium - Poster Papers*.

Kozek, M. & Benatzky, C. (2008). Ein maßstäbliches Experiment zur aktiven Schwingungsdämpfung eines Eisenbahn-Wagenkastens, *at - Automatisierungstechnik* 10(56): 504–512.

Kozek, M., Benatzky, C., Schirrer, A. & Stribersky, A. (2011). Vibration damping of a flexible car body structure using piezo-stack actuators, *Control Engineering Practice* 19(3): 298 – 310. Special Section: IFAC World Congress Application Paper Prize Papers.
URL: *http://www.sciencedirect.com/science/article/B6V2H-4X3MR4Y-2/2/3ef1d868e70c2b6f10fd9412f9c8c1de*

Luenberger, D. G. (1964). Observing the state of a linear system, *IEEE Transactions on Military Electronics* 8(2): 74–80.

Mohinder, S. & Angus, P. (2001). *Kalman Filtering: Theory and Practice Using MATLAB*, Wiley Interscience, John Wiley & Sons, USA.

Popprath, S., Benatzky, C., Bilik, C., Kozek, M., Stribersky, A. & Wassermann, J. (2006). Experimental modal analysis of a scaled car body for metro vehicles, *Proceedings of the 13th International Congress on Sound and Vibration (ICSV13), Vienna, Austria*.

Popprath, S., Schirrer, A., Benatzky, C., Kozek, M. & Wassermann, J. (2007). Experimental modal analysis of an actively controlled scaled metro vehicle car body, *Proceedings of the 14th International Congress on Sound and Vibration (ICSV14), Cairns, Australia*.

Preumont, A. (2006). *Mechatronics: Dynamics of Electromechanical and Piezoelectric Systems*, Springer.

Schandl, G. (2005). *Methodenuntersuchung zur aktiven Schwingungsreduktion eines Schienenfahrzeugwagenkastens*, PhD thesis, Vienna University of Technology, Vienna.

Schandl, G., Lugner, P., Benatzky, C., Kozek, M. & Stribersky, A. (2007). Comfort enhancement by an active vibration reduction system for a flexible railway car body, *Vehicle System Dynamics* 45(9): 835–847.

Schöftner, J. (2006). *Aktive Schwingungsdämpfung eines Schienenfahrzeugwagenkastens durch $\mathcal{H}_2$-Regelung*, Master's thesis, Institute for Mechanics and Mechatronics, Division of Control and Process Automation, Vienna University of Technology.

Schirrer, A. (2010). *Co-Simulation of Rail Car Body Vibration Control with SimPACK®*, VDM Verlag Dr. Müller, Saarbrücken, Germany.

Schirrer, A. & Kozek, M. (2008). Co-simulation as effective method for flexible structure vibration control design validation and optimization, *Control and Automation, 2008 16th Mediterranean Conference on*, pp. 481 –486.

Schirrer, A., Kozek, M. & Benatzky, C. (2008). Piezo stack actuators in flexible structures: Experimental verification of a nonlinear modeling and identification approach, *6th EUROMECH Nonlinear Dynamics Conference (ENOC 2008), OPEN-ACCESS library*. URL: *http://lib.physcon.ru*

Schirrer, A., Kozek, M., Plank, A., Neumann, M., Badshah, S. & Wassermann, J. (2008). Vibration analysis of an actively controlled flexible structure using speckle interferometry, *Proceedings of 15th International Congress on Sound and Vibration (ICSV15)*.

Skogestad, S. & Postlethwaite, I. (1996). *Multivariable feedback control*, John Wiley & Sons.

Stribersky, A., Müller, H. & Rath, B. (1998). The development of an integrated suspension control technology for passenger trains, *Proceedings of the Institution of Mechanical Engineers, Part F: Journal of Rail and Rapid Transit*, Vol. 212, pp. 33–42.

ADVANCES ON ANALYSIS AND CONTROL OF VIBRATIONS – THEORY AND APPLICATIONS

Edited by **Mauricio Zapateiro de la Hoz**
and **Francesc Pozo**

LPV Gain-Scheduled Observer-Based State Feedback for Active Control of Harmonic Disturbances with Time-Varying Frequencies

Wiebke Heins, Pablo Ballesteros, Xinyu Shu and Christian Bohn

Additional information is available at the end of the chapter

http://dx.doi.org/10.5772/50293

1. Introduction

The design of controllers for the rejection of multisine disturbances with time-varying frequencies is considered. The frequencies are assumed to be known. Such a control problem frequently arises in active noise and vibration control (ANC/AVC) applications where the disturbances are caused by imbalances due to rotating or oscillating masses or periodically fluctuating excitations, for example the torque of a combustion engine, and the rotational speed is measured. Application examples are automobiles and aircrafts.

For the rejection of disturbances with time-varying frequencies, time-varying controllers that are automatically adjusted to the disturbance frequencies are usually used. Although time-invariant controllers might be sufficient in some applications [22], time-varying controllers usually result in a much better performance, particularly if the disturbance frequencies vary over fairly wide ranges. Such a controller can be constructed in several ways (see Sec. 2.1). Two observer-based state-feedback controllers are presented in this chapter (see Sec. 3). General output-feedback controllers are treated in the next chapter. The approaches presented in this chapter use state augmentation in order to achieve disturbance rejection. One consists of a time-invariant plant observer and a time-varying state-feedback gain for the state-augmented plant, where the state augmenation is based on a time-varying error filter, as proposed by Kinney & de Callafon [19]. The other controller approach is based on the disturbance observer of Bohn et al. [7], where the plant is augmented with a time-varying disturbance model. A time-varying observer for the overall system and a time-invariant state-feedback gain are used to track and reject the disturbance.

The remainder of this chapter is organized as follows. In Sec. 2, existing approaches to the problem are classified and some general control design considerations are discussed. The state-augmented observer-based state-feedback approaches are described in Sec. 3. In Sec. 4, the calculation of stabilizing state-feedback gains for time-varying systems in polytopic linear parameter-varying (pLPV) form is discussed. The application of this method for the rejection of harmonic disturbances is treated in Sec. 5. Real-time results are presented in Sec. 6. A discussion and conclusions are given in Sec. 7.

2. Controller design: Overview, stability and implementation aspects

In this section, an overview of control approaches for the rejection of harmonic disturbances with time-varying frequencies is given (Sec. 2.1) and stability and implementation aspects are discussed (Secs. 2.2 and 2.3, respectively).

2.1. Overview and classification of control approaches

A common approach in ANC/AVC is adaptive filtering, where the filter weights are usually updated with the FxLMS algorithm [24]. In most cases, disturbance feedforward is used, although it is possible to use an adaptive filter in feedback control with a technique called "secondary path neutralization" [24], which is equivalent to internal model control [25].

Another approach is to use gain scheduling, where the scheduling parameters are calculated from the disturbance frequencies. This can be further subdivided into indirect and direct scheduling methods. In indirect scheduling, the controller, or part of it, for example a state-feedback or observer gain, is determined from a set of pre-computed quantities through interpolation or switching. For example, for linear parameter-varying (LPV) systems, where the uncertain parameters are contained in a polytope, one controller is calculated for each vertex and the resulting controller is obtained from interpolation [2]. For the rejection of harmonic disturbances, continuous-time LPV approaches have been suggested by Darengosse & Chevrel [10], Du & Shi [11], Du et al. [12], Witte et al. [28], Balini et al. [6] and tested for a single sinusoidal disturbance by Darengosse & Chevrel [10], Du et al. [12], Witte et al. [28] and Balini et al. [6]. Methods based on observer-based state-feedback controllers are presented by Bohn et al. [7, 8], Kinney & de Callafon [19, 20, 21] and Heins et al. [16, 17]. In the approach of Bohn et al. [7, 8], the observer gain is selected from a set of pre-computed gains by switching. In the other approaches of Kinney & de Callafon [21], Heins et al. [17] and in this chapter, the observer gain is calculated by interpolation. In the other approach presented here, which is also used by Kinney & de Callafon [19, 20] and Heins et al. [16], the state-feedback gain is scheduled using interpolation.

In direct scheduling, the dependence of the controller on the scheduling parameter does not correspond to a simple interpolation or switching law [2, 4, 5, 21, 23, 26]. For example, for LPV systems where the parameter dependence is expressed as a linear fractional transformation (LFT), the uncertain parameters also enter the controller through an LFT [2]. For harmonic disturbances, an LPV-LFT approach has been suggested and applied in real time by Ballesteros & Bohn [4, 5] and Shu et al. [26]. Another example for direct scheduling is a controller based on a time-varying state estimator, for example a Kalman filter, where the scheduling parameters enter the controller through the recursive equations for the state estimate and the error covariance matrix. Such a controller is presented and compared to an indirect (interpolation) approach by Kinney & de Callafon [21].

2.2. Stability considerations

The existing design approaches can be classified as approaches that take stability into consideration and such that do not. In indirect scheduling, for example, the controllers or gains are sometimes pre-computed for fixed operating points and then interpolated in an ad-hoc fashion [7, 8, 20]. Stability is then not guaranteed, although it might be expected that the system is stable for slow variations of the scheduling parameter. For the adaptive filtering approaches, only approximate stability results seem available to date [13, 24]. To take stability into consideration, it is attractive to model the control problem as an LPV system and then use suitable gain-scheduling techniques [4–6, 10, 16, 17, 19, 23, 26, 28]. If parameter-independent Lyapunov functions are used, this guarantees closed-loop stability even for arbitrarily fast changes of the scheduling parameters at the expense of conservatism. The methods presented in this chapter guarantee closed-loop stability. In designs based on parameter-dependent Lyapunov functions, limits on the rate of change of the parameters can be incorporated.

2.3. Implementation aspects

For a practical application, the resulting controller has to be implemented in discrete time. In applications of ANC/AVC, the plant model is often obtained through system identification. This usually gives a discrete-time plant model. It is therefore most natural to carry out the whole design in discrete time. If a continuous-time controller is computed, the controller has to be discretized. Since the controllers considered here are time-varying, the discretization would have to be carried out at each sampling instant. An exact discretization involves the calculation of a matrix exponential, which is computationally too expensive. Particularly in LPV gain-scheduling control, an approximate discretization is proposed by Apkarian [3]. However, this leads to a distortion of the frequency scale. Usually, this can be tolerated, but not for the suppression of harmonic disturbances. It is therefore not surprising that the continuous-time controllers of Darengosse & Chevrel [10], Du et al. [12], Kinney & de Callafon [19] and Köroğlu & Scherer [23] are only tested in simulation studies with a very simple system as a plant and a single frequency in the disturbance signal. The design methods that are tested in real time are usually formulated in discrete time [4, 5, 7, 8, 16, 17, 20, 21, 26]. Exceptions are Witte et al. [28] and Balini et al. [6], who designed continuous-time controllers which then are approximately discretized. However, Witte et al. [28] use a very high sampling frequency of 40 kHz to reject a harmonic disturbance with a frequency up to 48 Hz (in fact, the authors state that they chose "the smallest [sampling time] available by the hardware") and Balini et al. [6] use a maximal sampling frequency of 50 kHz. It seems more natural to directly carry out the design in discrete time to avoid discretization issues.

3. State-augmented observer-based state-feedback control

Usually, disturbances act somewhere on the plant under consideration. The control objective is then to reject this disturbance. For the design methods considered in this paper, it is assumed that the disturbance acts on the plant input. For linear systems, a disturbance acting somewhere in the plant can be represented as a disturbance acting on the plant input under very mild assumptions. Therefore, assuming an input disturbance does not mean that the true disturbance has to act on the plant input (in fact, it usually does not). The design method used in this chapter is based on the internal model principle [14]. Applying this principle results in controllers with high gain in the frequency ranges where disturbances are assumed. These disturbances are then rejected no matter where they act upon the plant.

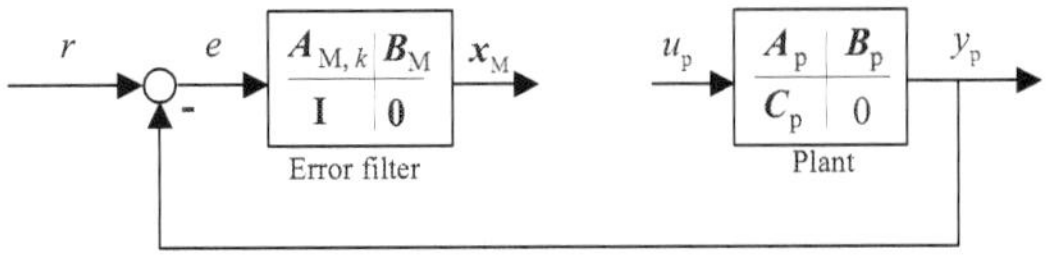

Figure 1. Interconnection of plant model and error filter

A discrete-time model for the plant with a disturbance $y_{d,k}$ acting on the input is given by

$$\begin{aligned} x_{p,k+1} &= A_p x_{p,k} + B_p(u_{p,k} + y_{d,k}), \\ y_{p,k} &= C_p x_{p,k}. \end{aligned} \tag{1}$$

In the approaches presented in the following, the disturbance rejection will be achieved with observer-based state-feedback controllers. The design methods presented here use state augmentation to add certain desired dynamics to the controller. In the time-invariant case, such a state augmentation can be used to prescribe controller poles. It can easily be verified that controller poles show up as zeros in the closed-loop disturbance transfer functions. Therefore, controller poles can be chosen to correspond to disturbances that are to be suppressed.

This argumentation, of course, only holds in the time-invariant case for which transfer functions and poles are defined (and only for controller poles that are not cancelled by plant zeros). If a time-varying internal model is used, it is not easy to interpret what happens when the internal model changes. Conceptually, the argument that the controller has high (infinite) gain at the disturbance frequencies that are to be rejected should still hold. It is confirmed by experiments (see Sec. 6) that even for fairly fast changes of the disturbance frequencies, excellent disturbance rejection is achieved.

3.1. State augmentation through an error filter

In this approach, disturbance rejection is achieved by including additional dynamics in a state-feedback controller through error filtering. A general error filter in state-space representation is given by

$$x_{M,k+1} = A_M x_{M,k} + B_M \left(r_k - y_{p,k}\right) \tag{2}$$

and interconnected with the plant model (without disturbance input)

$$\begin{aligned} x_{p,k+1} &= A_p x_{p,k} + B_p u_{p,k}, \\ y_{p,k} &= C_p x_{p,k}. \end{aligned} \tag{3}$$

as shown in Fig. 1. Usually, the controller error $e = r - y_p$ is used as input for the error filter. In the application considered here, however, no set-point r will be given and the error filter basically acts as an output filter with input $-y_p$. Nonetheless it will be referred to as error filter in the following.

The dynamics of the error filter can be used to include additional desired dynamics in a state-feedback controller for the overall system given by the plant and the filter. The additional dynamics can be chosen such that they describe the disturbances that the controller should reject, which corresponds to the internal model principle [14]. The additional dynamics can therefore be referred to as the "internal (disturbance) model." If, as in the control problem considered in this chapter, the disturbance characteristics change over time, a time-varying internal model

$$x_{M,k+1} = A_{M,k} x_{M,k} - B_M y_{p,k} \tag{4}$$

can be used.

In order to design a state-feedback controller for the overall system, first the overall system model is formed by combining the plant model (3) and the error filter dynamics (4) through introduction of an augmented state vector. This yields

$$\begin{aligned} \begin{bmatrix} x_{M,k+1} \\ x_{p,k+1} \end{bmatrix} &= \begin{bmatrix} A_{M,k} & -B_M C_p \\ 0 & A_p \end{bmatrix} \begin{bmatrix} x_{M,k} \\ x_{p,k} \end{bmatrix} + \begin{bmatrix} 0 \\ B_p \end{bmatrix} u_{p,k}, \\ y_{p,k} &= \begin{bmatrix} 0 & C_p \end{bmatrix} \begin{bmatrix} x_{M,k} \\ x_{p,k} \end{bmatrix}. \end{aligned} \tag{5}$$

Then, a state-feedback gain K_k for the overall system (5) can be designed. Due to the time-varying dynamics of the error filter, the state-feedback gain may be time-varying.

Usually the plant states are not available for feedback. Therefore, an identity observer

$$\widehat{x}_{p,k+1} = (A_p - L_{p,k} C_p)\widehat{x}_{p,k} + B_p u_{p,k} + L_{p,k} y_{p,k} \tag{6}$$

is used to obtain an estimate $\widehat{x}_p$ of the plant states x_p. The estimate is then used for feedback instead of the unknown plant states.

For an overall state-feedback gain

$$K_k = \begin{bmatrix} K_{M,k} & K_{p,k} \end{bmatrix}, \tag{7}$$

where $K_{M,k}$ and $K_{p,k}$ denote the parts which feed back the error filter states x_M and the plant states x_p (or in this case, the estimated plant states $\widehat{x}_p$), respectively, the state-feedback law is given by

$$u_{p,k} = -K_{M,k} x_{M,k} - K_{p,k} \widehat{x}_{p,k}. \tag{8}$$

The structure of the overall closed-loop system is shown in Fig. 2. The plant is affected by a disturbance y_d at the plant input. The plant output y_p is used as input for the error filter (with a sign reversal), which outputs are its states x_M, and for the identity observer for the plant. The latter gives an estimate $\widehat{x}_p$ of the plant states computed from y_p and u_p. The states are then combined and fed back with the overall state-feedback gain consisting of $K_{M,k}$ and $K_{p,k}$.

For stability analysis of the closed-loop system it is convenient to separate the state-space representation of the overall closed-loop dynamics including plant, observer, error filter, state-feedback and disturbance input in several parts as derived in the following. First, the controlled plant under disturbance at the input can be derived by combining (1), (6) and (8) to

$$x_{p,k+1} = (A_p - B_p K_{p,k}) x_{p,k} - B_p K_{M,k} x_{M,k} - B_p K_{p,k} (\widehat{x}_{p,k} - x_{p,k}) + B_p y_{d,k}. \tag{9}$$

It is driven by the error filter states x_M, the observer error $\widehat{x}_p - x_p$ and the real unknown disturbance y_d.

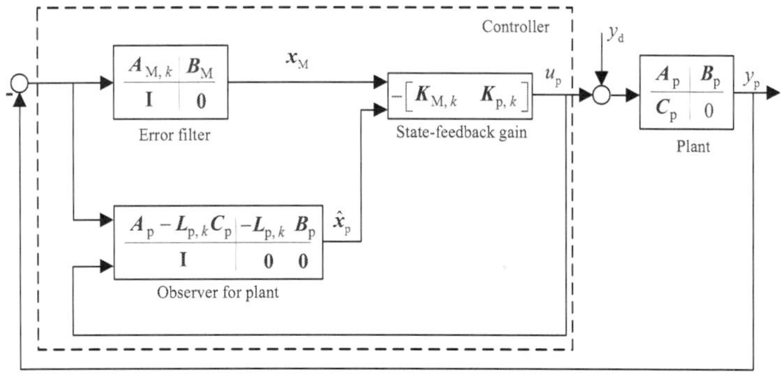

Figure 2. Structure of the resulting controller based on error filtering

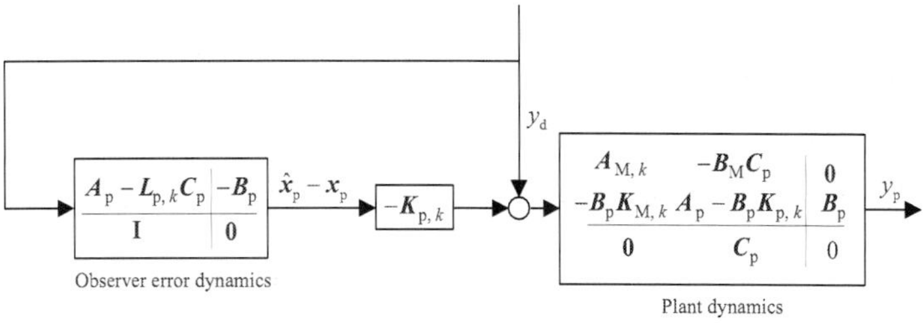

Figure 3. Dynamics of the overall closed-loop system for stability analysis

The dynamics of the identity observer for the plant states are described by (6). Since the observer does not include a model of the disturbance, the dynamics of the observer error, from now on denoted by $\tilde{x}_p$, are given as

$$\tilde{x}_{p,k+1} = (A_p - L_{p,k}C_p)\tilde{x}_{p,k} - B_p y_{d,k}, \tag{10}$$

which means that the observer is driven by the unknown disturbance y_d. With the error filter dynamics given by (4), the overall system can be described as

$$\begin{bmatrix} \tilde{x}_{p,k+1} \\ x_{M,k+1} \\ x_{p,k+1} \end{bmatrix} = \begin{bmatrix} A_p - L_{p,k}C_p & 0 & 0 \\ 0 & A_{M,k} & -B_M C_p \\ -B_p K_{p,k} & -B_p K_{M,k} & A_p - B_p K_{p,k} \end{bmatrix} \begin{bmatrix} \tilde{x}_{p,k} \\ x_{M,k} \\ x_{p,k} \end{bmatrix} + \begin{bmatrix} -B_p \\ 0 \\ B_p \end{bmatrix} y_{d,k}. \tag{11}$$

The overall system behavior is shown in Fig. 3. From this representation it can be seen that the time-varying dynamics of the augmented plant under state feedback are driven by the time-varying dynamics of the observer error for the plant states. The augmented plant under state feedback and the dynamics of the observer error for the plant states can therefore be interpreted as two distinct systems connected in series. From this it follows that choosing an observer gain $L_{p,k}$ such that $A_p - L_{p,k}C_p$ is stable and a state-feedback gain such that the time-varying dynamics of the augmented plant under state feedback are stable guarantees overall stability of the closed-loop system.

3.2. State augmentation through a disturbance model

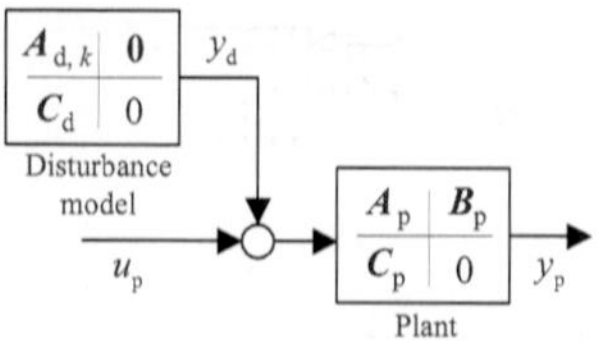

Figure 4. System model for the disturbance observer approach

In this approach, the disturbance y_d acting on the plant input u_p is explicitly modeled as the output of an unforced linear time-varying system

$$\begin{aligned} x_{d,k+1} &= A_{d,k} x_{d,k}, \\ y_{d,k} &= C_d x_{d,k} \end{aligned} \tag{12}$$

as shown in Fig. 4. An overall linear time-varying system model is then given by the combination of the plant model (1), which is affected by the disturbance $y_{d,k}$ at the plant input, and disturbance through

$$\begin{aligned} \begin{bmatrix} x_{d,k+1} \\ x_{p,k+1} \end{bmatrix} &= \begin{bmatrix} A_{d,k} & 0 \\ B_p C_d & A_p \end{bmatrix} \begin{bmatrix} x_{d,k} \\ x_{p,k} \end{bmatrix} + \begin{bmatrix} 0 \\ B_p \end{bmatrix} u_{p,k}, \\ y_{p,k} &= \begin{bmatrix} 0 & C_p \end{bmatrix} \begin{bmatrix} x_{d,k} \\ x_{p,k} \end{bmatrix}. \end{aligned} \tag{13}$$

With

$$x_k = \begin{bmatrix} x_{d,k} \\ x_{p,k} \end{bmatrix}, A_k = \begin{bmatrix} A_{d,k} & 0 \\ B_p C_d & A_p \end{bmatrix}, B = \begin{bmatrix} 0 \\ B_p \end{bmatrix}, C = \begin{bmatrix} 0 & C_p \end{bmatrix}, \tag{14}$$

the overall system can be written as

$$\begin{aligned} x_{k+1} &= A_k x_k + B u_{p,k}, \\ y_{p,k} &= C x_k. \end{aligned} \tag{15}$$

Estimates for the states of the disturbance and the plant model can then be obtained by application of a linear time-varying identity observer with time-varying observer gain L_k

$$\widehat{x}_{k+1} = (A_k - L_k C)\widehat{x}_k + B u_{p,k} + L_k y_{p,k}. \tag{16}$$

A state-feedback control for the overall system is given by

$$u_{p,k} = -K_k \widehat{x}_k. \tag{17}$$

This leads to the typical structure of an observer-based state-feedback controller, but due to the time-varying disturbance model, the observer gain as well as state-feedback gain might be time-varying. Fig. 5 shows the general structure. The idea behind the disturbance observer is to use the estimate of the disturbance states (with a sign reversal) as control input. This should

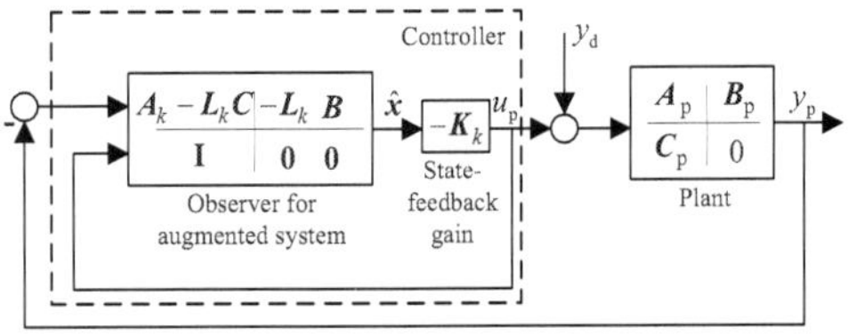

Figure 5. Observer-based state-feedback control in the disturbance observer approach

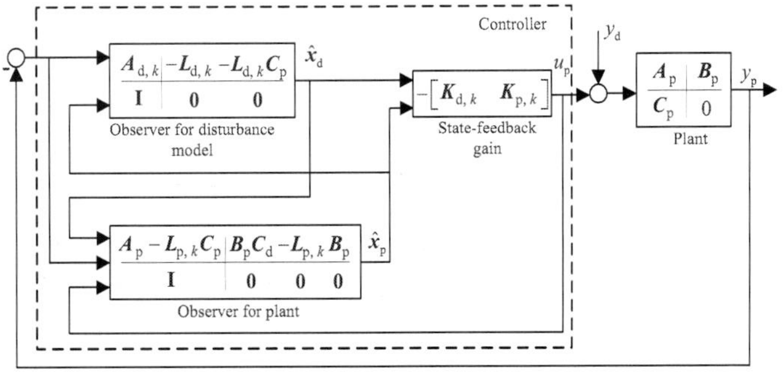

Figure 6. Detailed illustration of observer structure

lead to perfect disturbance cancellation if the disturbance model is chosen appropriately. For a more detailed analysis of this aspect, in the following observer gains and state-feedback gains are seperated into a part that feeds back the disturbance states via $L_{d,k}$ and $K_{d,k}$ and a part that feeds back the plant states or their estimates via $L_{p,k}$ and $K_{p,k}$. The estimated disturbance states can be described by

$$\hat{x}_{d,k+1} = A_{d,k}\hat{x}_{d,k} - L_{d,k}C_p\hat{x}_{p,k} + L_{d,k}y_{p,k}, \tag{18}$$

and the estimated plant states are given through

$$\hat{x}_{p,k+1} = (A_p - L_{p,k}C_p)\hat{x}_{p,k} + B_pC_d\hat{x}_{d,k} + L_{p,k}y_{p,k} + B_pu_{p,k}, \tag{19}$$

where

$$L_k = \begin{bmatrix} L_{d,k} & L_{p,k} \end{bmatrix}. \tag{20}$$

In Fig. 6, this more detailed representation is illustrated.

For stability analysis of the overall system including overall observer dynamics and state feedback, the dynamics of the observer error for the plant states

$$\hat{x}_{p,k+1} - x_{p,k+1} = (A_p - L_{p,k}C_p)(\hat{x}_{p,k} - x_{p,k}) + B_pC_d\hat{x}_{d,k} - B_py_{d,k} \tag{21}$$

are considered.

The dynamics of the states of the plant under state feedback are given by

$$x_{p,k+1} = (A_p - B_pK_{p,k})x_{p,k} - B_pK_{d,k}\hat{x}_{d,k} + B_pK_{p,k}(\hat{x}_{p,k} - x_{p,k}) + B_py_{d,k}. \tag{22}$$

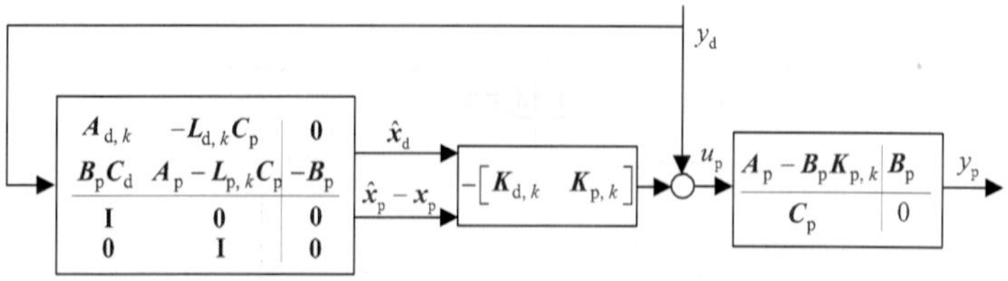

Figure 7. Dynamics of the overall closed-loop system for stability analysis

This yields the overall closed-loop system dynamics given by

$$\begin{bmatrix} \hat{x}_{d,k+1} \\ \tilde{x}_{p,k+1} \\ x_{p,k+1} \end{bmatrix} = \begin{bmatrix} A_{d,k} & -L_{d,k}C_p & 0 \\ B_pC_d & A_p - L_{p,k}C_p & 0 \\ -B_pK_{d,k} & B_pK_{p,k} & A_p - B_pK_{p,k} \end{bmatrix} \begin{bmatrix} \hat{x}_{d,k} \\ \tilde{x}_{p,k} \\ x_{p,k} \end{bmatrix} + \begin{bmatrix} 0 \\ -B_p \\ B_p \end{bmatrix} y_{d,k}. \tag{23}$$

Fig. 7 illustrates the structure of the overall system dynamics. This representation shows, similiar to the error filtering approach, that the dynamics of the plant under state feedback are driven by the time-varying dynamics of the observer for the augmented system, where in this case the outputs are given by the estimate of the disturbance states and the observer error for the plant states. The plant under state feedback and the observer of the augmented system can therefore be considered as two distinct systems in series connection. From this it follows that as long as a stabilizing state-feedback gain $K_{p,k}$ for the linear time-invariant plant and a stabilizing observer gain for the augmented system model is chosen, the overall closed-loop system is stable. It is also evident that the choice of the disturbance model does not influence the overall system stability, as long as the disturbance model itself is stable. Also, the choice of $K_{d,k}$ has no effect on the overall system stability (as long as $K_{d,k}\hat{x}_{d,k}$ remains bounded). Therefore, the overall state-feedback gain K does not have to be time-varying since only the linear time-invariant plant must be stabilized by K_p and the choice of K_d is not important for system stability. Furthermore, it can be seen that, assuming a perfect disturbance model, complete disturbance cancellation will be achieved after some transient if $K_d = C_d$.

4. Gain-scheduled state-feedback and observer design for pLPV systems

In this section, design methods for time-varying state-feedback and observer gains for pLPV systems based on quadratic stability theory are reviewed.

An LPV system is a linear system described by

$$\begin{aligned} x_{k+1} &= A(\theta)x_k + B(\theta)u_k \\ y_k &= C(\theta)x_k + D(\theta)u_k \end{aligned} \tag{24}$$

with $x_k \in \mathbb{R}^n$, where the state-space matrices depend on a possibly time-varying parameter vector

$$\theta = \begin{bmatrix} \theta_1 \ \theta_2 \cdots \theta_N \end{bmatrix}^T \in \Theta \subset \mathbb{R}^N. \tag{25}$$

In the following, only unforced systems of the form

$$x_{k+1} = A(\theta)x_k \tag{26}$$

are considered. In general, an LPV system is called a pLPV system, if the matrices depend affinely on the varying parameter. For the unforced system given in (26), this means that it is a pLPV system, if the system matrix $A(\cdot)$ depends affinely on $\boldsymbol{\theta}$, that is

$$A(\boldsymbol{\theta}) = \mathcal{A}_0 + \theta_1\mathcal{A}_1 + \theta_2\mathcal{A}_2 + \cdots + \theta_N\mathcal{A}_N, \tag{27}$$

where $\mathcal{A}_i \in \mathbb{R}^{n\times n}$ for all $i = 0, 1, ..., N$ are constant matrices and $\boldsymbol{\Theta}$ is a convex polytope in $\mathbb{R}^N$ with a finite set of vertices $V = \{v_1, v_2, ..., v_M\} \subset \mathbb{R}^N$. A point $\boldsymbol{\theta} \in \boldsymbol{\Theta}$ can be written as a convex combination of vertices, which means there exists a coordinate vector $\boldsymbol{\lambda} = [\lambda_1\,\lambda_2\,\ldots\,\lambda_M]^{\mathrm{T}} \in \mathbb{R}^M$ such that $\boldsymbol{\theta}$ can be written as

$$\boldsymbol{\theta} = \sum_{j=1}^{M} \lambda_j v_j, \quad \lambda_j \geq 0 \quad \forall j = 1, ..., M, \quad \sum_{j=1}^{M} \lambda_j = 1. \tag{28}$$

Defining $A_{\mathrm{v},j} = A(v_j)$ for $j = 1, \ldots, M,$ $A(\boldsymbol{\theta})$ can be represented as

$$A(\boldsymbol{\theta}) = A(\boldsymbol{\lambda}) = \lambda_1 A_{\mathrm{v},1} + \lambda_2 A_{\mathrm{v},2} + \cdots + \lambda_M A_{\mathrm{v},M}. \tag{29}$$

The matrices $A_{\mathrm{v},j}$ can be considered as system matrices of linear time-invariant vertex systems

$$x_{k+1} = A_{\mathrm{v},j} x_k, \quad j = 1, \ldots, M, \tag{30}$$

and the current system (26) is a convex combination of the vertex systems in (30).

The pLPV system (26) is called *quadratically stable* in $\boldsymbol{\Theta}$ if and only if there exists a positive definite matrix $\boldsymbol{P}$ such that for all $\boldsymbol{\theta} \in \boldsymbol{\Theta}$

$$A^{\mathrm{T}}(\boldsymbol{\theta})\boldsymbol{P}A(\boldsymbol{\theta}) - \boldsymbol{P} < 0, \tag{31}$$

or equivalently via the Schur complement [18]

$$\begin{bmatrix} \boldsymbol{P} & \boldsymbol{P}A(\boldsymbol{\theta}) \\ A^{\mathrm{T}}(\boldsymbol{\theta})\boldsymbol{P} & \boldsymbol{P} \end{bmatrix} > 0, \tag{32}$$

where "> 0" and "< 0" indicate positive and negative definiteness, respectively.

Since (32) has to hold for every $\boldsymbol{\theta}$, this constitutes an infinite set of linear matrix inequalities (LMIs), which is computationally intractable. By the following result, it is possible to use a finite set of inequalities. It holds that a pLPV system in the form of (26) is quadratically stable for every $\boldsymbol{\theta} \in \boldsymbol{\Theta}$ if and only if there exists a symmetric positive definite matrix $\boldsymbol{P}$ such that

$$A_{\mathrm{v},j}^{\mathrm{T}}\boldsymbol{P}A_{\mathrm{v},j} - \boldsymbol{P} < 0, \tag{33}$$

or equivalently

$$\begin{bmatrix} \boldsymbol{P} & \boldsymbol{P}A_{\mathrm{v},j} \\ A_{\mathrm{v},j}^{\mathrm{T}}\boldsymbol{P} & \boldsymbol{P} \end{bmatrix} > 0, \tag{34}$$

for all $j = 1, \ldots, M$. This result is based on one single quadratic Lyapunov function that assures stability for the whole parameter space $\boldsymbol{\Theta}$. Therefore, $\boldsymbol{\theta}$ is allowed to change arbitrarily fast with time and $\boldsymbol{\theta}$ and $\boldsymbol{\lambda}$ can explicitly be assumed time-varying and denoted as $\boldsymbol{\theta}_k$ and $\boldsymbol{\lambda}_k$, respectively. A proof is given in Amato [1].

These results can be applied to the design of time-varying state-feedback and observer gains, as reviewed in the following sections.

4.1. State-feedback design for pLPV systems based on quadratic stability

A system of the form

$$x_{k+1} = A(\theta_k)x_k + Bu_k \tag{35}$$

is considered and the objective is to find a state-feedback gain $K(\theta_k)$ that quadratically stabilizes the closed-loop system

$$x_{k+1} = (A(\theta_k) - BK(\theta_k))x_k. \tag{36}$$

From the above result it follows that it suffices to find a single symmetric positive definite matrix P and a finite set of matrices $K_{v,j}$ such that

$$\begin{bmatrix} P & P(A_{v,j} - BK_{v,j}) \\ (A_{v,j} - BK_{v,j})^T P & P \end{bmatrix} > 0, \quad j = 1, \ldots, M. \tag{37}$$

Then, $K(\theta_k)$ has to be chosen as

$$K(\theta_k) = \sum_{j=1}^{M} \lambda_{j,k} K_{v,j} = K(\lambda_k), \tag{38}$$

because then it holds that

$$A(\theta_k) - BK(\theta_k) = \sum_{j=1}^{M} \lambda_{j,k} A_{v,j} - B \sum_{j=1}^{M} \lambda_{j,k} K_{v,j} = \sum_{j=1}^{M} \lambda_{j,k}(A_{v,j} - BK_{v,j}), \tag{39}$$

and therefore quadratic stability of (36) is implied due to the results presented above.

4.2. Computation of state-feedback gains for the vertex systems

In this section, sufficient conditions that guarantee closed-loop stability and a certain H_2 performance level for the controlled vertex systems that can be used to calculate state-feedback gains for the vertex systems are derived. To use an H_2 performance level, an additional performance input is introduced that enters the state update equation for each vertex system of the pLPV system in (35). This input can be interpreted as process noise. For every $j = 1, \ldots, M$, this yields

$$x_{k+1} = A_{v,j}x_k + Bu_k + w_k. \tag{40}$$

As a performance output, the artificial signal

$$z_k = \begin{bmatrix} Q^{\frac{1}{2}} x_k \\ R^{\frac{1}{2}} u_k \end{bmatrix} \tag{41}$$

is defined that weighs the states and the control signal with matrices $Q^{\frac{1}{2}}$ and $R^{\frac{1}{2}}$, respectively. The objective then is to find state-feedback gains $K_{v,j}$ that stabilize the system and minimize the H_2 norm of the transfer path from w_k to z_k, when the control signal u_k in (40) and (41) is chosen as

$$u_k = -K_{v,j} x_k. \tag{42}$$

This objective can be stated as the minimization of the H_2 norm of the system

$$z = Gw \tag{43}$$

which has the state-space representation

$$\begin{aligned} x_{k+1} &= \widetilde{A} x_k + \widetilde{B} w_k, \\ z_k &= \widetilde{C} x_k, \end{aligned} \tag{44}$$

with

$$\widetilde{A} = A_{v,j} - BK_{v,j}, \quad \widetilde{B} = \mathrm{I}, \quad \widetilde{C} = \begin{bmatrix} Q^{\frac{1}{2}} \\ -R^{\frac{1}{2}} K_{v,j} \end{bmatrix}. \tag{45}$$

If the system is stable, the H_2 norm of this discrete-time linear time-invariant system is given by

$$\|G\|_2^2 = \text{trace}\left(\widetilde{C} W_c \widetilde{C}^{\mathrm{T}}\right), \tag{46}$$

where the controllability gramian W_c satisfies the discrete-time Lyapunov equation

$$\widetilde{A} W_c \widetilde{A}^{\mathrm{T}} - W_c + \widetilde{B}\widetilde{B}^{\mathrm{T}} = 0. \tag{47}$$

Therefore, if there exist symmetric positive definite matrices P and W such that

$$\widetilde{A} P \widetilde{A}^{\mathrm{T}} - P + \widetilde{B}\widetilde{B}^{\mathrm{T}} < 0, \tag{48}$$

$$W - \widetilde{C} P \widetilde{C}^{\mathrm{T}} > 0, \tag{49}$$

$$\text{trace}\,(W) < \gamma^2, \tag{50}$$

then it follows that $\|G\|_2 < \gamma$ for a $\gamma > 0$. Through the Schur complement, (48) and (49) can be transformed to

$$\begin{bmatrix} P & P\widetilde{A}^{\mathrm{T}} \\ \widetilde{A}P & P - \widetilde{B}\widetilde{B}^{\mathrm{T}} \end{bmatrix} > 0, \tag{51}$$

$$\begin{bmatrix} W & \widetilde{C}P \\ P\widetilde{C}^{\mathrm{T}} & P \end{bmatrix} > 0. \tag{52}$$

Introducing

$$\widetilde{Q} = \begin{bmatrix} Q^{\frac{1}{2}} \\ 0 \end{bmatrix}, \quad \widetilde{R} = \begin{bmatrix} 0 \\ R^{\frac{1}{2}} \end{bmatrix}, \quad Y_{v,j} = K_{v,j} P, \tag{53}$$

it follows that if solutions for the matrix variables P, W and $Y_{v,j}$, $j = 1, \ldots, M$, can be found that satisfy the $2M + 1$ LMIs

$$\begin{bmatrix} P & \left(A_{v,j} P - BY_{v,j}\right)^{\mathrm{T}} \\ A_{v,j} P - BY_{v,j} & P - \mathrm{I} \end{bmatrix} > 0, \; j = 1, \ldots, M, \tag{54}$$

$$\begin{bmatrix} W & \tilde{Q}P - \tilde{R}Y_{\mathrm{v},j} \\ \left(\tilde{Q}P - \tilde{R}Y_{\mathrm{v},j}\right)^{\mathrm{T}} & P \end{bmatrix} > 0, \; j = 1, \ldots, M, \tag{55}$$

$$\mathrm{trace}\,(W) < \gamma^2, \tag{56}$$

then the system G has an H_2 norm bounded by γ. From the solutions P and $Y_{\mathrm{v},j}$, the state-feedback gain for each vertex system can be calculated as

$$K_{\mathrm{v},j} = Y_{\mathrm{v},j}P^{-1}. \tag{57}$$

Quadratic stability is then implied for each closed-loop vertex system because of (48). In order to guarantee quadratic stability for the whole parameter space, solutions for the matrix variables P and W in (54)-(56) have to be the same for all vertex systems. Therefore, if solutions are found, also the performance bound γ is guaranteed for every fixed $\boldsymbol{\theta}$ in the parameter space $\boldsymbol{\Theta}$, since it depends only on P and W.

Once state-feedback gains are found for the vertex systems, in any instant of time of a realization of the pLPV system (35) with $\boldsymbol{\theta}_k \in \boldsymbol{\Theta}$, a state-feedback gain can be found via interpolation according to (38). Quadratic stability of the closed-loop system (36) is guaranteed. The question of how to compute the coordinates $\boldsymbol{\lambda}_k$ with the properties described in (28) depends on the specific polytope $\boldsymbol{\Theta}$. The case of an N-dimensional hyper box will be considered in detail in Sec. 4.4.

4.3. Observer design for pLPV systems based on quadratic stability

Although the observer design is dual to the controller design, the LMIs for observer design are shortly presented here for completeness.

For observer design, a system of the form

$$\begin{aligned} x_{k+1} &= A(\boldsymbol{\theta}_k)x_k + Bu_k, \\ y_k &= Cx_k \end{aligned} \tag{58}$$

is considered and the objective is to find an observer gain $L(\boldsymbol{\theta}_k)$ that quadratically (and therefore, asymptotically) stabilizes the dynamics of the observer error

$$\tilde{x}_{k+1} = (A(\boldsymbol{\theta}_k) - L(\boldsymbol{\theta}_k)C)\tilde{x}_k. \tag{59}$$

From the above result it follows that it suffices to find a single symmetric positive definite matrix P and a finite set of matrices $L_{\mathrm{v},j}$ such that

$$\begin{bmatrix} P & P(A_{\mathrm{v},j} - L_{\mathrm{v},j}C) \\ (A_{\mathrm{v},j} - L_{\mathrm{v},j}C)^{\mathrm{T}}P & P \end{bmatrix} > 0, j = 1, \ldots, M, \tag{60}$$

and $L(\boldsymbol{\theta}_k)$ has to be chosen as

$$L(\boldsymbol{\theta}_k) = \sum_{j=1}^{M} \lambda_{j,k} L_{\mathrm{v},j} = L(\boldsymbol{\lambda}_k). \tag{61}$$

As in the design of the state-feedback gain in the previous section, sufficient conditions for quadratic stability of (59) and a certain H_2 performance level can be derived. It is assumed that white noise of unit covariance, given by w_1 and w_2 and weighted by matrices $Q^{\frac{1}{2}}$ and $R^{\frac{1}{2}}$ respectively, affects the states and the outputs of the vertex systems of (58). This gives

$$\begin{aligned} x_{k+1} &= A(\theta_k)x_k + Cu_k + Q^{\frac{1}{2}} w_{1,k} \\ y_k &= Cx_k + R^{\frac{1}{2}} w_{2,k}. \end{aligned} \tag{62}$$

The H_2-norm of the transfer path from this noise input to the observer error is chosen as the performance measure. Defining

$$w_k = \begin{bmatrix} w_{1,k} \\ w_{2,k} \end{bmatrix}, \tag{63}$$

this transfer path is described by

$$\tilde{x}_{k+1} = \left(A_{\mathrm{v},j} - L_{\mathrm{v},k}C\right)\tilde{x}_k + \left[\, Q^{\frac{1}{2}} \;\; -L_{\mathrm{v},k}R^{\frac{1}{2}} \,\right] w_k. \tag{64}$$

Therefore the objective is to minimize the H_2-norm of the system $z = Gw$ with state-space representation

$$\begin{aligned} x_{k+1} &= \widetilde{A}x_k + \widetilde{B}w_k, \\ z_k &= \widetilde{C}x_k, \end{aligned} \tag{65}$$

with

$$\widetilde{A} = A_{\mathrm{v},j} - L_{\mathrm{v},j}C, \quad \widetilde{B} = \left[\, Q^{\frac{1}{2}} \;\; -L_{\mathrm{v},k}R^{\frac{1}{2}} \,\right], \quad \widetilde{C} = \mathrm{I}. \tag{66}$$

If the system is stable, the H_2 norm of this discrete-time linear time-invariant system is given by

$$\|G\|_2^2 = \mathrm{trace}\left(\widetilde{B}^{\mathrm{T}} W_{\mathrm{o}} \widetilde{B}\right), \tag{67}$$

where the observability gramian W_{o} satisfies the discrete-time Lyapunov equation

$$\widetilde{A} W_{\mathrm{o}} \widetilde{A}^{\mathrm{T}} - W_{\mathrm{o}} + \widetilde{C}^{\mathrm{T}}\widetilde{C} = 0. \tag{68}$$

For the same arguments as in Sec. 4.1 and with the introduction of

$$\widetilde{Q} = \begin{bmatrix} Q^{\frac{\mathrm{T}}{2}} \\ 0 \end{bmatrix}, \quad \widetilde{R} = \begin{bmatrix} 0 \\ R^{\frac{\mathrm{T}}{2}} \end{bmatrix}, \quad Y_{\mathrm{v},j} = L_{\mathrm{v},j}^{\mathrm{T}}P, \;\; j = 1, \ldots, M, \tag{69}$$

it follows that if solutions for the matrix variables P, W and $Y_{\mathrm{v},j}$, $j = 1, \ldots, M$, can be found that satisfy the $2M + 1$ LMIs

$$\begin{bmatrix} P & PA_{\mathrm{v},j} - Y_{\mathrm{v},j}^{\mathrm{T}}C \\ (PA_{\mathrm{v},j} - Y_{\mathrm{v},j}^{\mathrm{T}}C)^{\mathrm{T}} & P - \mathrm{I} \end{bmatrix} > 0, \;\; j = 1, \ldots, M, \tag{70}$$

$$\begin{bmatrix} W & \widetilde{Q}P - \widetilde{R}Y_{\mathrm{v},j} \\ (\widetilde{Q}P - \widetilde{R}Y_{\mathrm{v},j})^{\mathrm{T}} & P \end{bmatrix} > 0, \;\; j = 1, \ldots, M, \tag{71}$$

$$\operatorname{trace}(\boldsymbol{W}) < \gamma^2, \tag{72}$$

then the system $\boldsymbol{G}$ has an H_2 norm bounded by γ. From the solutions $\boldsymbol{P}$ and $\boldsymbol{Y}_{\mathrm{v},j}$, the observer gain for each vertex system can be calculated as

$$\boldsymbol{L}_{\mathrm{v},j} = \boldsymbol{P}^{-\mathrm{T}}\boldsymbol{Y}_{\mathrm{v},j}^{\mathrm{T}}. \tag{73}$$

In order to guarantee quadratic stability for the whole parameter space, solutions for the matrix variables $\boldsymbol{P}$ and $\boldsymbol{W}$ in (70)-(72) have to be the same for all vertex systems. Therefore, if solutions are found, also the performance bound γ is guaranteed for every fixed $\boldsymbol{\theta}$ in the parameter space $\boldsymbol{\Theta}$.

Once observer gains are found for the vertex systems, in any instant of time an observer gain for the realization of the pLPV system (58) with $\boldsymbol{\theta}_k \in \boldsymbol{\Theta}$ can be found via interpolation according to (61).

4.4. Computation of the coordinates for interpolation

The computation of the coordinates $\boldsymbol{\lambda}_k \in \mathbb{R}^M$ required for the interpolation depends on the specific polytope $\boldsymbol{\Theta} \subset \mathbb{R}^N$ that is used to cover the parameter space. In general, for a parameter vector $\boldsymbol{\theta}_k \in \boldsymbol{\Theta}$, where $\boldsymbol{\Theta}$ denotes a polytope with M vertices $\boldsymbol{v}_1, \boldsymbol{v}_2, \ldots, \boldsymbol{v}_M \in \mathbb{R}^N$, the coordinate vector $\boldsymbol{\lambda}_k$ as required for the interpolation is given by the solution to the constrained system of linear equations

$$\begin{bmatrix} \boldsymbol{v}_1 & \boldsymbol{v}_2 & \ldots & \boldsymbol{v}_M \\ 1 & 1 & \ldots & 1 \end{bmatrix} \boldsymbol{\lambda}_k = \begin{bmatrix} \boldsymbol{\theta}_k \\ 1 \end{bmatrix}, \; 0 \le \lambda_{j,k} \le 1, \; j = 1, \ldots, M. \tag{74}$$

Existence of the solution is guaranteed as long as $\boldsymbol{\theta}_k \in \boldsymbol{\Theta}$. If the polytope considered has less than or exactly $N+1$ vertices, the constrained linear equation system admits exactly one solution. If the polytope is given by more than $N+1$ vertices, the system is underdetermined and more than one solution is possible. Since the controller design is not a linear operation, different choices of coordinates might lead to different properties of the resulting controller.

In most cases, ranges of components of the parameter vector are given by closed intervals in $\mathbb{R}$ determined by upper and lower bounds, such that

$$\theta_{i,k} \in \left[\theta_{i,\min}, \theta_{i,\max}\right], i = 1, \ldots, N. \tag{75}$$

If no further information on the relations between parameters are known, the polytope $\boldsymbol{\Theta}$ has to be chosen as the N-dimensional hyper box, also referred to as the parameter box, given by

$$\boldsymbol{\Theta} = \left[\theta_{1,\min}, \theta_{1,\max}\right] \times \left[\theta_{2,\min}, \theta_{2,\max}\right] \times \ldots \times \left[\theta_{N,\min}, \theta_{N,\max}\right]. \tag{76}$$

One way to compute a set of coordinates $\boldsymbol{\lambda}_k$ for this case is introduced by Apkarian et al. [2] and generalized and implemented for an arbitrary number of vertices in the LMI Control Toolbox for Matlab [15]. Another implementation of this approach is proposed here that is suitable for real-time implementation purposes where variables have to be pre-initialized with fixed dimensions. Daafouz et al. [9] presented a compact way of writing the calculation scheme, on which the scheme proposed here and in [17] is based. If the order of vertices is not changed, any of the mentioned approaches leads to the same coordinates.

The i-th entry $v_{j,i}$ of a vertex j of the parameter box $\boldsymbol{\Theta}$ is either the lower bound $\theta_{i,\min}$ or the upper bound $\theta_{i,\max}$ of $\theta_{i,k}$. Now, $2N$ vectors

$$\boldsymbol{b}_{i_{\min}} = \begin{bmatrix} b_{i_{\min},1} \\ \vdots \\ b_{i_{\min},M} \end{bmatrix}, \boldsymbol{b}_{i_{\max}} = \begin{bmatrix} b_{i_{\max},1} \\ \vdots \\ b_{i_{\max},M} \end{bmatrix} \tag{77}$$

are pre-computed such that

$$b_{i_{\max},j} = \begin{cases} \frac{1}{\theta_{i,\max}-\theta_{i,\min}}, & \text{if } v_{j,i} = \theta_{i,\max}, \\ 0, & \text{if } v_{j,i} = \theta_{i,\min}, \end{cases} \tag{78}$$

$$b_{i_{\min},j} = \begin{cases} \frac{1}{\theta_{i,\max}-\theta_{i,\min}}, & \text{if } v_{j,i} = \theta_{i,\min}, \\ 0, & \text{if } v_{j,i} = \theta_{i,\max}. \end{cases} \tag{79}$$

The following steps are then carried out in every sampling instant:

1. $\theta_{i,k} = \cos(2\pi f_{i,k} T),\ i = 1, ..., N,$ (80)
2. $c_{i_{\max},k} = \theta_{i,k} - \theta_{i,\min},\ c_{i_{\min},k} = \theta_{i,\max} - \theta_{i,k},\ i = 1, ..., N,$ (81)
3. $\lambda_{j,k} = \prod_{i=1}^{N} (b_{i_{\max},j} c_{i_{\max},k} + b_{i_{\min},j} c_{i_{\min},k}),\ j = 1, ..., M.$ (82)

5. Application for the rejection of harmonic disturbances

In this section, the methods described in Sec. 4 are applied to the case of a harmonic multisine disturbance. Specific system models and the transformation that leads to a pLPV system are presented.

5.1. Internal model for harmonic disturbances

The disturbance is assumed to be a multisine. Let N be the number of components of the disturbance and $f_{i,k}$ the frequency of the i-th component at sampling instant k. The frequencies are assumed to vary in intervals $[f_{i,\min}, f_{i,\max}] \subset [0, 0.5 f_s]$, where f_s denotes the sampling frequency.

As discussed in Sec. 3, an error filter or a disturbance model can be used to include desired dynamics in the controller according to the internal model principle [14]. Since the following relations are used for either $A_{\mathrm{M},k}$ or $A_{\mathrm{d},k}$ depending on the control approach, the notation $\boldsymbol{A}_{\mathrm{M/d},k}$ is employed. The dynamics for a multisine disturbance with time-varying frequency are given by a discrete-time state-space model with system matrix

$$\boldsymbol{A}_{\mathrm{M/d},k} = \begin{bmatrix} \boldsymbol{A}_{\mathrm{M}_1/\mathrm{d}_1,k} & \cdots & 0 \\ \vdots & \ddots & \vdots \\ 0 & \cdots & \boldsymbol{A}_{\mathrm{M}_N/\mathrm{d}_N,k} \end{bmatrix}, \tag{83}$$

where

$$A_{\mathrm{M}_i/\mathrm{d}_i,k} = \begin{bmatrix} \cos(2\pi f_{i,k}T) & \sin(2\pi f_{i,k}T) \\ -\sin(2\pi f_{i,k}T) & \cos(2\pi f_{i,k}T) \end{bmatrix} \tag{84}$$

with sampling time $T = 1/f_{\mathrm{s}}$. This time-varying matrix might therefore be used as system matrix $A_{\mathrm{M},k}$ of the error filter described in Sec. 3.1 or $A_{\mathrm{d},k}$ of the disturbance model described in Sec. 3.2. In order to find a representation of $A_{\mathrm{M/d},k}$ that depends affinely on a parameter vector $\boldsymbol{\theta}$, one approach could be to choose

$$\theta_{i,k} = \cos(2\pi f_{i,k}T), \ i = 1,...,N. \tag{85}$$

Unfortunately, through

$$\sin(2\pi f_{i,k}T) = \sqrt{1-\theta_{i,k}^2}, \ i = 1,...,N, \tag{86}$$

a non-affine dependence is introduced into the model. To circumvent this, additional parameters

$$\theta_{i,k} = \sin(2\pi f_{i,k}T), i = N+1,...,2N \tag{87}$$

could be introduced. The use of this internal model leads to a large dimension of the parameter space and a polytope with many vertices. It can be expected that controllers based on this model are quite conservative, if there exists a solution to the LMIs at all. Therefore, a simplified disturbance model that leads to less conservative conditions is used for the controller design.

In the case of constant frequencies, instead of (84), the matrices

$$A_{\mathrm{M}_i/\mathrm{d}_i} = \begin{bmatrix} 0 & 1 \\ -1 & 2\cos(2\pi f_i T) \end{bmatrix}, i = 1,...,N \tag{88}$$

can be used in the system matrix of the internal model (83). This leads to a time-invariant overall system model. As discussed in Sec. 3, the internal model then is a way to determine controller poles (that show up as zeros in the disturbance transfer functions). With the above approach, for a frequency f, a complex conjugate controller pole pair is placed at $\exp(j2\pi fT)$ and $\exp(-j2\pi fT)$, which causes complete asymptotic rejection of harmonic disturbances with this frequency.

Although this model does not take into account the rate of change of the disturbance frequencies and the argumentation based on controller poles is valid only in the time-invariant case, this simpler model is used here even for the case of time-varying frequencies, since it reduces the dimension of the parameter space. Thus, the matrices

$$A_{\mathrm{M}_i/\mathrm{d}_i,k} = \begin{bmatrix} 0 & 1 \\ -1 & 2\cos(2\pi f_{i,k} T) \end{bmatrix}, i = 1,...,N \tag{89}$$

are used for the time-varying disturbance model. This might have an effect on the disturbance attenuation for fast changes of the frequencies, but it is expected that other effects are more dominant than the "incorrect" disturbance model. There will always be delay between the measured frequency used in the controller and the true frequency. Also, if the disturbance and the control signal do not enter at exactly the same point and the disturbance frequency varies, it is unclear which frequency "is present" at the point where the control signal enters the plant at a certain time. As shown in Sec. 3, closed-loop stability is not affected by the choice of the internal model.

5.2. Transformation to pLPV form

If the internal model can be written in the form of a pLPV system, it is possible to find a representation of the system matrix A_k of the overall system of the error filter approach (5) and the disturbance observer approach (15) in the form of a pLPV system as well. Then, the controller design and the gain-scheduling procedure can be carried out based on quadratic stability theory of pLPV systems, as has been presented in Sec. 4.

If the sampling theorem applies (which should be the case for any practical application), it holds that $0 \leq 2\pi f_{i,k}T \leq \pi$ for all $i = 1,...,N$. Since the function $f \mapsto \cos(2\pi fT)$ is monotonically decreasing on $[0, \pi]$, the frequencies under consideration have to fulfill $0 \leq 2\pi f_{i,k}T \leq \pi$ for all $i = 1,...,N$ in order to guarantee a unique mapping between frequencies and parameters. This can usually be achieved by choosing a sufficiently small value for T. Then, parameters $\boldsymbol{\theta}_i$ as defined in (85) are bounded by

$$\theta_{i,\min} = \cos(2\pi f_{i,\max}T),\ \theta_{i,\max} = \cos(2\pi f_{i,\min}T),\ i = 1,...,N, \tag{90}$$

respectively. Defining

$$\boldsymbol{\Theta} = [\theta_{1,\min}, \theta_{1,\max}] \times [\theta_{2,\min}, \theta_{2,\max}] \times ... \times [\theta_{N,\min}, \theta_{N,\max}] \subset \mathbb{R}^N \tag{91}$$

with vertices $v_1, v_2, ..., v_M \in \mathbb{R}^N$ it follows that the parameter vector

$$\boldsymbol{\theta}_k = \begin{bmatrix} \theta_{1,k} \\ \theta_{2,k} \\ \vdots \\ \theta_{N,k} \end{bmatrix} \in \mathbb{R}^N \tag{92}$$

is always inside the hyper box $\boldsymbol{\Theta}$. Then, the system matrix $\boldsymbol{A}_{\mathrm{M/d},k}$ of the internal model can be written as

$$\boldsymbol{A}_{\mathrm{M/d},k} = \boldsymbol{A}_{\mathrm{M/d}}(\boldsymbol{\theta}_k) = \mathcal{A}_{\mathrm{M_0/d_0}} + \theta_{1,k}\mathcal{A}_{\mathrm{M_1/d_1}} + \theta_{2,k}\mathcal{A}_{\mathrm{M_2/d_2}} + ... + \theta_{N,k}\mathcal{A}_{\mathrm{M_N/d_N}} \tag{93}$$

with

$$\mathcal{A}_{\mathrm{M_0/d_0}} = \begin{bmatrix} \mathcal{A}_{\mathrm{M_0/d_0},1} & \cdots & 0 \\ \vdots & \ddots & \vdots \\ 0 & \cdots & \mathcal{A}_{\mathrm{M_0/d_0},N} \end{bmatrix}, \tag{94}$$

where

$$\mathcal{A}_{\mathrm{M_0/d_0},i} = \begin{bmatrix} 0 & 1 \\ -1 & 0 \end{bmatrix},\ i = 1,...,N, \tag{95}$$

and $\mathcal{A}_{\mathrm{M}_i/\mathrm{d}_i}$ matrices with zero entries only except for

$$\mathcal{A}_{\mathrm{M}_i/\mathrm{d}_i}(2i, 2i) = 2,\ i = 1,...,N. \tag{96}$$

Therefore, the system matrix $\boldsymbol{A}_{\mathrm{M/d},k}$ of the internal model used for the error filter from Sec. 3.1 and the disturbance model in Sec. 3.2 can be written in pLPV form based on the

N-dimensional hyper box Θ with vertices $v_1, v_2, ..., v_M \in \mathbb{R}^N$. The system matrices A_k of (5) in the error filter approach can then be written as

$$A_k = A(\theta_k) = \mathcal{A}_0 + \theta_{1,k}\mathcal{A}_1 + ... + \theta_{N,k}\mathcal{A}_N \tag{97}$$

with

$$\mathcal{A}_0 = \begin{bmatrix} \mathcal{A}_{\mathrm{M}_0} & B_{\mathrm{M}}C_{\mathrm{p}} \\ 0 & A_{\mathrm{p}} \end{bmatrix}, \tag{98}$$

$$\mathcal{A}_i = \begin{bmatrix} \mathcal{A}_{\mathrm{M}_i} & 0 \\ 0 & 0 \end{bmatrix}, i = 1, ..., N, \tag{99}$$

and analogously in (15) for the disturbance observer approach

$$A_k = A(\theta_k) = \mathcal{A}_0 + \theta_{1,k}\mathcal{A}_1 + ... + \theta_{N,k}\mathcal{A}_N \tag{100}$$

with

$$\mathcal{A}_0 = \begin{bmatrix} A_{\mathrm{p}} & B_{\mathrm{p}}C_{\mathrm{d}} \\ 0 & \mathcal{A}_{\mathrm{d}_0} \end{bmatrix}, \tag{101}$$

$$\mathcal{A}_i = \begin{bmatrix} 0 & 0 \\ 0 & \mathcal{A}_{\mathrm{d}_i} \end{bmatrix}, i = 1, ..., N. \tag{102}$$

Since Θ is a convex polytope, in every instant there exist coordinates $\lambda_{j,k}$, $j = 1, ..., M$, such that

$$\theta_k = \sum_{j=1}^{M} \lambda_{j,k} v_j. \tag{103}$$

Since A_k depends affinely on θ_k in both approaches, it follows that the system matrix can be expressed as

$$A(\theta_k) = \sum_{j=1}^{M} \lambda_{j,k} A_{\mathrm{v},j} = A(\lambda_k), \tag{104}$$

where

$$A_{\mathrm{v},j} = A(v_j), j = 1, \ldots, M. \tag{105}$$

The required time-varying state-feedback and observer gains for the overall systems (5) and (15), respectively, can therefore be obtained by application of the design methods based on quadratic stability theory as reviewed in Sec. 4.

5.3. Controller implementation

Once the pLPV representation of the considered system is found, the required time-varying state-feedback and observer gains for the overall systems (5) and (15), respectively, can be obtained by application of the design methods based on quadratic stability theory as reviewed in Sec. 4.

The time-invariant vertex state-feedback or vertex observer gains are calculated in advance. Since the plant itself is linear time-invariant, also the time-invariant observer required for the estimation of the plant states in the error filter approach as well as the time-invariant state-feedback gain of the disturbance observer approach can be calculated off-line with

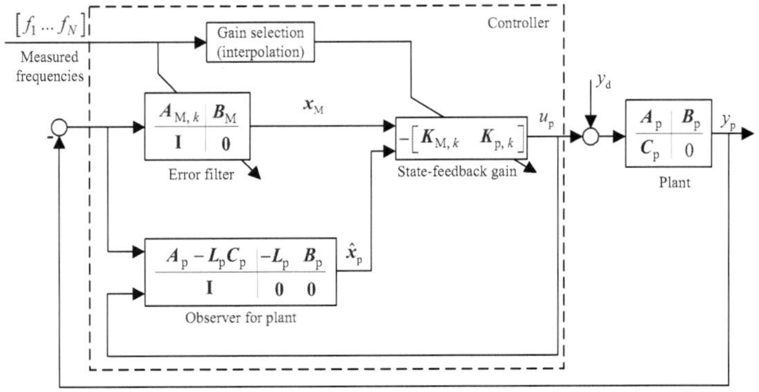

Figure 8. Control structure and interpolation for the error filter approach

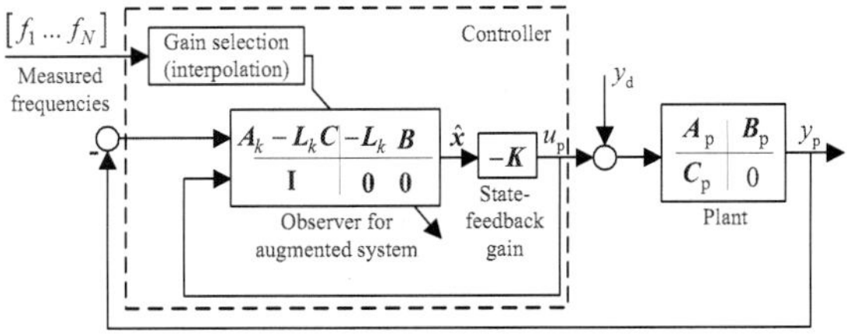

Figure 9. Control structure and interpolation the disturbance model approach

a standard method (e.g. LQR or pole placement). The controllers are then implemented according to the structures given in Sec. 3 and shown in Figs. 8 and 9 with the interpolation. Wherever possible, system matrices are updated in every sampling instant directly with the measured frequencies, while the time-varying state-feedback and observer gains are obtained by interpolation with the coordinates obtained by the scheme given in Sec. 4.4.

6. Real-time results

The gain-scheduled observer-based state-feedback controllers obtained through the design procedures presented in this chapter are validated with experimental results. Both controllers have been tested on an AVC and an ANC system and were found to work well. Results are presented for the test of the error filter with time-varying state-feedback gain on an ANC headset and the controller based on a disturbance observer with time-varying observer gain on an AVC test bed.

6.1. Real-time results for state-feedback gain-scheduling (error filtering) approach

The controller based on a disturbance observer with time-varying observer gain is tested experimentally on a Sennheiser PXC 300 headset. The experimental setup is shown schematically in Fig. 10. An external loudspeaker is used to generate a harmonic disturbance. The headset has one microphone on each loudspeaker. The objective is to cancel the

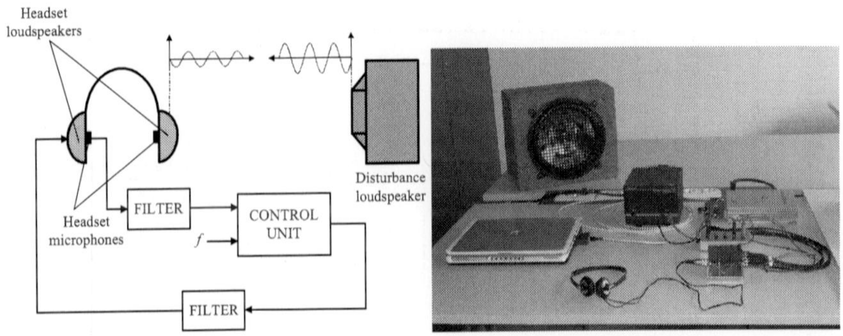

Figure 10. Block diagramm (left) and photograph (right) of the ANC system

disturbance with the loudspeakers of the headset. An anti-aliasing filter is applied to the output signal and a reconstruction filter to the control input. The control algorithm is implemented on a rapid control prototyping unit (dSpace MicroAutoBox). A sampling frequency of 1 kHz was used.

Standard black-box system identification techniques were used to obtain the transfer function between output and input of the control unit. The identified transfer function is of 12th order and the resulting controller of 20th order. As a disturbance signal, a sum of four harmonically related sine signals with fundamental frequency between 90 Hz and 100 Hz is used. The control algorithm has been implemented on both sides and results for the right side are shown.

Fig. 11 shows the amplitude frequency responses for constant fundamental frequencies of 90 Hz and 100 Hz. The amplitude frequency response plots show that amplification takes place in some frequency ranges, which is the known "waterbed" effect. It might cause problems in practical applications where significant background noise, e.g. broadband stochastic disturbances, is present. In Fig. 12, results are shown for a case where the fundamental frequency suddenly jumps from 90 Hz to 100 Hz. The controller remains stable even for such drastic variations of the fundamental frequency. The transient spike might be undesirable, particularly in ANC applications where it could be audible. However, a step change in the frequency does not commonly occur in real applications.

In Fig. 13, results for a gradually changing fundamental frequency are shown. The measurements show that excellent disturbance attenuation is achieved. The performance of the controller is tested with fast variations of the disturbance frequencies. Results for this experiment are shown in Fig. 14. It can be observed that the attenuation performance decreases, but the system remains stable.

6.2. Real-time results for observer gain-scheduling (disturbance model) approach

The scheme and a photograph of the AVC test bed are shown in Fig. 15. Two shakers (inertia mass actuators) are attached to a steel cantilever beam. One shaker acts as the disturbance source and the other shaker is driven by the control signal to counteract this disturbance. An accelerometer is used to measure the output signal. An anti-aliasing filter is applied to the output signal and a reconstruction filter to the control input.

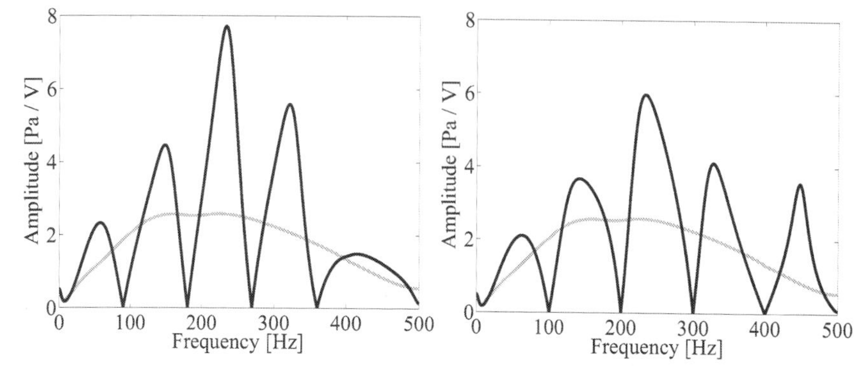

Figure 11. Open-loop (gray) and closed-loop (black) amplitude frequency responses for fixed disturbance frequencies of 90, 180, 270 and 360 Hz (left) and of 100, 200, 300 and 400 Hz (right)

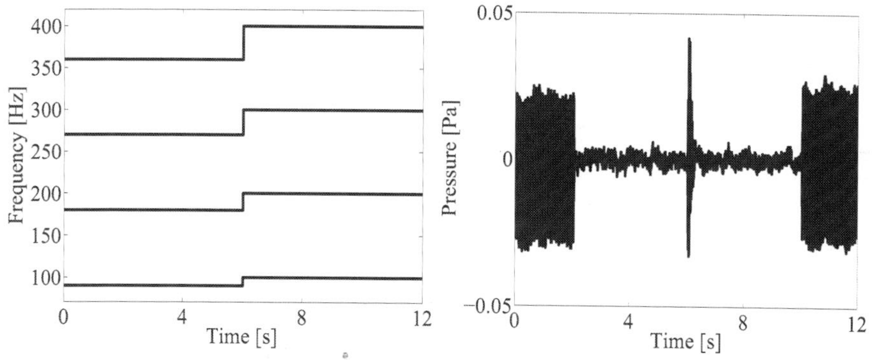

Figure 12. Results for a disturbance with time-varying frequencies. Variation of the frequencies (left) and measured sound pressure (right). The control sequence is off/on/off

A multisine test signal is used to identify the transfer function between output and input of the control unit using standard black-box system identification techniques. The controller is implemented on a rapid control prototyping unit (dSpace MicroAutoBox). A sampling frequency of 1 kHz was chosen. The identified system is of 10th order and the controller of 16th order. As a disturbance signal, a sum of three harmonically related sine signals with fundamental frequency between 110 and 120 Hz is used. The amplitude frequency responses of the open-loop and closed-loop systems for fixed disturbance frequencies of 110 Hz, 220 Hz and 330 Hz and 120 Hz, 240 Hz and 360 Hz are shown in Fig. 16. The resonance frequencies are damped by a factor 0.9 using pole placement with the state-feedback gain K_p. In Fig. 17, results are shown for a (rather unrealistic) case where the fundamental frequency suddenly jumps from 110 Hz to 120 Hz. After a short transient spike the disturbance rejection resumes quickly. The transient spike might be undesirable but a step change in the frequency does not commonly occur in real applications. The measurements confirm the excellent disturbance rejection that can be expected from the amplitude frequency responses. The amplitude response plots show that amplification takes place in frequency ranges between the rejected

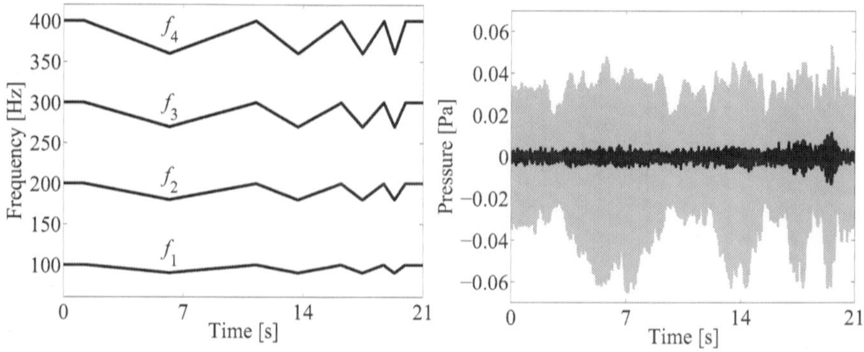

Figure 13. Results for a disturbance with time-varying frequencies. Variation of the frequencies (left) and measured sound pressure (right) in open loop (gray) and closed loop (black)

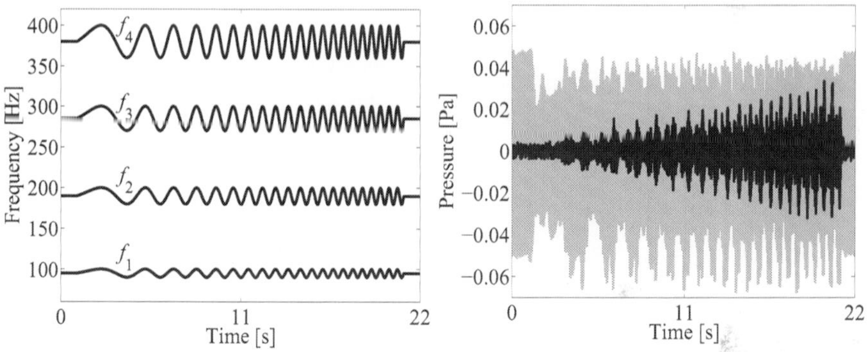

Figure 14. Results for a disturbance with time-varying frequencies. Variation of the frequencies (left) and measured sound pressure (right) in open loop (gray) and closed loop (black)

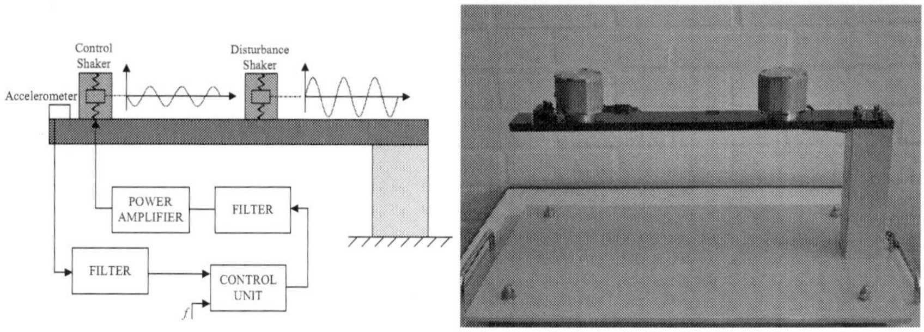

Figure 15. Schematic representation (left) and photograph (right) of the AVC system

frequencies. This is due to Bode's sensitivity integral ("waterbed" effect). Whether this is tolerable or not depends on the application. It is possible to limit the maximum disturbance amplification by a suitable overall controller design (which need not be observer based), but at the expense of either not fully suppressing the harmonic disturbances (that is, the notches in the frequency response would not get to zero) or requiring the frequency measurements to be very exact (that is, making the notches narrower) or worsening the transient behavior (that is, it would take longer before a harmonic disturbance is suppressed to a certain level). However, in this chapter only harmonic disturbances are considered, so these issues are not addressed in the design.

In Fig. 18, results for a gradually changing fundamental frequency are shown. It decreases linearly from 120 Hz to 110 Hz, remains constant for a while and then rises back to 120 Hz. This is repeated four times, every time in a shorter time interval. At the end the fourth harmonic rises from 330 Hz to 360 Hz and decreases back to 330 Hz in less than two seconds. The effect of fast variations of the disturbance frequencies has been further investigated in another experiment. The results are shown in Fig. 19. The disturbance frequency varies

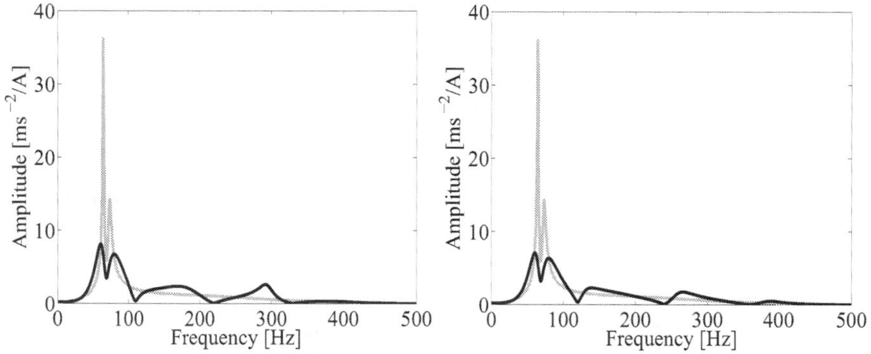

Figure 16. Open-loop (gray) and closed-loop (black) amplitude frequency responses for fixed disturbance frequencies of 110, 220 and 330 Hz (left) and of 120, 240 and 360 Hz (right)

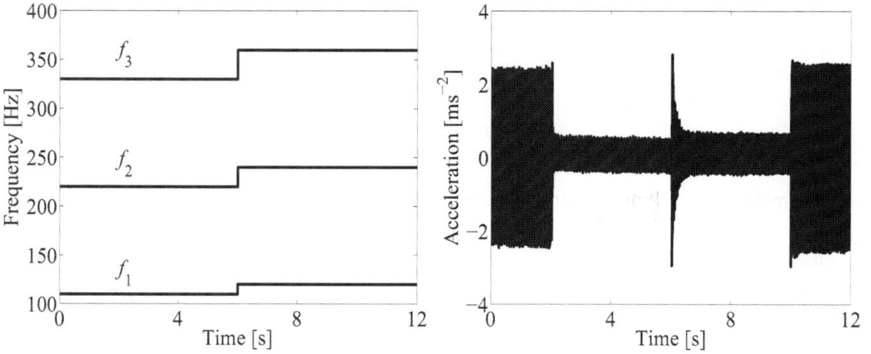

Figure 17. Results for a disturbance with time-varying frequencies. Variation of the frequencies (left) and measured acceleration (right). The control sequence is off/on/off

sinusoidally between the minimum and the maximum value with a period that decays from 10 to 0.5 seconds. It is seen that for very fast frequency variations, the attenuation performance decreases slightly (but the system remains stable).

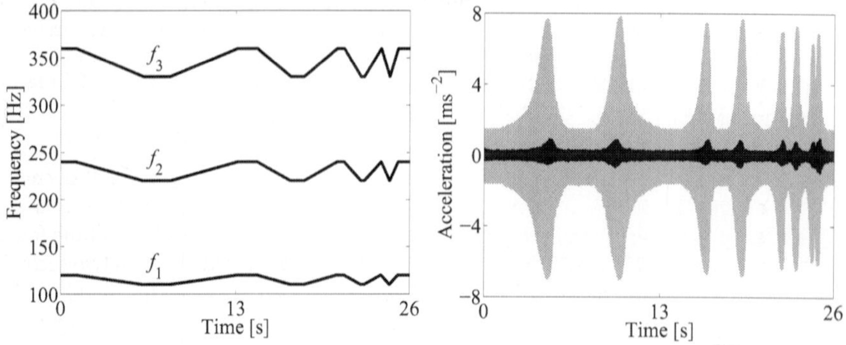

Figure 18. Results for a disturbance with time-varying frequencies. Variation of the frequencies (left) and measured acceleration (right) in open loop (gray) and closed loop (black)

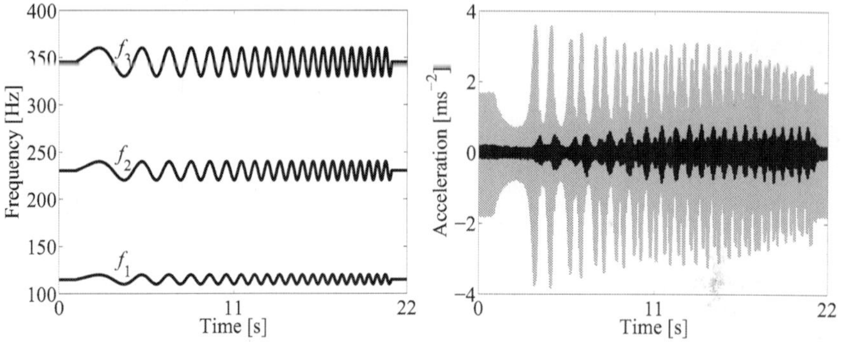

Figure 19. Results for a disturbance with time-varying frequencies. Variation of the frequencies (left) and measured acceleration (right) in open loop (gray) and closed loop (black)

7. Discussion and conclusion

Two discrete-time LPV observer-based state-feedback controllers for the rejection of harmonic disturbances with time-varying frequencies are presented. The control design methods are based on quadratic stability theory for pLPV systems. They guarantee stability of the closed-loop system also for arbitrarily fast changes of the disturbance frequencies. This is an advantage over other approaches such as adaptive filtering or heuristic gain scheduling. The experimental results show that an excellent disturbance rejection is achieved and that the proposed controllers can be applied in a real-time setup.

However, some degree of conservatism is present in these approaches. Using a single parameter independent Lyapunov function limits the range of admissible disturbance frequencies that can be covered with the resulting controller. Also, the polytope that contains the uncertain parameters could be chosen much smaller and with fewer vertices than the cuboid applied here, if information on the relations between the disturbance frequencies is given (as, for example, in the case of harmonically related frequencies). Thus, feasibility of the LMIs and the upper bound on the system performance could be improved as well as the computation time for the coordinates required for the on-line interpolation. This might be important for applications where many harmonics have to be cancelled, a wide frequency range has to be covered and the computational resources are limited, for example in automotive applications [7]. This aspect as well as a direct comparison of the proposed methods with conventional algorithms (such as the FxLMS) will be subject of future research.

To the best of the authors' knowledge, industrial applications of LPV controllers are rather limited. The results of this and the following chapter show that the implementation of even high-order LPV controllers can be quite straightforward.

Nomenclature

Acronyms

ANC	Active noise control.
AVC	Active vibration control.
FxLMS	Filtered-x least mean squares.
LFT	Linear fractional transformation.
LMI	Linear matrix inequality.
LPV	Linear parameter varying.
pLPV	Polytopic linear parameter varying.

Variables

(in order of appearance)

$\boldsymbol{A}_{\mathrm{p}}, \boldsymbol{B}_{\mathrm{p}}, \boldsymbol{C}_{\mathrm{p}}$	State-space matrices of the plant.
$\boldsymbol{x}_{\mathrm{p},k}$	State vector of the plant.
$y_{\mathrm{p},k}, u_{\mathrm{p},k}$	Output signal and control input of the plant.
$y_{\mathrm{d},k}$	Disturbance signal.
$\boldsymbol{A}_{\mathrm{M}}, \boldsymbol{B}_{\mathrm{M}}$	State-space matrices of the error filter.
$\boldsymbol{x}_{\mathrm{M},k}$	State-space vector of the error filter.
r_k, e_k	Reference and error signal.
$\boldsymbol{A}_{\mathrm{M},k}$	Time-varying system matrix of the error filter.

$\mathbf{0}$	Zero matrix.
$\mathbf{I}$	Identity matrix.
$\boldsymbol{L}_{\mathrm{p}}$	Observer gain for the plant.
$\widehat{x}_{\mathrm{p},k}$	Estimated state vector of the plant.
$\boldsymbol{K}_k$	Time-varying state-feedback gain for the overall system.
$\boldsymbol{K}_{\mathrm{M},k}, \boldsymbol{K}_{\mathrm{p},k}$	Time-varying state-feedback gains for the error filter and the plant.
$\widetilde{x}_{\mathrm{p},k}$	Observer error of the plant.
$\boldsymbol{A}_{\mathrm{d},k}, \boldsymbol{C}_{\mathrm{d}}$	State-space matrices of the disturbance model.
$x_{\mathrm{d},k}$	State vector of the disturbance model.
$\boldsymbol{A}_k, \boldsymbol{B}, \boldsymbol{C}$	State-space matrices of the overall system.
$\boldsymbol{L}_k$	Time-varying observer gain of the overall system.
$\boldsymbol{L}_{\mathrm{d},k}, \boldsymbol{L}_{\mathrm{p},k}$	Time-varying observer gains for the disturbance model and the plant.
$\widehat{x}_{\mathrm{d},k}$	Estimated state vector of the disturbance model.
$\boldsymbol{K}_{\mathrm{d},k}, \boldsymbol{K}_{\mathrm{p},k}$	Time-varying state-feedback gains for the disturbance model and the plant.
$\boldsymbol{A}(\boldsymbol{\theta}), \boldsymbol{B}(\boldsymbol{\theta}),$ $\boldsymbol{C}(\boldsymbol{\theta}), \boldsymbol{D}(\boldsymbol{\theta})$	State-space matrices of an LPV system.
$\boldsymbol{\theta} = [\theta_1\, \theta_2\, \ldots \theta_N]^{\mathrm{T}}$	Parameter vector.
N	Size of the parameter vector.
$\boldsymbol{\Theta}$	Parameter polytope.
$\boldsymbol{\mathcal{A}}_i$	Constant matrices of a polytopic representation.
V	Set of vertices of the parameter polytope.
v_j	Vertex j of the parameter polytope.
$\boldsymbol{\lambda} = [\lambda_1\, \lambda_2\, \ldots \lambda_M]^{\mathrm{T}}$	Coordinate vector.
M	Size of the coordinate vector.
$\boldsymbol{A}_{\mathrm{v},j}$	System matrices for the j-th vertex.
$\boldsymbol{P}$	Symmetric positive definite matrix.
$\boldsymbol{\theta}_k$	Vector of time-varying parameters.
$\boldsymbol{\lambda}_k = \left[\lambda_{1,k}\, \lambda_{2,k}\, \ldots \lambda_{M,k}\right]^{\mathrm{T}}$	Time-varying coordinate vector.
$\boldsymbol{K}_{\mathrm{v},j}$	State-feedback gain for the j-th vertex.

w_k, z_k, u_k	Process noise, performance output and control input.
$\boldsymbol{Q}, \boldsymbol{R}$	Weighting matrices.
G	Transfer function between w_k and z_k.
$\widetilde{A}, \widetilde{B}, \widetilde{C}$	State-space matrices of the transfer path from w_k to z_k.
$\widetilde{Q}, \widetilde{R}$	Matrices to build the LMIs.
W_c	Controllability gramian.
W_o	Observability gramian.
W	Symmetric positive definite matrix.
$Y_{\mathrm{v},j}$	Solution of the LMIs for the j-th vertex.
$L_{\mathrm{v},j}$	Observer gain for the j-th vertex.
$\boldsymbol{b}_{i_\mathrm{min}}, \boldsymbol{b}_{i_\mathrm{max}}$	Vectors for the computation of the coordinate vector.
$c_{i_\mathrm{min},k}, c_{i_\mathrm{max},k}$,	Scalars for the computation of the coordinate vector.
$b_{i_\mathrm{min},j}, b_{i_\mathrm{max},j}$	
$A_{\mathrm{M/d},k}$	System matrix of a general time-varying internal disturbance model.

Author details

Pablo Ballesteros, Christian Bohn, Wiebke Heins and Xinyu Shu
Institute of Electrical Information Technology, Clausthal University of Technology, Clausthal-Zellerfeld, Germany

8. References

[1] Amato, F. 2006. *Robust control of linear systems subject to uncertain time-varying parameters.* Berlin: Springer.

[2] Apkarian, P., P. Gahinet and G. Becker. 1995. Self-scheduled control of linear parameter-varying systems: A design example. *Automatica* 31:1251-61.

[3] Apkarian, P. 1997. On the discretization of LMI-synthesized linear parameter-varying controllers. *Automatica* 33:655-61.

[4] Ballesteros, P. and C. Bohn. 2011a. A frequency-tunable LPV controller for narrowband active noise and vibration control. *Proceedings of the American Control Conference.* San Francisco, June 2011. 1340-45.

[5] Ballesteros, P. and C. Bohn. 2011b. Disturbance rejection through LPV gain-scheduling control with application to active noise cancellation. *Proceedings of the IFAC World Congress.* Milan, August 2011. 7897-902.

[6] Balini, H. M. N. K., C. W. Scherer and J. Witte. 2011. Performance enhancement for AMB systems using unstable H_∞ controllers. *IEEE Transactions on Control Systems Technology* 19:1479-92.

[7] Bohn, C., A. Cortabarria, V. Härtel and K. Kowalczyk. 2003. Disturbance-observer-based active control of engine-induced vibrations in automotive vehicles. *Proceedings of the*

SPIE's 10th Annual International Symposium on Smart Structures and Materials. San Diego, March 2003. Paper No. 5049-68.

[8] Bohn, C., A. Cortabarria, V. Härtel and K. Kowalczyk. 2004. Active control of engine-induced vibrations in automotive vehicles using disturbance observer gain scheduling. *Control Engineering Practice* 12:1029-39.

[9] Daafouz, J., G. I. Bara, F. Kratz and J. Ragot. 2000: State observers for discrete-time LPV systems: an interpolation based approach. *Proceedings of the 39th IEEE Conference on Decision and Control* 5: 4571-72.

[10] Darengosse C. and P. Chevrel. 2000. Linear parameter-varying controller design for active power filters. *Proceedings of the IFAC Control Systems Design*. Bratislava, June 2000. 65-70.

[11] Du, H. and X. Shi. 2002. Gain-scheduled control for use in vibration suppression of system with harmonic excitation. *Proceedings of the American Control Conference*. Anchorage, May 2002. 4668-69.

[12] Du, H., L. Zhang and X. Shi. 2003. LPV technique for the rejection of sinusoidal disturbance with time-varying frequency. *IEE Proceedings on Control Theory and Applications* 150:132-38.

[13] Feintuch, P. L., N. J. Bershad and A. K. Lo. 1993. A frequency-domain model for filtered LMS algorithms - Stability analysis, design, and elimination of the training mode. *IEEE Transactions on Signal Processing* 41:1518-31.

[14] Francis, B. and W. Wonham. 1976. The internal model principle of control theory. *Automatica* 12:457-65.

[15] Gahinet, P., A. Nemirovskii, A. J. Laub and M. Chilali. 1995. LMI Control Toolbox. The Mathworks Inc.

[16] Heins, W., P. Ballesteros and C. Bohn. 2011. Gain-scheduled state-feedback control for active cancellation of multisine disturbances with time-varying frequencies. Presented at the *10th MARDiH Conference on Active Noise and Vibration Control Methods*. Krakow-Wojanow, Poland, June 2011.

[17] Heins, W., P. Ballesteros and C. Bohn. 2012. Experimental evaluation of an LPV-gain-scheduled observer for rejecting multisine disturbances with time-varying frequencies. *Proceedings of the American Control Conference*. Montreal, June 2012. Accepted for publication.

[18] Horn, R. A. and C. R. Johnson. 1985. *Matrix Analysis*. Cambridge: Cambridge UP.

[19] Kinney, C. E. and R. A. de Callafon. 2006a. Scheduling control for periodic disturbance attenuation. *Proceedings of the American Control Conference*. Minneapolis, June 2006. 4788-93.

[20] Kinney, C. E. and R. A. de Callafon. 2006b. An adaptive internal model-based controller for periodic disturbance rejection. *Proceedings of the 14th IFAC Symposium on System Identification*. Newcastle, Australia, March 2006. 273-78.

[21] Kinney, C. E. and R. A. de Callafon. 2007. A comparison of fixed point designs and time-varying observers for scheduling repetitive controllers. *Proceedings of the 46th IEEE Conference on Decision and Control*. New Orleans, December 2007. 2844-49.

[22] Köroğlu , H. and C. W. Scherer. 2008. Robust generalized asymptotic regulation against non-stationary sinusoidal disturbances. *Proceedings of the 47th IEEE Conference on Decision and Control*. Cancun, December 2008. 5426-31.

[23] Köroğlu, H. and C. W. Scherer. 2008. LPV control for robust attenuation of non-stationary sinusoidal disturbances with measurable frequencies. *Proceedings of the 17th IFAC World Congress*. Korea, July 2008. 4928-33.

[24] Kuo, S. M. and D. R. Morgan. 1996. *Active Noise Control Systems*. New York: Wiley.

[25] Morari, M. and E. Zafiriou 1989. *Robust Process Control*. Englewood Cliffs: Prentice Hall.

[26] Shu, X., P. Ballesteros and C. Bohn. 2011. Active vibration control for harmonic disturbances with time-varying frequencies through LPV gain scheduling. *Proceedings of the 23rd Chinese Control and Decision Conference*. Mianyang, China, May 2011. 728-33.

[27] Stilwell, D. J. and W. J. Rugh. 1998. Interpolation of observer state feedback controllers for gain scheduling. *Proceedings of the American Control Conference* 2:1215-19.

[28] Witte, J., H. M. N. K. Balini and C. W. Scherer. 2010. Experimental results with stable and unstable LPV controllers for active magnetic bearing systems. *Proceedings of the IEEE International Conference on Control Applications*. Yokohama, September 2010. 950-55.

Transverse Vibration Control for Cable Stayed Bridge Under Construction Using Active Mass Damper

Hao Chen, Zhi Sun and Limin Sun

Additional information is available at the end of the chapter

http://dx.doi.org/10.5772/45732

1. Introduction

Since the erection of the Stromaund Bridge of Sweden in 1956, cable-stayed bridge, as an efficient and economic bridge type to surmount a long-distance obstacle, has attracted more and more interests both from bridge engineering community and from the society and government. Nowadays, the cable stayed bridge is the most competitive type for the bridge with the span of 300-1000 meters. For a cable supported bridge, which is generally quite flexible and of low damping, its vibration under ambient excitation (such as the wind and ground motion excitation) and operational loading (such as the vehicle and train loads) is quite critical for its safety, serviceability and durability. Vibration control countermeasures, such as the installation of the energy dissipating devices, are thus required [1, 2]. Structural active control, which applies a counter-force induced by a control device to mitigate structural vibration, has been widely proposed for the vibration control of cable stayed bridges and proven to be efficient by many researchers [3-6].

Although the vibration response of a fully erected cable-stayed bridge should be controlled, a cable-stayed bridge under construction, which is of low damping and not as stable as the completed structure, is generally more vulnerable to dynamic loadings. During the construction stage, the cable pylons were generally erected firstly and the cable and main girders are then hang on the pylons symmetrically in a double-cantilever way. With the increase of the cantilever length, the bridge is more and more flexible. When the girder is on its longest double-cantilever state, the bridge is the most vulnerable to the external disturbance (such as the ambient wind fluctuation and ground motions). Moreover, if the cables, pylons and main girder of the bridge are all steel components and thus the damping of the bridge is very low, its vibration under ambient excitations will be quite large. The vibration reduction countermeasures are thus in great demand. Frederic conducted several mock-up tests representing a cable-stayed bridge during the construction stage [7]. Since the control objective was set to reduce the girder vertical vibration response or cable parametric vibration response, the active tendon was installed as the control device. While for a cable stayed bridge under uncontrollable ambient excitations, structure will vibrate not only in the vertical direction but also in the transverse direction. Moreover, since the bridges are generally designed to carry the vertical loads, the unexpected transverse loads, especially the transverse dynamic loads, will induce structural safety and durability problems. It is thus of crucial importance to install some vibration reduction devices to control bridge transverse vibration. For this type of structure vibration control problem, the active mass damper (AMD) or the active tuned mass damper (ATMD) will be a competent candidate [8, 9].

In this chapter, the general procedure and key issues on adopting an active control device, the active mass damper (AMD), for vibration control of cable stayed bridges under construction are presented. Taking a typical cable stayed bridge as the prototype structure; a lab-scale test structure was designed and fabricated firstly. A baseline FEM model was then setup and updated according to the modal frequencies measured from structural vibration test. A numerical study to simulate the bridge-AMD control system was conducted and an efficient LQG-based controller is designed. Based on that, an experimental implementation of AMD control of the transverse vibration of the bridge model was performed.

2. Model structure description and vibration test

The lab-scale bridge model studied in this chapter is designed according to a prototype cable-stayed bridge, the Third Nanjing Yangtze River Bridge located in Jiangsu Province of China. Since the prototype bridge is of all the characteristics of a modern cable-stayed bridge, the test model is assumed to be a good test bed to study the feasibility of active structural control applied to cable-stayed bridges under construction. The test model was designed and fabricated to simulate the longest double cantilever state during the construction stage of the prototype bridge. Since structural dynamic response control is the main focus of this chapter, the test model is preliminarily designed according to the dynamic scaling laws. However, concerning the restriction of test conditions, some modifications were made during the detailed design of the model bridge [10]. Fig. 1 shows the dimension of the designed model bridge. The bridge is composed of a 1.433 meters high cable pylon, a 3.08 meters long main girder, and six couples of stay cables. The cross section of the main girder is a rectangular of 16 mm wide and 10 mm high. At two ends of the main girder, the 3.6 kg and 3.8 kg weight AMD orbits were installed on the side span and main span respectively. The stay cables are made of steel wire with the diameter of 1 mm. At the upper end of each cable, an original 30 cm long spring was installed and adjusted to simulate the cable force. The cable forces were adjusted to provide supporting force to the main girder and to force it to match the designed layout of the test bridge. Table 1 shows the length, Young's Modulus and computed cable force for the cables. All of the components were made of steel. Since the model consists of only one cable pylon and no other piers, it is symmetric with respect to the cable pylon.

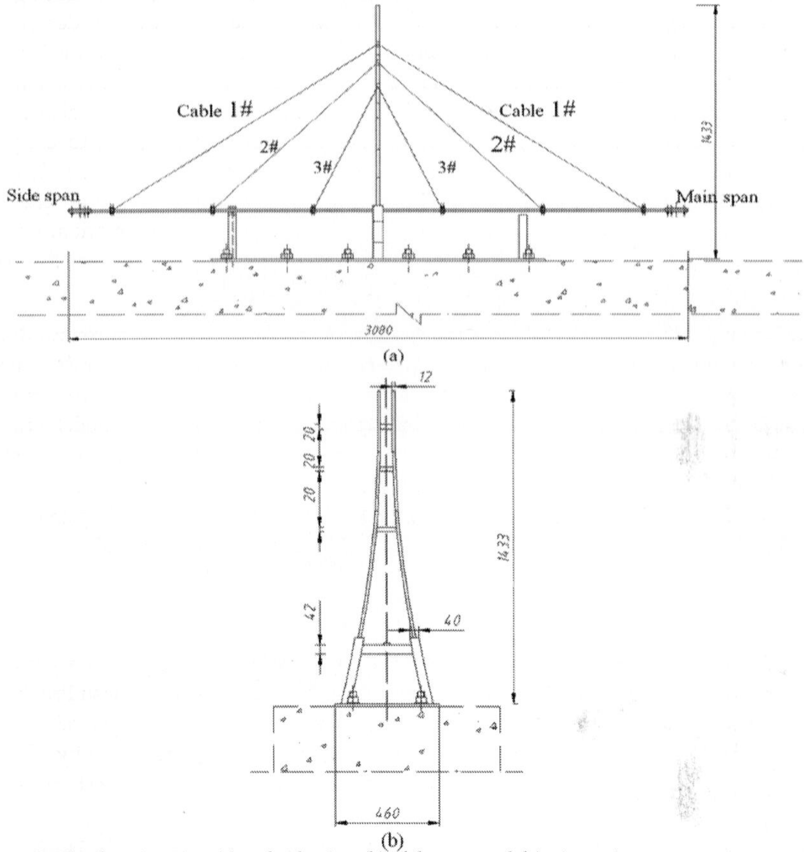

Figure 1. The elevation view (a) and side view (b) of the test model (unit: mm)

Cable No.	1#	2#	3#
Young's Modulus (Mpa)	451.7	327.0	216.0
Length with spring (m)	1.62	1.17	0.77
Spring Force (N)	32.0	13.9	7.2

Table 1. Cable parameters

Num.	1	2	3	4	5	6	7	8
Sensitivity (mv/g)	134.6	140.8	127.4	134.6	149.2	138.3	138.5	130.8
Location (m)	0.06	0.32	0.62	1.02	2.06	2.46	2.76	3.02

Table 2. Sensor sensitivity and location distances from the tip end of the side span

Vibration tests under forced excitations were conducted to identify the dynamic characteristics of the bridge model. Eight accelerometers (as described in Table 2) were distributed along the main girder both on the side span and the main span to collect structural acceleration responses at a sampling frequency of 50 Hz (as shown in Fig. 2a). A transverse impulse force was acted on the cantilever end of the side span to excite the structure. Concerning that the bridge model is symmetrical about the cable pylon and thus of repeated or close frequency modes, a modal identification algorithm of the capacity to identify the close modes, the wavelet based modal identification method developed by the research group, is used [11]. During the analysis, the mother wavelet function adopted is the complex Morlet function with the central frequency of 300 Hz and the scale increment is set to be 0.25 during the analysis. Fig. 2b shows the wavelet scalgram of a set of response measurement on the tip end of the side span. As shown in the figure, structural transverse vibration responses were dominated by two modes at the scale of 168.0 and 176.5, which correspond to the vibration modes of the natural frequencies of 1.701 Hz and 1.786 Hz, respectively. This figure also shows that the adopted modal identification method can separate these two close modes successfully. Structural modal parameters can then be estimated and the results are shown in Table 3.

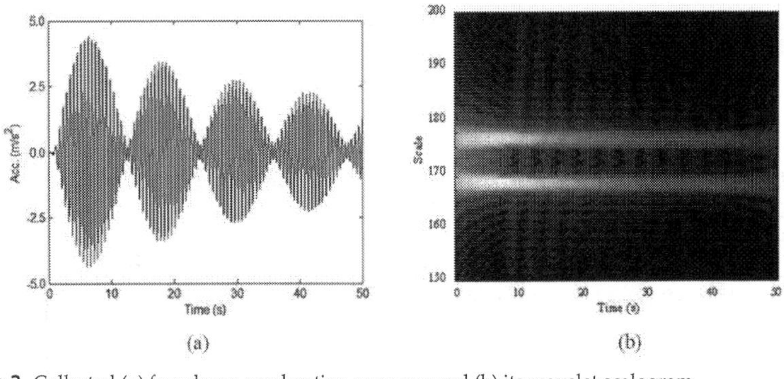

Figure 2. Collected (a) free decay acceleration response and (b) its wavelet scalogram

3. Numerical modeling and model updating

For structural active control, a baseline numerical model is generally required for controller design. A FEM bridge model is thus setup in ANSYS according to the design diagram of the bridge model. The cable is modeled using a 3D uniaxial tension-only truss element. Equivalent modulus for the cables without spring are computed using Ernst formula and then series wound equivalent modulus for the cable with spring can be established. Other structural elements are all modeled using 3D elastic beam element. Cables are connected to the main girder using rigid beam element. Additional masses were modeled using isotropic mass element. The cable pylons and the main girder are linked by coupling the horizontal projective intersection points of the lowest transverse beam of the pylon with the main girder. The six DOFs at the feet of the pylon are fixed.

		The first transverse symmetrical bending (TSB) mode	The first transverse anti-symmetrical bending (TAB) mode
Frequency	Test	1.701 Hz	1.786 Hz
	Initial Model	1.848 Hz	1.856 Hz
	Updated Model	1.787 Hz	1.786 Hz
Mode Shape	Test	[1.00 0.86 0.70 0.32 0.00 0.30 0.70 0.90 0.99]	[1.00 0.84 0.72 0.28 0.00 -0.32 -0.70 -0.86 -0.99]
	Initial Model	[1.00 0.74 0.46 0.17 0.00 0.13 0.38 0.60 0.81]	[1.00 0.74 0.47 0.17 0.00 -0.14 -0.38 -0.60 -0.81]
	Updated Model	[1.00 0.74 0.46 0.17 0.00 0.17 0.46 0.74 1.00	[1.00 0.74 0.46 0.17 0.00 -0.17 -0.46 -0.74 -1.00]
MAC*	Initial Model	0.9620	0.9620
	Updated Model	0.9722	0.9718

* MAC (modal assurance criteria) is defined to be a correlation coefficient of two mode shape vectors.

Table 3. The computed modal parameters before and after model updating compared with the tested mode parameters

Taken eigenvalue analysis of the numerical model, structural natural frequencies and mode shape vectors can be computed. Table 4 shows the computed modal parameters of the transverse bending modes. As shown in the table, since the sum of the effective mass of the first transverse anti-symmetric bending mode (TAB) and the first transverse symmetric bending mode (TSB) is 80.1% of the sum of the effective mass of all transverse modes, these 2 transversal modes dominant the transverse vibration of the bridge. Fig. 3 shows the mode shape of these two transverse bending modes.

No.	Frequency (Hz)	Mode description	Effective mass (kg)
1	1.848	The first TAB mode	6.00
2	1.856	The first TSB mode	6.00
3	19.802	The second TAB mode	0.94
4	19.848	The second TSB mode	0.94

Table 4. The computed natural frequencies for the first 4 transverse modes of the main girder

Comparing the natural frequencies obtained from the eigenvalue analysis on the numerical model and the vibration test on the test bridge, some differences can be observed (as shown in Table 3). A model updating process is thus conducted to get an accurate baseline numerical model. The results of the sensitivity analysis show that the vertical and transverse vibration modes are sensitive to the tip mass magnitude, while the Young's module of the cable is critical for vertical bending modes. So these two parameters were updated: the tip masses of two spans were updated from 3.8kg and 3.6kg to 4.16kg, respectively, and the spring modulus was updated from 220N/m to 203N/m. Table 4 shows the natural frequencies and mode shapes of the updated model. As shown, the modal parameters of the first TSB and TAB modes have a good match with the modal test results.

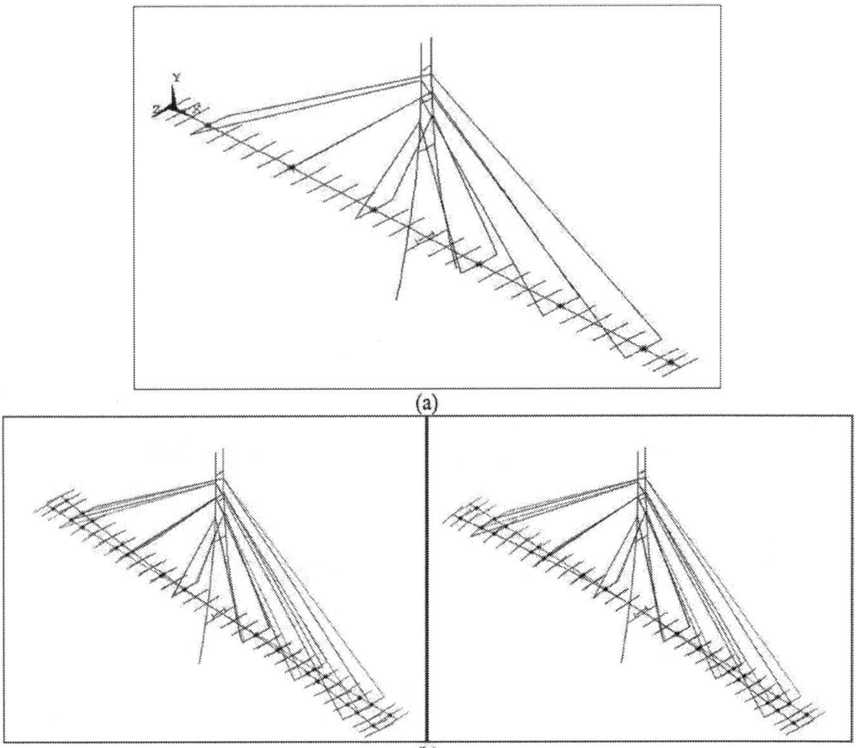

Figure 3. The baseline FEM model (a) setup in ANSYS and the computed mode shape (b) of the first transverse anti-symmetric and symmetric bending mode

4. Control system simulation

Based on the baseline numerical model updated according to dynamic measured structural modal parameters, a system simulation study is conducted. For a bridge-AMD system, its governing equation of motion is

$$M_s\ddot{X}(t)+C_s\dot{X}(t)+K_sX(t)=F_e(t)+D^TT^Tf_d(t) \tag{1}$$

$$m_d\ddot{x}_d(t)+f_d(t)=0 \tag{2}$$

$$f_d(t)=-c_d[\dot{x}_d(t)-TD\dot{X}]+b_du(t) \tag{3}$$

where, M_s, C_s, and K_s are the mass, damping and stiffness matrices of the bridge structure; $X(t)=\left[x_1\cdots x_i\cdots x_n\right]'$ is a n-dimensional vector, in which x_i is the displacement response vector of the ith DOF of the structure and n is the number of the DOF of the structure; $F_e(t)$

is the excitation force vector acting on the structure; $D=\begin{bmatrix}0 & I & 0\end{bmatrix}$ is the AMD location matrix, in which 0 and I are zero and identity matrix with appropriate sizes; $f_d(t)$ is the actuation force on the structure applied by the AMD ; T is the transfer matrix from the affiliated nodes of AMD to the principle nodes of the structure ; $m_d = diag\left[m_{d1}\cdots m_{di}\cdots m_{dl}\right]$ is the mass matrix of the AMDs, in which m_{di} is the mass of the ith AMD; $x_d(t)$ are absolute displacement of the AMD to the bridge structure, respectively; c_d is the viscous coefficient of the AMD; b_d is the force-voltage coefficient of the AMD and $u(t)$ is the control input voltage.

Since the numerical model of a complex structure is generally of a large amount of DOFs, for example in this study the FEM model obtained from the last section is of 1188 DOFs, this will induce great computation difficulty to design the controller according to this so called full order model. A reduced order model is thus required. In this study, the critical modal reduction method is adopted for this purpose because this method can greatly reduce structural DOFs and the reduced order model is of clear physical meaning [12, 13]. Concerning a structure whose vibration is dominated by the first r modes, its dynamic response can be approximately expressed as $X(t) \cong \sum_{i=1}^{r} \phi_i Y_i(t)$ and the full order model can thus be reduced to an r-order reduced order model. The governing equation of motion for this reduced order model is

$$\hat{M}_s \ddot{Y}(t) + (\hat{C}_s - c_d \phi^T D^T T^T T D \phi)\dot{Y}(t) + \hat{K}_s Y(t) + \phi^T D^T T^T c_d \dot{x}_d(t) = \hat{F}_e(t) + \phi^T D^T T^T b_d u(t) \quad (4)$$

$$m_d \ddot{x}_d(t) - c_d \dot{x}_d(t) + c_d T D \phi \dot{Y} = -b_d u(t) \quad (5)$$

where, $Y(t)$ is the modal response vector superposed by the modal responses of the rth principle modes; ϕ is the generalized mode shape matrix composed of the mode shape vectors of the first r principle modes; $\hat{M}_s = \phi^T M_s \phi$, $\hat{C}_s = \phi^T C_s \phi$ and $\hat{K}_s = \phi^T K_s \phi$ are structural generalized mass, damping and stiffness matrices, respectively; $\hat{F}_e = \phi^T F_e$ is the generalized excitation vectors. In this study, since the vibration responses of the first two transversal vibration modes are the most accountable for structural transverse vibration, structural modal mass, stiffness and damping matrices are computed using the modal parameters of these two modes. Correspondingly, the governing equation of motion for the reduced order bridge-AMD system can be derived.

For this reduced order bridge-AMD system, its state space equation is

$$\dot{x}_r = A_r x_r + B_r u + E_r w \quad (6)$$

$$z_r = C_r^z x_r + D_r^z u + E_r^z w \quad (7)$$

$$y_r = C_r^y x_r + D_r^y u + E_r^y w \quad (8)$$

where, x_r is an a dimensional state-space vector, $a = 2r + l$, r and l are the number of DOF for the reduced order structure and the number of the installed AMD, respectively; A_r is an $a \times a$ system matrix; u is the control input vector for the l AMDs; B_r is an $a \times l$ AMD location matrix; w is the generalized modal excitation vector; E_r is an $a \times r$ excitation matrix ; z_r is an l dimensional control output vector; y_r is a q dimensional observer output vector. The system matrices can be expressed as:

$$x = \begin{Bmatrix} Y \\ \dot{Y} \\ \dot{x}_d \end{Bmatrix};\ A_r = \begin{bmatrix} 0 & I & 0 \\ -\hat{M}_s^{-1}\hat{K}_s & -\hat{M}_s^{-1}\hat{C}^* & -\hat{M}_s^{-1}\phi^T D^T T^T c_d \\ 0 & m_d^{-1} c_d H & m_d^{-1} c_d \end{bmatrix};\ B_r = \begin{bmatrix} 0 \\ \hat{M}_s^{-1}\hat{H} b_d \\ -m_d^{-1} b_d \end{bmatrix};\ E_r = \begin{bmatrix} 0 \\ \hat{M}_s^{-1} \\ 0 \end{bmatrix};$$

$$C_r^z = \alpha \begin{bmatrix} I \\ \begin{bmatrix} 0 & I \end{bmatrix} A_r \end{bmatrix};\ D_r^z = \alpha \begin{bmatrix} I \\ \begin{bmatrix} 0 & I \end{bmatrix} B_r \end{bmatrix};\ E_r^z = \alpha \begin{bmatrix} 0 \\ \begin{bmatrix} 0 & I \end{bmatrix} E_r \end{bmatrix};$$

$$C_r^y = \beta \begin{bmatrix} I \\ \begin{bmatrix} 0 & I \end{bmatrix} A_r \end{bmatrix};\ D_r^y = \beta \begin{bmatrix} I \\ \begin{bmatrix} 0 & I \end{bmatrix} B_r \end{bmatrix};\ E_r^y = \beta \begin{bmatrix} 0 \\ \begin{bmatrix} 0 & I \end{bmatrix} E_r \end{bmatrix};$$

where $\hat{C}^* = \hat{C}_S - \hat{H} c_d \hat{H}^T$; $\hat{H} = \begin{bmatrix} -TD\phi & I \end{bmatrix}^T$; 0 and I are zero and identity matrix with appropriate sizes; α and β are appropriately selected weighting matrix to adjust the optimize objective of the controller, respectively. The control output vector z_r and the observer output vector y_r are the displacement, velocity and acceleration responses of the bridge structure or the AMDs.

The control output and observer output matrices C_r^z, D_r^z, C_r^y, and D_r^y are determined according to the sensor and actuator placement. The sensor placement should basically satisfy the following observability criteria $rank \begin{bmatrix} A_r - \lambda I \\ C_r^y \end{bmatrix} = a$, where λ is an arbitrary complex number [14]. For a system with n modes of repeated or close frequency, n sensors are required for full state response measurement. In this study, to provide redundant channels to collect structural response, eight accelerometers, the same as aforementioned in the modal test, are installed along the main girder during the control process. The actuators are placed on the girder by checking the following controllability criteria $a = rank\begin{bmatrix} A_r - \lambda I, B_r \end{bmatrix}$. In this study, different schemes for AMD placement will be discussed in detail in the following sections.

5. Controller design

The design of a controller is very important for the success of structural active vibration control. In this study, the LQG control algorithm is adopted since this control algorithm can offer excellent control performance and is of good robustness as shown in some preliminary study [10]. During the controller design process, the excitation is assumed to be a stationary white noise, and the following cost function is set as the control objective:

$$J = \lim_{t\to\infty} E\left[\int_0^t (z_r{}^T Q z_r + u^T R u)dt\right] \tag{9}$$

where z_r, system output variables, are set to be the transverse displacement or acceleration response at the tip ends; Q, a square matrix of the same order as z_r, equals to the multiplication of an identity matrix with a parameter q; u is a control force variable; R, the active force weight matrix, is set to be an identity matrix of the same order as the number of AMD applied. The design of the *LQG* controller is to adjust the weight parameter q via optimizing the performance of the system with compensator under the limitation of energy supply. The design of the controller relies on the full state feedback vector X_r. Since limited sensors are mounted on the structure, this full state vector cannot be directly measured but be estimated from the sensor measurements. When the excitation forces w and the measurement noise v are uncorrelated white noise process, the Kalman-Bucy filter is employed to get an estimation of the state vector X_r [15].

To obtain a good controller for experimental implementation, a series of numerical analysis with different value of weight parameter q are conducted. During the numerical analysis, structural modal damping coefficients are set to be the same as the real measured modal damping ratios. A scaled *El Centro* earthquake time history, whose dominant frequency band covers the first 2 transverse modal frequencies of the bridge, is adopted to excite the bridge in the numerical study. Two AMD placement strategies are employed for the comparison of optimal actuator location. These two strategies are the strategy of one AMD placed at the tip end of the main span (named S1) and the strategy of two AMDs placed at the tip ends of both spans (named S2). The AMDs are simulated to be the two electric servo-type AMD carts provided by Quanser Inc. with the following expression of the actuation force

$$f_d^1(t) = 8.246\dot{x}_d(t) + 1.42u(t) \tag{10}$$

$$f_d^2(t) = 12.576\dot{x}_d(t) + 1.73u(t) \tag{11}$$

To simulate a more practical control condition during experimental implementation, the following constraint condition is adopted: 1) The discrete digital computation is employed for the controller computation with the sampling frequency of 500Hz; 2) The precision of the A/D converter is set to be 12-bits and the range of the input voltage is set to be ± 5 V; 3) The measurement noise of a root mean square (RMS) value of 0.015 V is added into each channel of the acceleration responses, which corresponds to the 0.3% of voltage range of the A/D converter; 4) The maximum actuation voltage is set to be ± 5 V with the corresponding RMS voltage of 1.67 V and the maximum actuation displacement is set to be ± 0.08 m with the corresponding RMS displacement of 0.027 m.

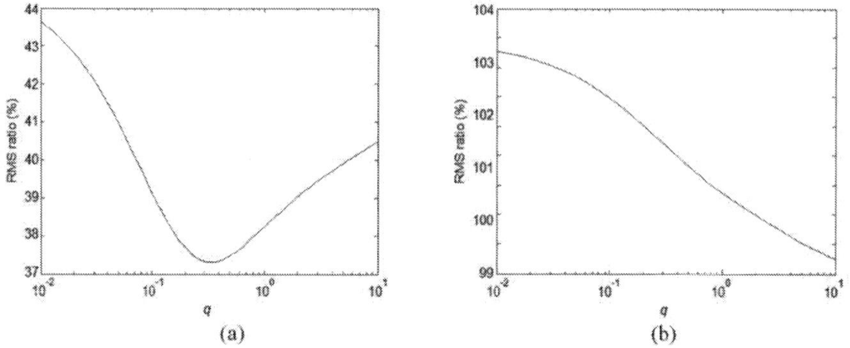

Figure 4. The ratio of the controlled acceleration RMS value to uncontrolled RMS value at the tip ends of the side span (a) and main span (b) with respect to the weighting parameter q for S1 control

For S1 control, a series of numerical studies simulating the control system with one AMD cart installed on the tip end of the main span of the bridge, whose actuation force is expressed as Eq. (10), are conducted when the weighting parameter q varies from 0.01 to 10. Fig. 4 shows the relative RMS ratio of the controlled acceleration response to the uncontrolled acceleration response at the tip end of both main span and side span with the varying of q. As shown in the figure, q = 0.398 are set for S1 to achieve an optimal control performances. This figure also tells that for S1 control, the tip acceleration response of the main span can be well controlled; however, a good control performance for the tip acceleration at the side span cannot be achieved by adjusting the weight parameter. Fig. 5 shows the controlled and uncontrolled tip acceleration response of the side span and the main span when q is set to be 0.398. This figure also tells that the well designed controller can greatly reduce the acceleration response of the tip end of the main span but cannot mitigate the vibration of the side span.

For S2 control, the control system with two AMD carts, whose actuation force expressions are shown as Eq. (10) and (11), installed on the tip ends of both spans of the bridge, is simulated. Numerical studies are conducted when the weighting parameter q varies from 0.01 to 100. Fig. 6 shows the relative RMS ratio of the controlled acceleration response to the uncontrolled acceleration response at the tip end of both spans with the varying of q. As shown in the figure, when q = 9.1, the control system provides the most optimal control performances. This figure also tells that for S2 control, the tip acceleration response of both the main span and the side span can be well reduced. Fig. 7 shows the controlled and uncontrolled tip acceleration time response of the side span and the main span when q is set to be 9.1. This figure also tells that the well designed controller can greatly reduce the acceleration response at the tip end of both spans.

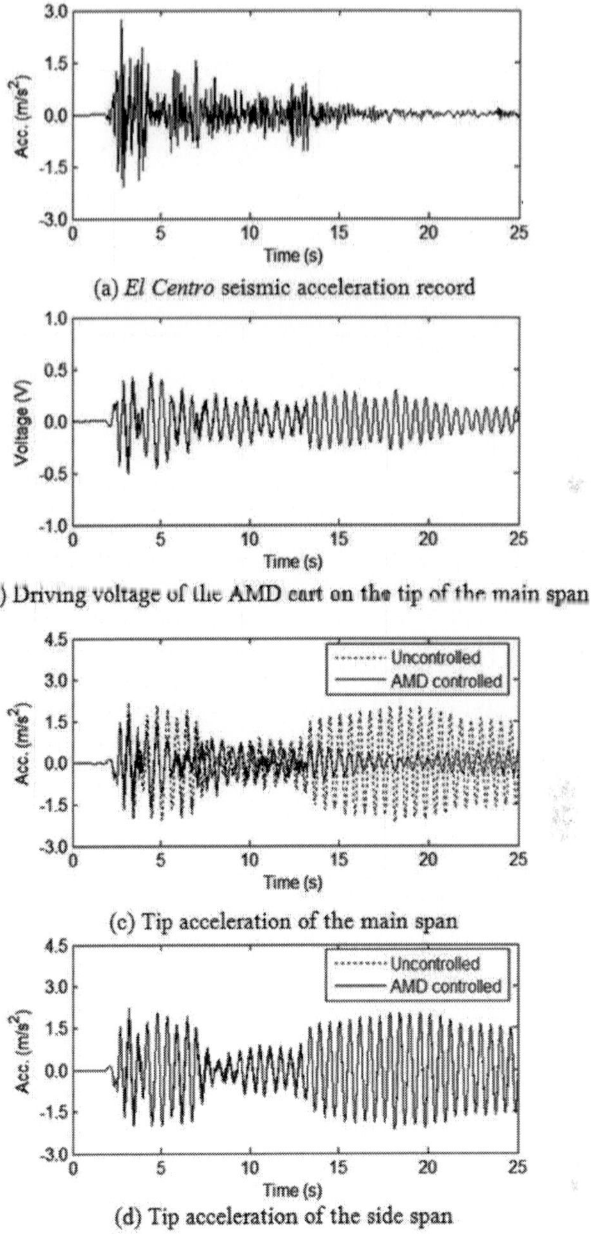

Figure 5. Excitation (a), driving voltage (b), main span tip acceleration (c), and side span tip acceleration (d) time histories of the bridge under El Centro seismic excitation for S1 control

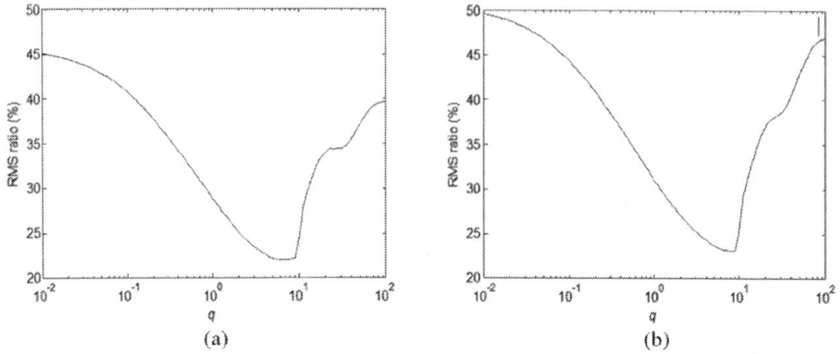

Figure 6. The ratio of the controlled acceleration RMS value to uncontrolled RMS value at the tip ends of the side span (a) and main span (b) with respect to the weighting parameter q for S2 control

The control performance comparison of these two AMD placing strategies tell that for the cable-stayed bridge studied, which is of two dominant transverse vibration modes with close frequencies, the single AMD control strategy (S1) can only reduce structural vibration response of the AMD-instrumented span, and the double AMD control strategy (S2) can achieve a good control performance for reducing structural response of both spans. These observations are verified by checking the controllability criteria. If the control system is of two close eigenvalues, at least two actuators are required to ensure the system is controllable. Moreover, since structural dominant transverse modes are anti-symmetric and symmetric bending modes, the shift of the AMD position along one span of the bridge will only proportionally vary the coefficients of B_r. If two AMDs are placed at one span of the bridge, the corresponding two columns of B_r are linear dependant. Therefore, the two AMDs must be placed at both the side span and the main span respectively to ensure the controllability of the bridge.

6. Experimental implementation

To verify the feasibility of the AMD control for transverse vibration reduction of cable-stayed bridge under construction, an experimental study on the fabricated test model is conducted in the Bridge Testing Laboratory of Tongji University. During the experiment, the S2 control strategy was adopted according to the conclusion obtained from the above numerical simulation study. Fig. 8 shows the layout of the experimental setup. As shown in the figure, the control system includes a data acquisition system, a central control computer, and two AMD carts. The data acquisition system consists of eight accelerometers, whose sensitivity is checked using dynamic calibration method; Dspace signal amplifier and filter; a general purpose data acquisition and control board MultiQ-3, which has 8 single ended analog inputs, 8 analog outputs, 16 bits of digital input, 16 bits of digital output, 3 programmable timers and up to 8 encoder inputs decoded in quadrature (option 2E to 8E). The central control computer is of 512 Mb memory and 1.0 GHz Intel Celeron processor. The

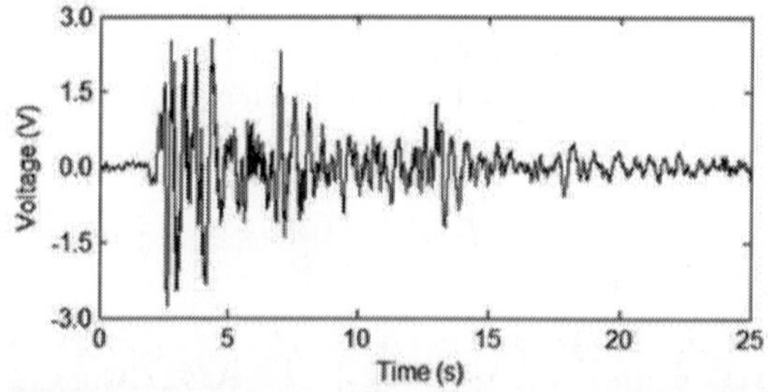

(a) Driving voltage of the AMD cart on the tip of the main span

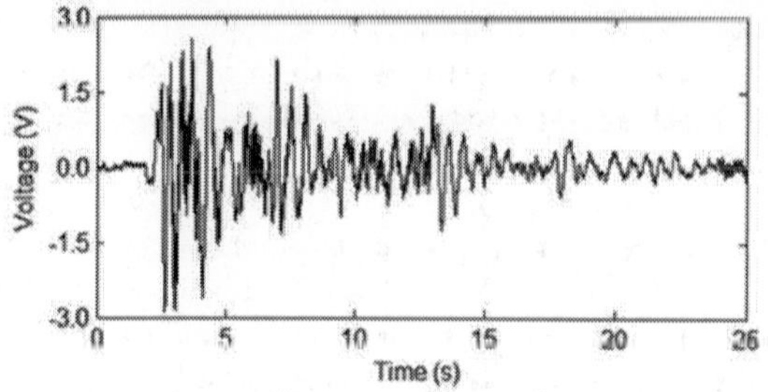

(b) Driving voltage of the AMD cart on the tip of the side span

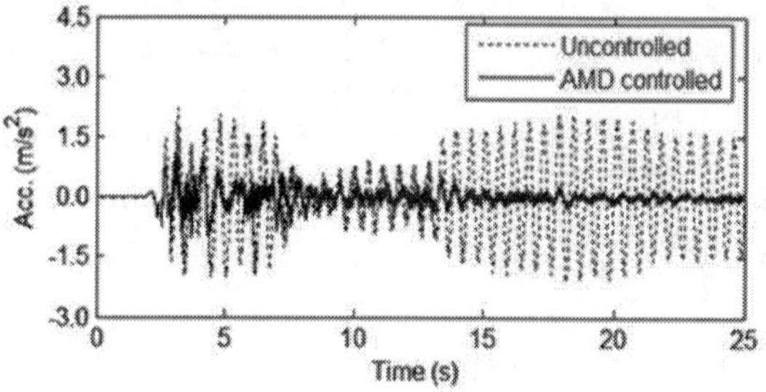

(c) Tip acceleration of the main span

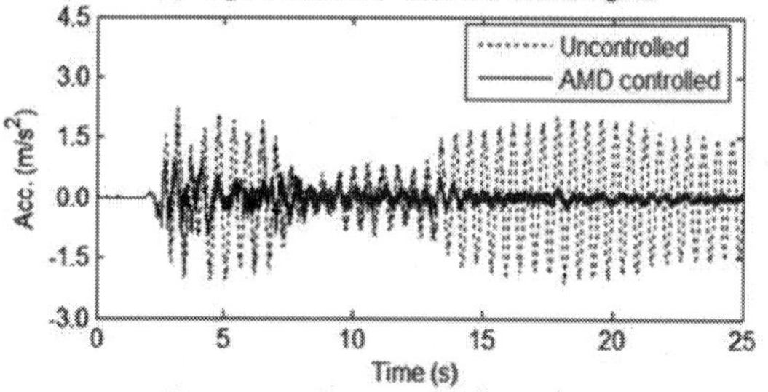

(d) Tip acceleration of the side span

Figure 7. Driving voltage of the main span (a) and side span (b), and tip acceleration time histories of main span (c) and side span (d) of the bridge under *El Centro* seismic excitation for S2 control

AMD carts (as shown in Fig. 8) are electric servo type [16]. They are 0.645 kg weight. Their track length is 32 cm and the max safe input voltage is $\pm 5V$. The system is designed under the constraint of avoiding the AMD to knock the baffle. If the AMD position exceeds the safe range of the orbit, the system will be shut down.

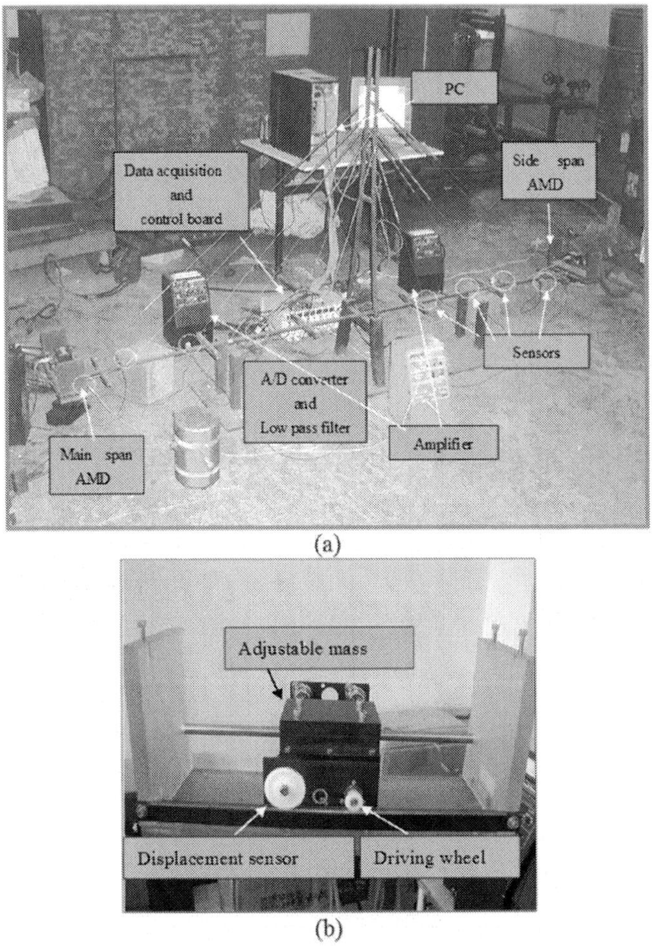

Figure 8. The experimental setup (a) and the AMD devices (b)

During the experiment, the tip ends of the main girder were pulled transversely using steel wires firstly to generate an initial displacement in this direction. The wires were then cut suddenly and the bridge started to vibrate due to this initial potential energy imported. Several seconds later, the power of the AMD control system was turned on and structural vibration response was recorded. Three case studies were conducted to check the performance of the AMD control system under different excitation schemes. For the first case, case C1, the tip ends of both the main span and the side span were pulled in the same direction. After the steel wires were cut, the transverse symmetric bending mode of the bridge was excited. For case C2, the tip ends of the two spans were pulled in the opposite direction to excite the transverse anti-symmetric bending mode of the bridge. Case C3 simulated bridge free vibration under an initial displacement of the main span. Both anti-symmetric and symmetric transverse bending modes of the bridge would thus be excited.

For test case C1, Fig. 9 shows the AMD driving voltage, recorded tip acceleration responses with or without AMDs, and their Fourier spectrums. As shown in these figures, when the power of the AMD carts was still turned off, structural acceleration responses were already reduced. When the power is turned on, the free decay ratios of structural responses were further increased. That verifies the performance of the active control system on structural response reduction. The comparison of the uncontrolled and AMD controlled Fourier spectrum magnitude of structural responses also verifies this statement. Moreover, as shown, when the AMD carts were installed, the peaks of Fourier spectrum were shifted to the left-hand-side, which meant that the natural frequencies of the system were decreased. Table 5 shows the peak and RMS acceleration response of the structure with and without AMD. As shown in the table, after the AMD was mounted, the peak and RMS accelerations recorded at the tip point of the side span were reduced 44.1 % and 82.1 %, respectively. For the tip point of the main span, the recorded peak and RMS accelerations were reduced 31.1 % and 81.4 %, respectively. For test cases C2 and C3, the peak and RMS acceleration responses of the structure with and without AMD were recorded. As shown in Table 5, the transverse vibration responses for these sensor-mounted points were also efficiently reduced: For case C2, structural peak and RMS acceleration responses were respectively reduced 28.3% and 65.4% for the tip point of the side span and 22.0% and 68.4% for the tip point of the main span; For case C3, structural peak and RMS acceleration responses were reduced 65.8% and 85.6% respectively for the tip point of the side span and 40.5% and 76.7% for the tip point of the main span. Moreover, concerning that the controller is designed via numerical studies on a reduced order model, the good control performances obtained on the bridge model tells that the control algorithm adopted in this study is of good robustness.

Case	Peak acc. of sensor 1		RMS acc. of sensor 1		Peak acc. of sensor 8		RMS acc. of sensor 8	
	No AMD	With AMD	No AMD	With AMD	No AMD	With AMD	No AMD	With AMD
C1	4.431	2.479	1.826	0.326	4.312	2.970	1.810	0.336
C2	4.198	3.012	1.341	0.464	3.563	2.779	1.327	0.419
C3	5.186	1.773	1.800	0.260	5.833	3.473	1.820	0.424

Table 5. Peak and RMS acceleration at the tip ends of the bridge with and without AMDs for the experimental cases

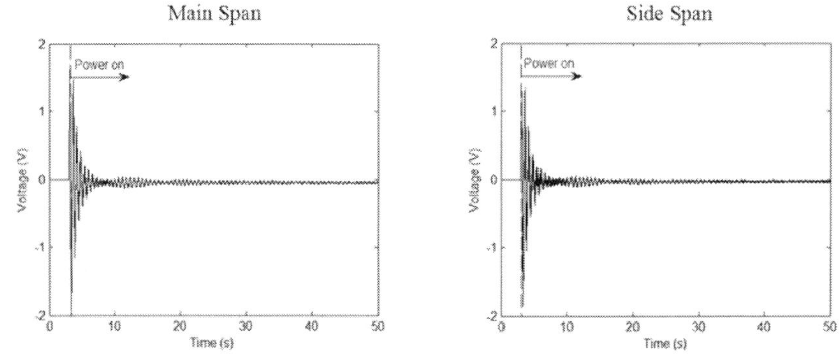

(a) Driving voltage of the AMD carts

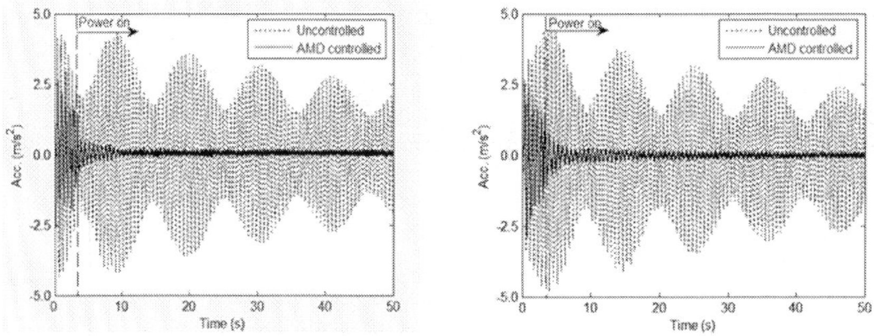

(b) Tip acceleration response

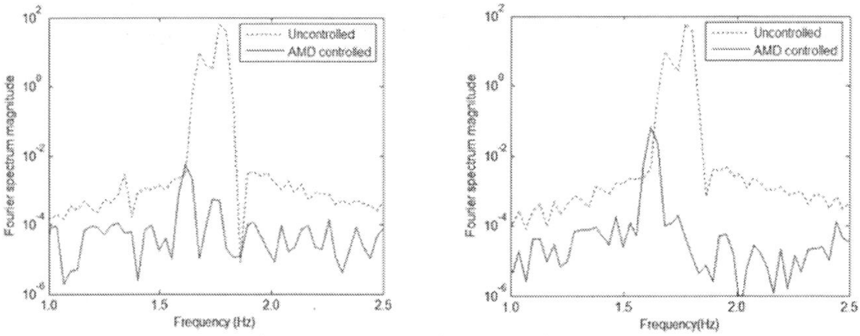

(c) Fourier spectrum of the tip acceleration responses

Figure 9. The driving voltage (a) and tip acceleration (b) time histories and Fourier spectrum (c) of responses for control case C1

7. Concluding remarks

In this chapter, the active mass dampers are implemented for vibration control of a lab-scale cable stayed bridge in double cantilever construction state. The results of both numerical simulation and experimental study verified that the proposed AMD control technique is applicable and efficient for the control of transverse vibration of cable-stayed bridge under construction.

For the cable-stayed bridge studied in this chapter, structural vibration test showed that the bridge was of two dominant transverse bending modes with close frequencies. The numerical study verified that for the control of such a structure with repeated frequencies; at least two AMDs should be installed for a good control performance. Moreover, the placement of those two AMDs should be carefully studied.

For the control of a linear, time-invariant system, an accurate and complete system models are generally required. However, for the bridge-AMD system studied in this chapter, it is very difficult, if not impossible, to set up such a numerical model due to the complex layout of the bridge structure. Since structural vibration responses are generally governed by some dominant vibration modes and the objective of structural vibration control is to reduce but not to completely restrain structural vibration responses, this study verified that a reduced order model constructed from the critical modes is good enough for the controller design to achieve an excellent vibration reduction performance. Considering the differences between the numerical model and the real structure, the control algorithm adopted in this study must be robust to the vibration property change of the controlled structure. The results of experimental studies show that the vibration of the test model can be well controlled using the controller designed from the numerical studies. That means the *LQG* control algorithm is of good robustness for real application.

This study is an initial work on AMD control of transverse vibration of a cable stayed bridge under construction before it can be used for real applications. Although the experimental study verified the efficiency of the adopted AMD control for structural response reduction under given excitation, some primary issues for real application of the AMD control technique, such as how to deal with time delay, how to reduce the requirement on power supply of the control system, and et al., are not addressed in this study. Further laboratory studies or field applications on some real bridges will be conducted in the coming future to discuss these issues and make the technique more efficient and practical for real applications.

Author details

Hao Chen

Institute of Engineering Mechanics, China Earthquake Administration, Sanhe, Hebei, China

Zhi Sun and Limin Sun

State Key Laboratory for Disaster Reduction in Civil Engineering, Tongji University, Shanghai, China

Acknowledgement

This research is partially supported by the National High-tech R&D Program of China (863 Program) (Grant No. 2006AA11Z109), and Shanghai Rising Star Tracking Program (Grant No. 09QH1402300). These supports are greatly appreciated.

8. References

[1] Boonyapinyo, V., Aksorn, A., and Lukkunaprasit, P. (2007). Suppression of aerodynamic response of suspension bridges during erection and after completion by using tuned mass dampers. *Wind and Structures, An Int\ 'l Journal*, 10(1): 1-22.

[2] Ubertini, F. (2010). Prevention of suspension bridge flutter using multiple tuned mass dampers. *Wind and Structures, An Int\ 'l Journal*, 13(3): 235-256.

[3] Warnitchai, P., Fujino, Y., Pacheco, B.M. and Agret, R. (1993). An experimental study on active tendon control of cable-stayed bridges. *Earthquake Engineering and Structural Dynamics*, 22: 93-111.

[4] Achkire, Y. and Preumont, A. (1996). Active tendon control of cable-stayed bridges. *Earthquake Engineering and Structural Dynamics*, 25(6): 585-597.

[5] Dyke, S.J., Caicedo, J.M., Turan, G., Bergman, L.A., Hague, S. (2003). Phase I benchmark control problem for seismic response of cable-stayed bridges. *Journal of Structural Engineering, ASCE*, 129(7): 857–872.

[6] Caicedo, J.M., Dyke, S.J., Moon, S.J., Bergman, L.A., Turan, G., Hague, S. (2003). Phase II benchmark control problem for seismic response of cable-stayed bridges. *Journal of Structural Control*, 10: 137–168.

[7] Frederic, B. and Andre, P. (2001). Active tendon control of cable-stayed bridges: a large-scale demonstration. *Earthquake Engineering & Structural Dynamics*, 30: 961–979.

[8] Shelley, S.J., Lee, K.L., Aksel, T. and Aktan, A.E. (1995). Active control and forced vibration studies on highway bridges. *Journal of Structural Engineering, ASCE*, 121(9): 1306-1312.

[9] Korlin, R. and Starossek, U. (2007). Wind tunnel test of an active mass damper for bridge decks. *Journal of Wind Engineering and Industrial Aerodynamics*, 95: 267-277.

[10] Chen, H. (2007). Experimental study on active control of cable-stayed bridge under construction. Master Thesis. Tongji University.

[11] Sun, Z., Hou, W., Chang, C. C. (2009). Structural system identification under random excitation based on asymptotic wavelet analysis. *Engineering Mechanics*, 26(6), 199-204 (in Chinese).

[12] Schemmann, A.G. (1997). Modeling and active control of cable-stayed bridges subject to multiple-support excitation. PhD thesis. Standford University.

[13] Xu, Y. and Chen, J. (2008). Modal-based model reduction and vibration control for uncertain piezoelectric flexible structures. *Structural Engineering and Mechanics*, 29(5): 489-504.

[14] Laub, A.J. and Arnold, W.F. (1984). Controllability and observability criteria for multivariable linear second-order models, *IEEE Transactions on Automatic Control*, AC-29(2).

[15] Wu, J.C., Yang, J.N., and Schmitendorf W.E. (1998). Reduced-order H_∞ and *LQR* control for wind-excited tall buildings. *Engineering Structures*. 20(3): 222-236.

[16] Quanser Consulting Inc. (2002). Active Mass Damper: Two-Floor (AMD-2), User Manual.

ADVANCES IN VIBRATION ANALYSIS RESEARCH

Edited by **Farzad Ebrahimi**

Vibration Analysis of Long Span Joist Floors Submitted to Human Rhythmic Activities

José Guilherme Santos da Silva, Sebastião Arthur Lopes de Andrade, Pedro Colmar Gonçalves da Silva Vellasco, Luciano Rodrigues Ornelas de Lima and Rogério Rosa de Almeida
State University of Rio de Janeiro (UERJ)
Rio de Janeiro/RJ,
Brazil

1. Introduction

In the last years, building structures are more and more becoming the modern landmarks of urban areas. Designers seem to continuously move the safety border, in order to increase slenderness and lightness of their structural systems. However, more and more steel and composite floors (steel-concrete) are carried out as light weight structures with low frequencies and low damping. These facts have generated very slender composite floors, sensitive to dynamic excitation, and consequently changed the serviceability and ultimate limit states associated to their design.

The increasing incidence of building vibration problems due to human rhythmic activities led to a specific design criterion for rhythmic excitations to be addressed in structural design (Allen et al. 1985); (Almeida, 2008); (Almeida et al., 2008); (Bachmann & Ammann, 1987); (Faisca, 2003); (Ji & Ellis, 1994); (Langer, 2009); (Murray et al., 2003); (Silva et al., 2008). This was the main motivation for the development of a design methodology centred on the structural system dynamical response submitted to dynamic loads due to human activities.

This paper investigated the dynamic behaviour of composite floors (steel-concrete) subjected to the human rhythmic activities. The dynamic loads were obtained through experimental tests conducted with individuals carrying out rhythmic and non-rhythmic activities such as stimulated and non-stimulated jumping and aerobics (Faisca, 2003).

The description of the loads generated by human activities is not a simple task. The individual characteristics in which each individual perform the same activity and the existence of external excitation are relevant factors when the dynamic action is defined. Numerous investigations were made aiming to establish parameters to describe such dynamic loads (Allen et al. 1985); (Bachmann & Ammann, 1987); (Faisca, 2003); (Murray et al., 2003).

The present investigation considered the dynamic loads, based on results achieved through a long series of experimental tests made with individuals carrying out rhythmic and non-rhythmic activities. This investigation described these dynamic loads, generated by human activities, such as jumps with and without stimulation, aerobics, soccer, rock concert audiences and dancing (Faisca, 2003).

The load modelling was able to simulate human activities like aerobic gymnastics, dancing and free jumps. In this paper, the Hanning function was used to represent the human dynamic actions since it was verified that this mathematical representation was very similar to the signal force obtained through experimental tests (Faisca, 2003). Based on the experimental results, human load functions due to rhythmic and non-rhythmic activities were proposed.

The computational model, developed for the composite floors dynamic analysis, adopted the usual mesh refinement techniques present in finite element method simulations implemented in the Ansys program (ANSYS, 2003). In the present computational model, the floor steel joists were represented by three-dimensional beam elements, considering flexural and torsion effects, while the composite slab was represented by shell finite elements.

The investigated structural model was associated to a floor composed by steel joists and a concrete slab. The structural system was a typical floor used as a restaurant with an adjacent dancing area (Almeida, 2008); (Almeida et al., 2008); (Murray et al., 2003); (Silva et al., 2008). The composite (steel-concrete) floor system consisted of long span (14m) joists supported by concrete block walls.

The floor effective weight was estimated to be equal to 3.6kPa, including 0.6kPa for people dancing and dining. The joists effective composite moment of inertia was selected based on its required strength, i.e., 1.1×10^6 mm^4. This structural system geometry was based on a typical example described in literature (Almeida, 2008); (Almeida et al., 2008); (Murray et al., 2003); (Silva et al., 2008).

The parametric study considered correlations between analytical and numerical results found in literature. The peak acceleration values were compared to the limits proposed by design codes and recommendations (ISO 2631-2, 1989); (Murray et al., 2003), based on human comfort criteria. The results indicated that the limits suggested by the design recommendations were not satisfied. This fact indicated that these rhythmic activities could generate peak accelerations that surpass design criteria limits developed for ensuring human comfort.

2. Human-induced dynamic loads

Floor vibrations induced by human rhythmic activities like: walking, running, jumping or even aerobics consist on a very complex problem. This is due to the fact that the dynamical excitation characteristics generated during these activities are directly related to the individual body adversities and to the specific way in which each human being executes a certain rhythmic task. All these aspects do not contribute for an easy mathematical or physical characterization of this phenomenon.

Human beings have always analysed the most apparent distinctions of the various activities they perform. However the fundamental mechanical analysis of these tasks was not possible before a significant development of the mechanical science. Initially the human motion received an incipient attention from researchers like Borelli in 1679 (Lehmkuhl & Smith, 1985) and the Weber brothers in 1836 (Lehmkuhl & Smith, 1985). The first pioneer on this field was Otto Fischer, a German mathematician that in 1895 made the first study containing a comprehensive evaluation of the forces involved in human motion.

In order to determine the dynamical behaviour of floor structural systems subjected to excitations from human activities, various studies have tried to evaluate the magnitude of these rhythmic loads. The following stage of this research line was the development of a loading platform by Elftman (Lehmkuhl & Smith, 1985), that enable the determination of the ground reactions to the foot forces associated to the human walk motion. The typical force platform is made by an approximate $1m^2$ steel plate supported by four small columns at the plate midsides. Load cells were installed at each of the columns to detect the magnitude of the load variation at these points. With these results in hand it was possible to determine the magnitude and direction of the forces transmitted to the supporting surface, denominated ground reaction forces.

Rainer also contributed in this investigation developing more sophisticated load platforms that recorded the ground reaction forces coming from the foot forces associated to the human motion (Rainer et al., 1987). Ebrahimpur developed a 14.2m length x 2m wide platform designed to record the actions from a single individual, or groups of two or four individual walk motion (Ebrahimpur, 1996).

Another load model used to represent the walk motion forces is expressed as a function of tests that recorded the heel impact over the floor. This load type, considered as the main excitation source during the human walk motion, produces a transient response, i.e., when the system is excited by an instantaneous force application. Its graphical representation was presented by Ohmart (Ohmart, 1968) in experiments denominated heel drop tests, where the individual drops its heel over the floor after elevating it to a height corresponding to its weight.

The heel drop test was also made by Murray and Hendrick in different building types (Murray & Hendrick, 1977). A 0.84kN impact force was measured by a seismograph in nine church ceremonial rooms, three slabs located at a shopping mall highest floor, two balcony slabs of a hotel and one slab located at a commercial building second floor. With these results in hand, the structural dynamic responses, in terms of the force amplitudes, frequencies and damping, associated to the investigated structural systems, could be determined.

Murray (Murray, 1975) classified the human vibration perception in four categories, i.e.: the vibration is not noticed by the occupants; the vibration is noticed but do not disturb the occupants; the vibration it is noticed and disturb the occupants; the vibration can compromise the security of the occupants. These categories were established based on 100 heel drop tests performed on composite floors made of steel beams and concrete slabs.

Allen et al. (Allen et al., 1985) proposed minimum values for the natural frequencies of structures evaluated according to the type of occupation and their main characteristics. These values were based on the dynamical load values produced by human rhythmic activities like dancing and aerobics and on the limit acceleration values associated to those activities.

A significant contribution to this field was made in Brazil by Alves (Alves, 1997) and Faisca (Faisca, 2003) based on experiments made with a group of volunteers acting on a concrete platform. These tests enabled the development of approximated descriptions of the loads induced by human activities such as: jumps, aerobics, soccer and rock show audience responses. These tests were executed over two concrete platforms, one rigid and the other flexible, both of them over movable supports. The experimental results analysis, allied to an analytical model, led to the development of load functions associated to synchronous and asynchronous activities that could be used in structural designs intended for stadiums and other related structures.

3. Loads generated by human activities

The description of the loads generated by human activities is not a simple task. The individual characteristics in which each individual perform the same activity and the existence of external excitation are relevant factors when the dynamic action is defined. Numerous investigations were made aiming to establish parameters to describe such loads (Allen et al. 1985); (Bachmann & Ammann, 1987); (Faisca, 2003); (Murray et al., 2003).

Several investigations described the loads generated by human activities as a Fourier series, which consider a static part due to the individual weight and another part due to the dynamic load. The dynamic analysis is performed equating one of the activity harmonics to the floor fundamental frequency, leading to resonance (Almeida, 2008); (Bachmann & Ammann, 1987); (Langer, 2009); (Murray et al., 2003); (Silva et al., 2008).

The present investigation considered the dynamic loads, based on results achieved through a long series of experimental tests made with individuals carrying out rhythmic and non-rhythmic activities (Faisca, 2003). These dynamic loads, generated by human activities, were described such as jumps with and without stimulation, aerobics, soccer, rock concert audiences and dancing.

The load modelling was able to simulate human activities like aerobic gymnastics, dancing and free jumps. In this paper, the Hanning function was used to represent the human dynamic actions since it was verified that this mathematical representation was very similar to the signal force obtained through experimental tests (Faisca, 2003).

The mathematical representation of the human dynamic loading is described by Equation (1). This expression requires some parameters like the activity period, T, contact period with the structure, T_c, period without contact with the model, T_s, impact coefficient, K_p, and phase coefficient, CD, see Fig. 1 and Table 1.

$$F(t) = CD\left\{K_p P\left[0.5 - 0.5\cos\left(\frac{2\pi}{T_c}t\right)\right]\right\}, \text{for } t \le T_c$$

$$F(t) = 0, \text{for } T_c < t \le T \tag{1}$$

Where:

F(t) : dynamic loading, in (N);
t : time, in (s);
T : activity period (s);
T_c : activity contact period (s);
P : weight of the individual (N);
K_p : impact coefficient;
CD : phase coefficient.

Figure 1 illustrates the phase coefficient variation, CD, for some human activities, initially, considering a few number of individuals and later extrapolating for a larger number of people (Faisca, 2003). Figure 2 presents an example of dynamic action related to human rhythmic activities using the following parameters: T = 0.53s, T_c = 0.43s, T_s = 0.10, K_p = 2.78 and CD = 1.0, see Table 1.

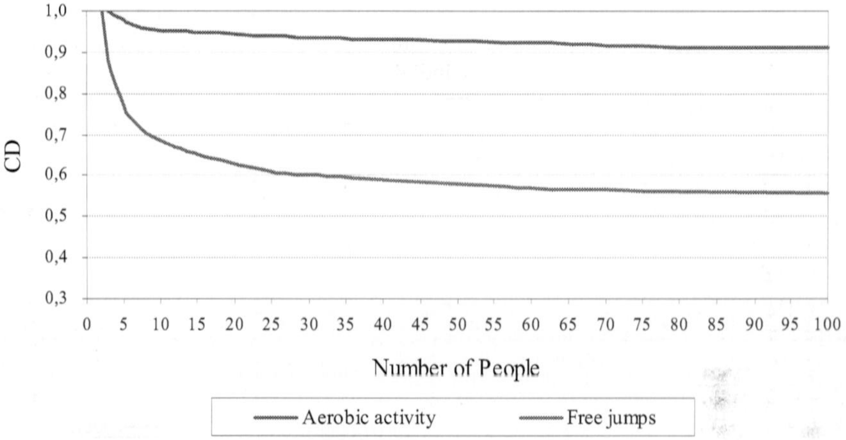

Fig. 1. Phase coefficients for the studied activities (Faisca, 2003)

Activity	T (s)	T_c (s)	K_p
Aerobics	0.44 ± 0.09	0.34 ± 0.09	2.78 ± 0.60
Free jumps	0.44 ± 0.15	0.32 ± 0.09	3.17 ± 0.58

Table 1. Parameters used for human rhythmic activities representation (Faisca, 2003)

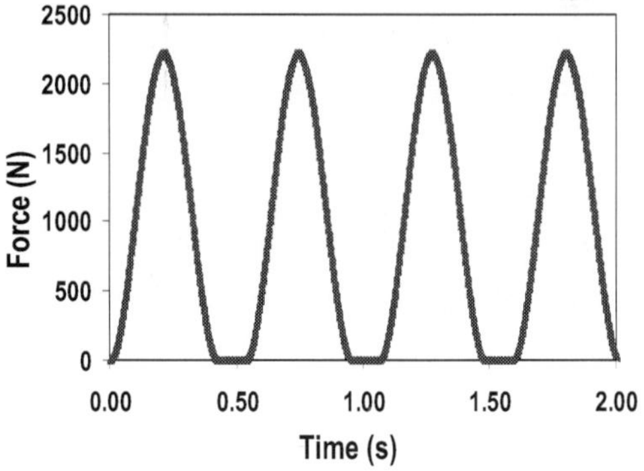

Fig. 2. Dynamic loads induced by dancing associated to the following parameters: T=0.53s, T_c=0.43s, T_s=0.10, K_p=2.78 and CD=1.0

4. Investigated structural model

The investigated structural model was associated to a floor composed by steel joists and a concrete slab, as presented in Figs. 3 to 6. The structural system was a typical floor used as a restaurant with an adjacent dancing area (Almeida, 2008); (Almeida et al., 2008); (Murray et al., 2003); (Silva et al., 2008).

The composite floor system consisted of long span (14m) joists supported by concrete block walls, see Figs. 3 to 6. The floor effective weight was estimated to be equal to 3.6kPa, including 0.6kPa for people dancing and dining. The joists effective composite moment of inertia was selected based on its required strength, i.e., 1.1×10^6 mm^4. This structural system geometry was based on a typical example described in literature (Almeida, 2008); (Almeida et al., 2008); (Murray et al., 2003); (Silva et al., 2008).
The adopted steel sections were made with a 300MPa yield stress steel grade. A 2.05×10^5 MPa Young's modulus was used for the steel joists. The concrete slab had a 30MPa specified compression strength and a 2.4×10^4 MPa Young's Modulus. The structural model geometrical characteristics are illustrated in Table 2.

Main Span	Bottom Chords	Top Chords	Vertical Members	Diagonals
14.0m	⌋⌊2x(1 ½″ x 1/8″)	⌉⌈2x(2″ x 1/8″)	L (½ ″x 1/8″)	L (½ ″x 1/8″)

Table 2. Structural model geometric properties

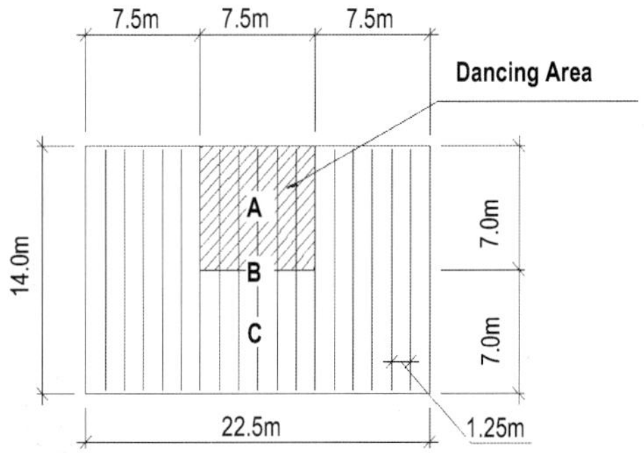

Fig. 3. Dancing floor layout (dimensions in m)

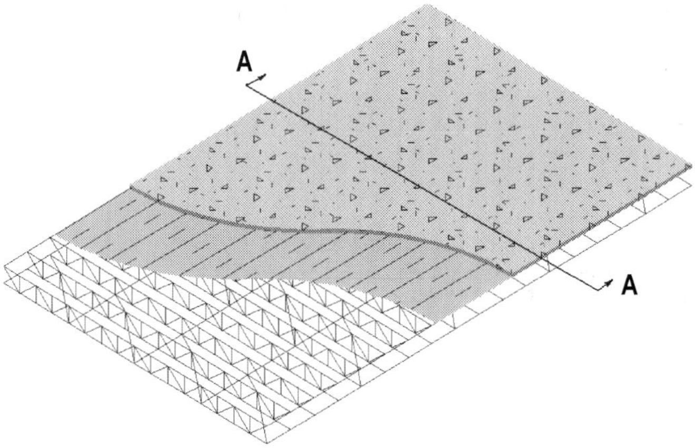

Fig. 4. Structural model three-dimensional view

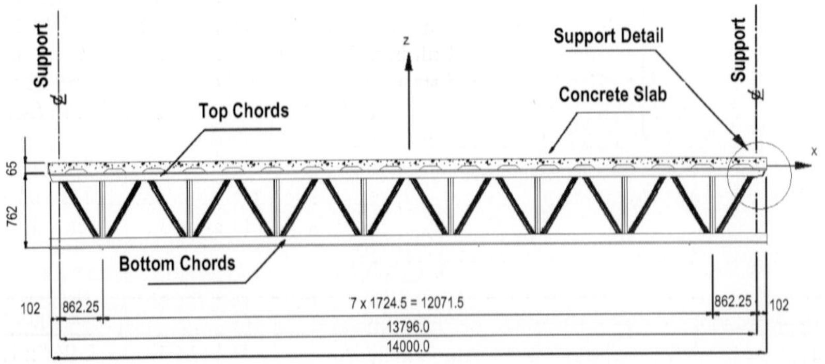

Fig. 5. Composite floor cross section - Section AA (dimensions in mm)

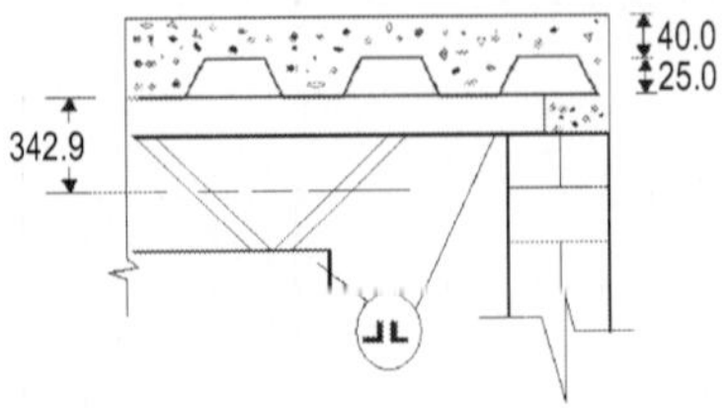

Fig 6. Support details (dimensions in mm)

The human-induced dynamic action was applied to the dancing area, see Figs. 3 and 7. The composite floor dynamical response, in terms of peak accelerations values, were obtained on the nodes A, B and C, to verify the influence of the dynamical loads on the adjacent slab floor, see Figs. 3 and 7. In the current investigation, the human rhythmic dynamic loads were applied to the structural model corresponding to the effect of 1, 3, 6, 9 and 12 individuals practicing aerobics or couples dancing.

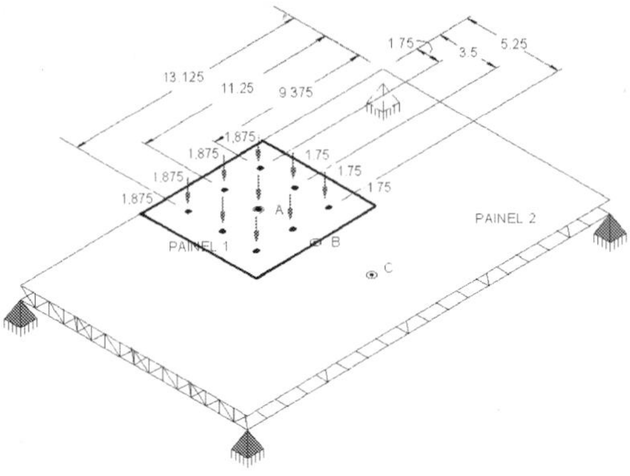

Fig. 7. Load distribution associated to nine individuals acting on the floor (dimensions in m)

The live load considered in this analysis corresponds to one individual for each 4.0m² (0.25 person/m²), (Bachmann & Ammann, 1987). The load distribution was considered symmetrically centred on the slab panel, as depicted in Fig. 7. The present investigation also assumed that an individual person weight was equal to 800N (0.8kN) (Bachmann & Ammann, 1987) and that the adopted damping ratio was equal to, ξ=3% (ξ = 0.03) in all studied cases (Almeida, 2008); (Almeida et al., 2008); (Murray et al., 2003); (Silva et al., 2008).

5. Finite element modelling

The proposed computational model, developed for the composite floors dynamic analysis, adopted the usual mesh refinement techniques present in finite element method simulations implemented in the ANSYS program (ANSYS, 2003). In the present computational model, the floor steel joists were represented by three-dimensional beam elements (BEAM44), with tension, compression, bending and torsion capabilities (ANSYS, 2003). The composite slab was represented by shell finite elements (SHELL63) (ANSYS, 2003), as illustrated in Fig. 8. In this investigation, it was considered that both materials (steel and concrete) presented total interaction and have an elastic behaviour. The finite element model has 11673 nodes, 5267 three-dimensional beam elements (BEAM44), 6912 shell elements (SHELL63) and 62568 degrees of freedom. The developed computational model is illustrated in Fig. 8.

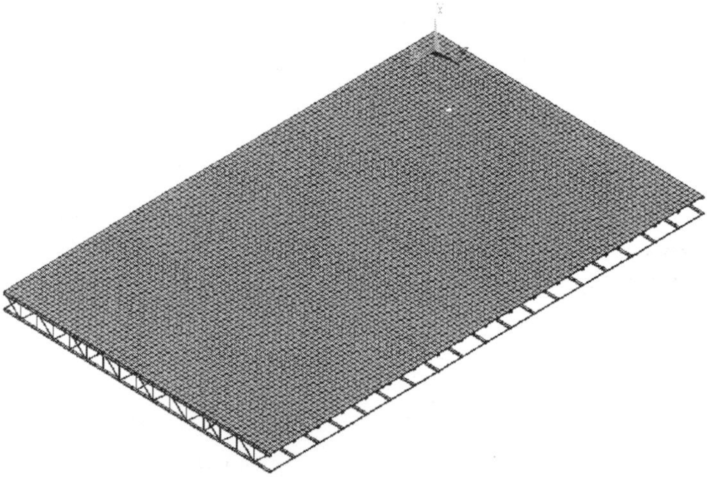

Fig. 8. Composite floor (joists and concrete slab) finite element model

6. Natural frequencies and vibration modes

The composite (steel-concrete) floor natural frequencies were determined with the aid of the numerical simulations, as illustrated in Table 3. The structural system vibration modes were illustrated in Fig. 9.

It can be clearly noticed from Table 3 results, that there is a very good agreement between the structural model fundamental frequency value calculated using finite element simulations and the AISC recommendation (Murray et al., 2003).

Such fact validates the numeric model here presented, as well as the results and conclusions obtained throughout this work. It must be emphasized that the structural model presented vibration modes with a predominance of flexural effects, as illustrated in Fig. 9.

Natural Frequencies (Hz)						AISC*	Error
f_{01}	f_{02}	f_{03}	f_{04}	f_{05}	f_{06}	f_{01}	%
5.70	5.91	6.13	6.42	6.56	7.89	5.80	2.0

*(Murray et al., 2003)

Table 3. Composite floor (steel-concrete) natural frequencies

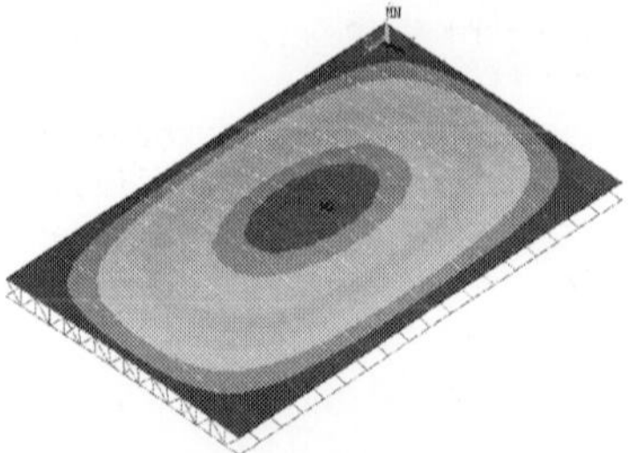

a) Vibration mode associated to the first natural frequency: f_{01}=5.80 Hz.

b) Vibration mode associated to the second natural frequency: f_{02}=5.91 Hz.

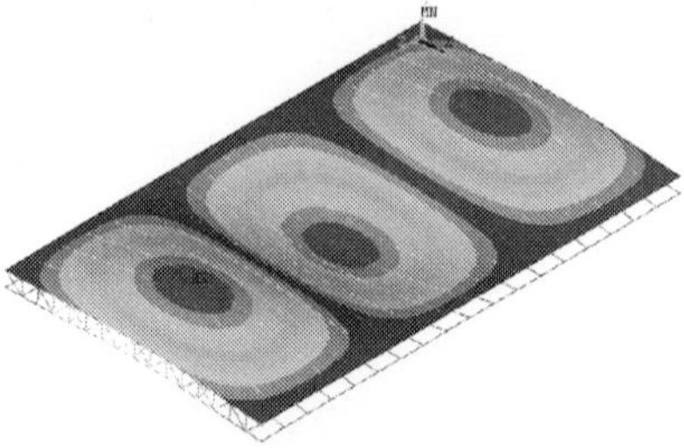

c) Vibration mode associated to the third natural frequency: f_{03}=6.13 Hz.

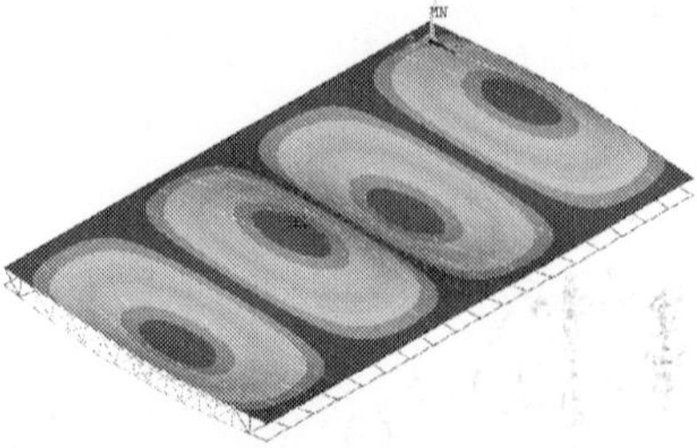

d) Vibration mode associated to the fourth natural frequency: f_{04}=6.42 Hz.

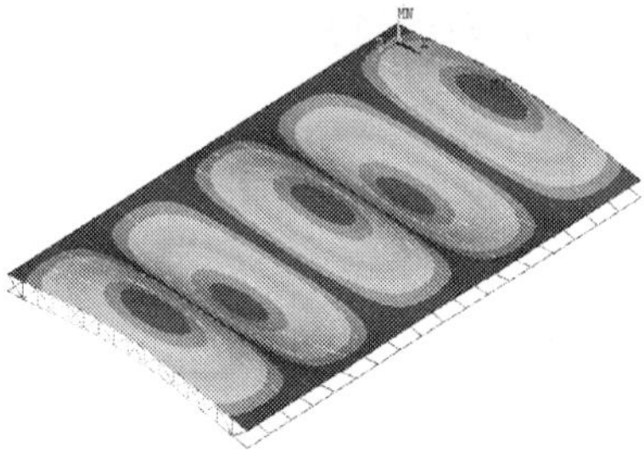

e) Vibration mode associated to the fifth natural frequency: f_{05}=6.56 Hz.

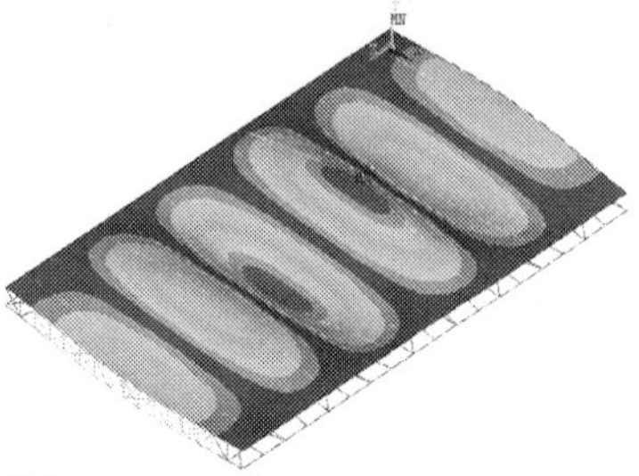

f) Vibration mode associated to the sixth natural frequency: f_{06}=7.89 Hz.

Fig. 9. Composite floor (steel-concrete) vibration modes

7. Time domain analysis

For practical purposes, a linear time-domain analysis was also performed throughout this study. This section presents the evaluation of the structural systems vibrations levels when submitted to dynamic excitations coming from human rhythmic activities (aerobics and dancing).

The composite floor (steel-concrete) dynamic responses were determined through an analysis of its displacements and accelerations. The results of the dynamic analysis were obtained from an extensive numerical analysis, based on the finite element method using the ANSYS program (ANSYS, 2003).

Figures 10 and 11present the vertical displacement and acceleration, respectively, versus time graphs for the analysed composite floor (steel-concrete) at point A (see Figs. 3 and 7), when only one individual is acting on the structural model (aerobics).

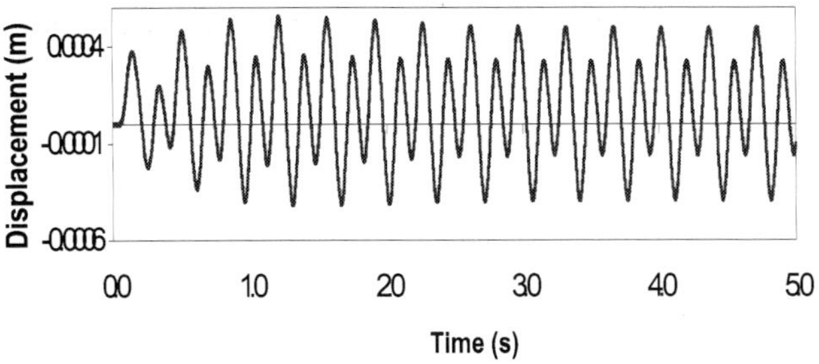

Fig. 10. Composite floor displacement response due to one individual practicing aerobics at Point A (see Figs. 3 and 7): T_c=0.25s, T_s=0.10s, K_p=2.78 and CD=1.

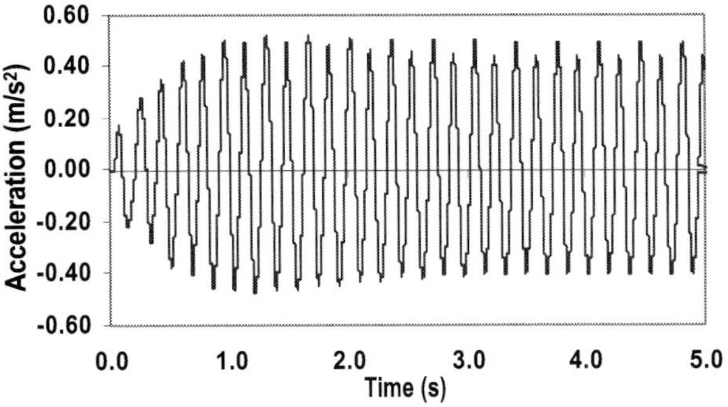

Fig. 11. Composite floor acceleration response due to one individual practicing aerobics at Point A (see Figs. 3 and 7): T_c=0.25s, T_s=0.10s, K_p=2.78 and CD=1.

Figures 10 and 11 show that, the vertical displacement and acceleration, at point A (see Figs. 3 and 7) of the structural model, gradually increase with time until the beginning of the composite floor steady state response, which occurred at the time of approximately 2.0 seconds. From this point (t = 2.0s) onwards, the maximum displacement and acceleration values were, respectively, equal to 0.051 cm and 0.55 m/s^2.
It must be emphasized that considering only one individual acting on the floor (aerobics) the calculated peak acceleration value (a_p = 0.55m/s^2), was higher than limits proposed by design recommendations (a_{lim} = 5%g = 0.50 m/s^2), violating the human comfort criteria (ISO 2631-2, 1989); (Murray et al., 2003).

8. Peak accelerations

The peak acceleration analysis was focused in dancing activities and considered a contact period carefully chosen to simulate dancing activities on the composite floor. The adopted parameters were: T_c, equal to 0.43s (T_c = 0.43s) and the period without contact to the structure, T_s, of 0.10s (T_s = 0.10s). Based on the experimental results (Faisca, 2003), the composite floors dynamic behaviour was evaluated keeping the impact coefficient value, K_p, equal to 2.78 (K_p = 2.78). Tables 4 and 5 depict the peak accelerations, a_p, corresponding to nodes A, B and C, see Figs. 3 and 7, when 1, 3, 6, 9 and 12 dynamical loads, simulating individual dancing, see Table 4, and couples dancing, see Table 5, were applied to the composite floor.

Nodes (see Fig. 3)	Number of individuals - a_p (m/s^2)					ISO 2631-2 and AISC* (m/s^2)
	1	3	6	9	12	
A	0.12	0.24	0.41	0.52	0.69	0.50
B	0.11	0.28	0.53	0.72	0.88	
C	0.07	0.17	0.31	0.42	0.54	
*(ISO 2631-2, 1989); (Murray et al., 2003)						

Table 4. Structural model peak accelerations corresponding to individuals dancing: T_c=0.43s; T_s=0.10s; K_p=2.78.

Nodes (See Fig. 3)	Number of couples - a_p (m/s^2)					ISO 2631-2 and AISC* (m/s^2)
	1	3	6	9	12	
A	0.23	0.47	0.83	1.05	1.36	0.50
B	0.23	0.57	1.04	1.45	1.76	
C	0.13	0.34	0.62	0.83	1.09	
*(ISO 2631-2, 1989); (Murray et al., 2003)						

Table 5. Structural model peak accelerations corresponding to couples dancing: T_c=0.43s; T_s=0.10s; K_p=2.78

It can be verified that the obtained peak acceleration values are proportional to an increase of the number of considered individuals, Tables 4 and 5. These values tend to decrease when the dynamical response obtained on the node C (see Figs. 3 and 7) was compared to the response of nodes A and B (see Figs. 3 and 7), as presented in Tables 4 and 5.

Based on the results presented in Table 4, it was possible to verify that dancing activities on the structural model, represented by Equation (1), led to peak accelerations higher than 0.50 m/s^2 (5%g) (ISO 2631-2, 1989); (Murray et al., 2003), when the composite floors was submitted to six individuals dancing, violating the human comfort criteria. The situation becomes even more significant when nine and twelve individuals were considered in the analysis, see Table 4.

On the other hand, when couples dancing were considered, the human comfort criterion was violated starting from cases associated with only three couples. It must be emphasized that in this situation the peak accelerations presented higher values when compared to individual dancing.

Observing the results illustrated in Tables 4 and 5, it was also possible to verify that the analyzed composite floor presented peak accelerations higher than 5.0% g (ISO 2631-2, 1989); (Murray et al., 2003) and the human comfort criteria was not satisfied even when an adjacent area, where no dancing actions are present (Node C, see Figs. 3 and 7), was investigated, Tables 4 and 5.

9. Final remarks

This paper investigated the dynamic behaviour of composite floors (steel-concrete) when subjected to the human rhythmic activities corresponding to aerobics and dancing effects. The dynamic loads were obtained through experimental tests conducted with individuals carrying out rhythmic and non-rhythmic activities such as stimulated and non-stimulated jumping and aerobics.

The proposed analysis methodology adopted the usual mesh refinement techniques present in the finite element method (FEM). Based on the experimental results (Faisca, 2003), human load functions due to rhythmic and non-rhythmic activities were proposed. The investigated structural system was a typical floor used as a restaurant with an adjacent dancing area. The composite floor system consisted of long span (14m) joists supported by concrete block walls.

The parametric analysis considered correlations between analytical and numerical results found in literature. The results, in terms of maximum accelerations, were compared to the limits proposed by design recommendations, focusing on human comfort considerations.

The results obtained throughout this study indicated that the limits recommended by design standards (ISO 2631-2, 1989); (Murray et al., 2003) were not satisfied for the investigated structural model when subjected dancing load actions. Such fact shows that these rhythmic activities may generate peak accelerations that violated design criteria related to human comfort.

The present investigation also indicated that these dynamic loads can even generated considerable perturbations on adjacent areas, where there is no human rhythmic activity of such kind present. Despite this fact there was still a surpassing of the associated human comfort criteria.

10. Acknowledgements

The authors gratefully acknowledge the support for this work provided by the Brazilian Science Foundations CAPES, CNPq and FAPERJ.

11. References

Almeida, R. R. de. (2008). Análise de vibrações em sistemas estruturais para pisos mistos com joists submetidos a atividades humanas rítmicas (Vibration analysis of composite joist floors subjected to human rhythmic activities), MSc Dissertation (in Portuguese), Civil Engineering Post-Graduate Programme, PGECIV, State University of Rio de Janeiro, UERJ, Rio de Janeiro, Brazil, pp. 1-154.

Almeida, R. R. de; Silva, J.G.S. da; Vellasco, P.C.G. da S.; Andrade, S.A.L. de; Lima, L.R.O de. (2008). Dynamical behaviour of long span joist floors submitted to human rhythmic activities. Proceedings of the 5th European Conference on Steel and Composite Structures, Eurosteel 2008, Graz, Austria, v. A. p. 735-740.

Allen, D.E.; Rainer, J.H.; Pernica, G. (1985). Vibration criteria for assembly occupancies, *Canadian Journal of Civil engineering*, Vol. 12, pp. 617-623.

Alves, N.K.C. (1997). Cargas dinâmicas devido a pessoas em movimento (Dynamic loading due to human movements), MSc Dissertation (in Portuguese), COPPE/UFRJ, Rio de Janeiro, RJ, Brazil, pp. 1-113.

ANSYS, Swanson Analysis Systems (2003). Inc. P.O. Box 65, Johnson Road, Houston, PA. 15342-0065, Version 10.0. Basic analysis procedures. Second edition.

Bachmann, H. & Ammann, W. (1987). Vibrations in structures induced by man and machines. Structural Engineering Document 3e, International Association for Bridges and Structural Engineering.

Ebrahimpur, A.; Haman, A.; Sack, R.L.; Patten, W.N. (1996). Measuring and modelling dynamic loads imposed by moving crowds, *Journal of Structural Engineering*, Vol. 122 (12), pp. 1468-1473.

Faisca, R. G. (2003). Caracterização de cargas dinâmicas geradas por atividades humanas (Characterization of Dynamic Loads due to Human Activities), PhD Thesis (in Portuguese), COPPE/UFRJ, Rio de Janeiro, RJ, Brazil, pp. 1-240.

International Standard Organization - ISO 2631-2 (1989). Evaluation of Human Exposure to Whole-Body Vibration, Part 2: Human Exposure to Continuous and Shock-Induced Vibrations in Buildings (1 to 80Hz), International Standard.

Ji, T. & Ellis, B.R. (1994). Floor vibration induced by dance-type loads: theory and verification, *The Structural Engineer*, Vol. 72, Nº 3, pp. 37-50.

Langer, N.A. dos S.; Silva, J.G.S. da; Vellasco, P.C.G. da S.; Lima, L.R.O de; Neves, L.F. da C. (2009). Vibration analysis of composite floors induced by human rhythmic activities. Proceedings of the 12th International Conference on Civil, Structural and Environmental Engineering Computing, CC 2009, Funchal, Ilha da Madeira, Portugal, CD-ROM, pp. 1-14.

Lehmkuhl, L. & Smith, L.K. (1985). Cinesealogia clínica de brunnstrom, ed. Manole, pp. 472-499.

Murray, T.M. (1975). Design to prevent floor vibration, *Engineering Journal*, Vol. 12(3), pp. 82-87.

Murray, T.M.; Hendrick, W.E. (1977). Floor vibrations and cantilevered construction, *Engineering Journal*, AISC.

Murray, T.M.; Allen, D.E.; Ungar, E.E. (2003). Floor vibration due to human activity. Steel Design Guide Series, AISC, Chicago, USA.

Ohmart, R.D. (1968). An approximate method for the response of stiffened plates to aperiodic excitation studies in engineering mechanics, Report n° 30, The University of Kansas, Center for Research in Engineering Science, Lawrence, Kansas.

Rainer, J.H.; Pernica, G.; Allen, D.E. (1987). Dynamic loading and response of footbridges. Structures Section, Institute for Research in Construction, National Research Council of Canada, Ottawa, Ont. Canada KIA OR6, pp 66-71.

Silva, J.G.S. da; Vellasco, P.C.G. da S.; Andrade, S.A.L. de; Lima, L.R.O de; Almeida, R.R. de. (2008). Vibration analysis of long span joist floors when submitted to dynamic loads due to human activities. Proceedings of the 9th International Conference on Computational Structures Technology, CST 2008, Athens, Greece, CD-ROM, pp. 1-11.

Progress and Recent Trends in the Torsional Vibration of Internal Combustion Engine

Liang Xingyu, Shu Gequn, Dong Lihui, Wang Bin and Yang Kang
State Key Laboratory of Engines, Tianjin University, 300072
P. R. China

1. Introduction

With modern machinery industry developing, the application of internal combustion engine is getting wider and research direction is towards high-power, high speed and strong loads. So the issue of torsional vibration of the engine is becoming more prominent. All kinds of work conditions of the engine may have great impacts on the shafting, leading to all sorts of torsional vibration and resonance, and many accidents which lead to much detriment have occurred at home and abroad due to torsional vibration.

As the problem of torsional vibration of the engine is becoming more and more prominent, broad research is made both at home and abroad. This article mainly refers to the literatures on torsional vibration issue published in recent years, summarizes on the modeling of torsional vibration, corresponding analysis methods, appropriate measures and torsional vibration control, and points out the problems to be solved in the study and some new research directions.

2. Modeling of engine crankshaft

2.1 Engine crankshaft modeling method

Crankshaft is the main component of internal combustion engine. Shaft vibration is one of the most important factors affecting engine operation safety. Crankshaft modeling is the base of crankshaft torsional vibration analysis, whose accuracy and simple practical applicability will greatly improve the efficiency and credibility of research results.

At present, there are 3 kinds of most basic shaft models used in analyzing torsional vibration: the first type is simple mass - spring model, the second is continuous mass model, and the third is multi-segment concentrated mass model.

2.1.1 Simple mass - spring model

Simple mass - spring model is the earliest mechanics model in the calculation of shaft vibration [1-6], which was also called lumped parameter model in some literatures. It disperses crankshaft onto the disk with concentration of inertia moment, elastic axis without mass, internal damping and external damping, as shown in figure 1. Each disk rotational inertia includes: the rotational inertia of the crank, the equivalent rotational inertia of connecting rod and piston , transmission system, shock absorber, the rotational inertia of the flywheel, etc.

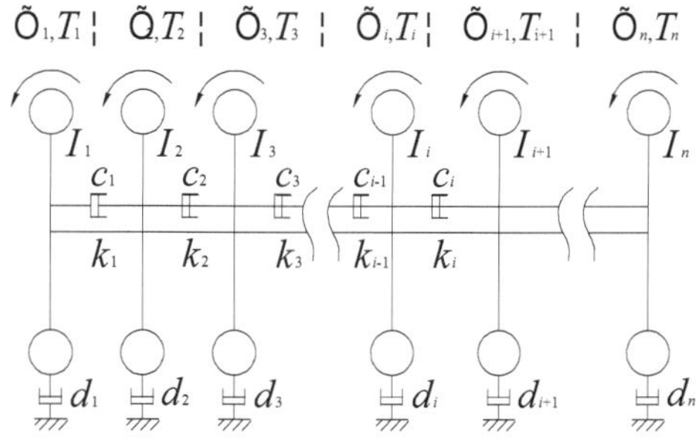

Fig. 1. Simple mass - spring model schematic diagram

This model has certain precision for lower frequency of torsional vibration modal and clear physical concept. It's simple to use and easy to calculate. But since this model is simplified, when precise calculation of the crankshaft is required, its precision is limited. This model is established completely for rigid shaft and rotation parts, so it can not simulate the actual shaft.

2.1.2 Continuous mass model

Continuous mass model is based on continuum theory, regarding shaft as elastomer, established in finite element method. It's also called distributed mass model in some literatures [7-8]. It adopts finite element method in general, dissecting the crankshaft entities directly into finite element calculation model of division. Hence, the mass of the shaft is distributed continuously along the shaft, closer to practice than that of simple mass - spring model. Partial differential equations can be used in this model, which can accurately calculate low frequency and vibration model of the shaft, as well as high frequency and vibration model, solve by numerical method, and also can calculate arbitrary section stress conveniently. But the model is complex and with low speed to calculate, and is easy to cause greater accumulative error. It is more difficult to use this model in system simulation and design. Due to the method of forced vibration calculation, it is hard to realize, thus it's mainly used in the calculation of free vibration.

Recently, two consecutive quality models also have derived from this model: framework model and multi-diameter model.

Framework model is a model, in which, circular cross section straight beam represents main journal and crank pin, and variable cross-section rectangular beam represents crank arm and counterbalance in finite element analysis [9]. For these analyses, circular cross section beam also can represent main journal and crank pin, but the crank arm and counterbalance should be treated as simple rectangular beam. Model schematic diagram is shown in figure 2. In framework model, different structural parts of the crankshaft are substituted by the continuous entities with regular shape, and the original basic shapes of crankshaft are kept. Thus this model has higher precision to analyze the crankshaft vibration.

Multi-diameter model is a model used in elastic wave propagation theory solving torsional vibration of internal combustion engine [10-12]. Assign piston-rod additional mass to two crank arms and simplify a unit crankcase into a group of concentric multi-diameter. Model schematic diagram is shown in figure 3. Because the model has continuous mass distribution, the effect of distribution parameters on shafting vibration characteristics can be considered. It also can adopt different mathematical methods to calculate and compare with simple mass - spring model. This model can have high precision.

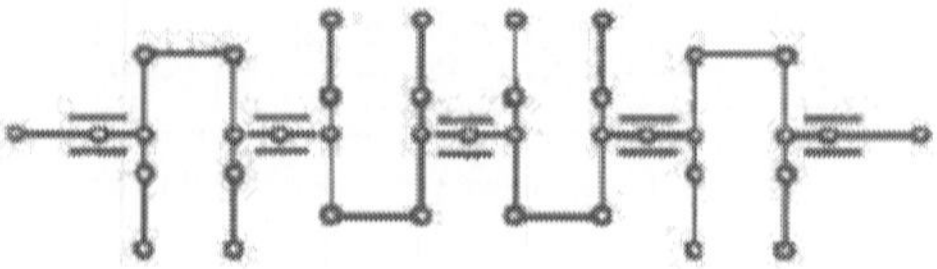

Fig. 2. Framework model schematic diagram

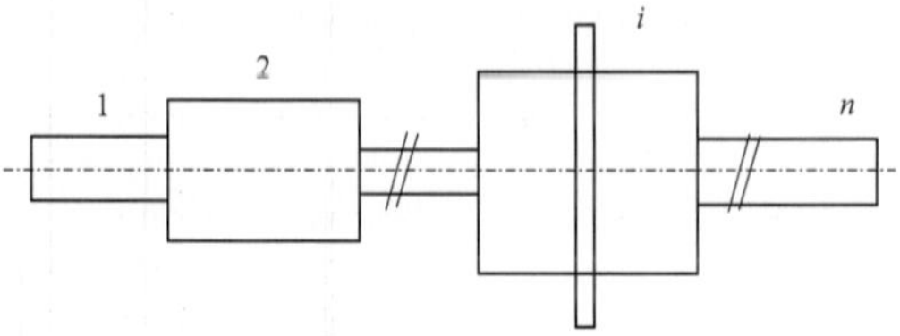

Fig. 3. Multi-diameter model schematic diagram

2.1.3 Multisegment concentrated mass model [13]

This model is similar to the simple mass model in essence. However, it can be separated into dozens to hundreds sections according to the structure characteristics upon analysis demand. It can calculate high order torsional vibration frequencies that can't be determined by simply mass model, and also avoids the large amount of computation that required in the calculation of continuous mass model. Thus it has been widely used.

2.1.4 Soft body dynamics model [14]

In the calculation of flexible multi-body dynamics, flexible body is described as modal flexible body. A flexible body contains a series of modals. In the breakdown steps, each model unit requires obtaining system state variables and calculating the relative amplitude of each characteristic vector, then using linear superposition principle to integrate node deformation of each time step to reflect total deformation of flexible body.

2.1.5 Other axis modeling methods

In recent years, with the further study of shafting vibration, many new modeling methods came up in the engine industry and other related industries.

Continuous beam model was used in the crankshaft load calculation by Li Renxian [15]. The crank and conrod were equalized to the concentrated force acting on a non-equal continuous beam, and all kinds of force were also equalized to simplify. The author analyzed various loads of crankshaft and its changes in an operation cycle comprehensively. This is a simplified model force shown in figure 4. Of course, in order to calculate simply, the author treats both gas load and centrifugal force function load as concentrated loads. If they were expressed as some forms of distributed loads, the calculation might be more accurate. Calculation model may also adopt continuous beam to make it more close to the actual situation of crankshaft.

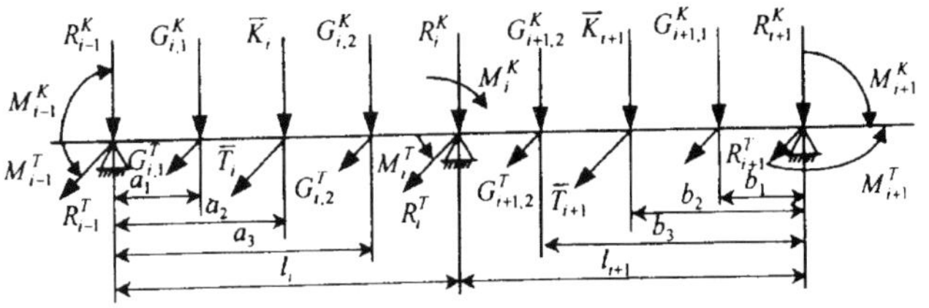

Fig. 4. Continuous beam model schematic diagram

Gu Yujiong[16] used the four-terminal network model in calculating torsional vibration. The author starts with motor control equations and its general solution, equalizing torsional vibration system to the four-terminal network model based on the principle of electromechanical analogy; then adopts mechanical impedance method to obtain frequency equation controlling torsional vibration according to the input impedance of the system and the resonant characteristics, and then obtains the natural frequency, vibration mode and stress distribution, etc. of each order. Four-terminal network model is an accurate low-order model, whose algorithm is convenient and fast. The physical significance is obvious to analyze the mechanical impedance of the system, thus it is a good attempt to model torsional vibration. He Shanghong and Duan Jian [17] also used network method in calculating torsional vibration, and based on dynamic in the process of modeling, which was also a kind of deepening of the method.

Through the analysis of different vibration mathematical models, Xiang Jianhua [18] proposed a graphical modeling method based on system matrix method solving the axis of torsional vibration. The modeling method only requires users providing original torsional vibration mechanics model, and is not restricted by axis branches and modeling scale. In actual implementation, torsional vibration module is divided into two kinds of module unit in this method, which can be used to build various kinds of torsional vibration mechanical model. Model topology relation can be generated through the module traverse and equalisation conversion of the torsional vibration model, and finally the system integration required for solution is integrated.

2.2 Axis modeling research direction

Thanks to the development of modern computer, the precise calculation for shaft can be easily realized by finite element method. So for the continuous mass model, the main development direction is how to make the model have better simulation with material object in computer modeling, thus to peer analysis of the model to material object.

, As the model parameters (especially rigidity parameters) of a simple mass - spring model are obtained by a large number of experience formula and approximate calculations, as its accuracy is hard to ensure, resulting in rather great error between calculated results and actual machine test results. The reason that theoretical calculation result has lager error is often not because of the calculation method of itself, but lies in the accuracy of the model. Since rotational inertia has a relatively accurate analytic calculation, and the torsional vibration damping of the axis is small, to improve the accuracy of the models, focus should be on the amendment of rigidity parameters. Multi-segment concentrated mass model also has similar problem.

The establishment and derivation of a new model should be based on the amendment with material object, adopting all kinds of similar models used in other industries to derive, thus further perfect engine crankshaft torsional vibration model. The influence of damping should not only be considered, the influence of bending-torsional mixture should also be taken into account. Now, many scholars apply model reduction method [19]used in dealing with torsion vibration of steam turbine and generator as well as the method [20]used in identifying parameters of experimental data to the calculation of the engine torsional vibration. This is also the evitable trend of torsional vibration integration.

3 Solving method of torsional vibration of internal combustion engine

3.1 Common method of Torsional vibration

Based on the above-mentioned several shaft models, the common methods and algorithms solving torsional vibration for multi-freedom free vibration calculation include Holzer method, system matrix method and transfer matrix method, etc. The methods for multi-freedom forced vibration calculation include energy method, amplification coefficient method and system matrix method, etc. With the development of computer technology, the traditional manual calculation has been replaced by computers gradually, while some common calculation methods of torsional vibration emerged, such as modal analysis method and finite element method, etc. Various methods are described below.

3.1.1 Holzer method.

Holzer method [1, 2] is always a classic and effective solution in "free- free" system of power machine. The Holzer form method or the Tolle form method derived from its basic principle are often used in engineering. The Holzer method, widely used today, is a numerical calculation method and corresponding calculation program derived from its principle. The advantage of this method is clear physical concept.

From its scientific name, this method can be called "method of sum of torsion moments". The basic idea is: the sum of inertia moment of each lumped mass (disc) should be zero when the shaft doing free vibration without damp, that is

$$\sum I_k \ddot{\varphi}_k = 0$$

Due to the characteristics of simple harmonic oscillator, the relation between the displacement α_k and the acceleration $\ddot{\varphi}_k$ of each inertia I_k is:

$\ddot{\varphi}_k = -p^2 \alpha_k$, namely, $\sum I_k p^2 \alpha_k = 0$. This is the foundation of Holzer method.

This method is effective in estimating low order torsional vibration frequency in initial design stage. This method has simple algorithm and is easy to use, thus is widely used in actual engineering. But its higher-order calculation has lower precision and is time consuming.

3.1.2 System matrix method

System matrix method [1, 2, 21] is a method using each parameter matrices of torsional vibration equation of the axis to solve characteristic root to calculate torsional vibration. All methods which can calculate the eigenvalue and eigenvector of the matrix can be used to calculate the free torsional vibration of multi-mass system.

The basic principle of this method is: for more freedom vibration equation: $\mathbf{I}\ddot{\varphi}+\mathbf{K}\varphi=0$ by assuming the form of solution and input them into equation, the following can be obtained: $\mathbf{KA}=\omega_n{}^2\mathbf{IA}$ let $\lambda=\omega_n^2$ and $\mathbf{H}=\mathbf{I}^{-1}\mathbf{K}$, now the system matrix of $W_M=W_C$ .can be obtained. So free vibration calculation can come down to the question of solving characteristic equations $|\mathbf{D}|=0$.
System matrix method is widely used, not only in free vibration solution, but also in the solution of forced vibration. However, it's generally only suitable for solving low frequencies and its accumulative error would be bigger for calculating high frequencies.

3.1.3 Transfer matrix method

The transfer matrix method [16,22] is a commonly used method for analyzing various vibration problems, which was first introduced by Holzer to analyze crankshaft vibration and calculate the inherent frequency of undamped-free vibration of the shafting.
The basic concept of transfer matrix method is: decomposing the studied system into several two-terminal elements with simple mechanical properties, and building relation between the state vectors of the two terminals of one component by transfer matrix. Then, connect all components one by one, and multiply them together to obtain and solve the transfer matrix. Internal combustion engine shafting, according to its composition configuration characteristics, can be divided into three kinds of components: inertial disks - viscous damper components, elastic elements and even elastomer shaft section components.
The advantage of vibration calculation by transfer matrix method is that the order of transfer matrix will be not affected by the increased unit number, namely, the dimension of matrix will not increase with the increase of the freedom degree of the system, and the calculation method of each order vibration mode is identical. So with simple calculation, convenient programming and less memory for calculation and less time consumption, this method is widely used in the analysis and research of crankshaft vibration. However, when analyzing complex shaft with many freedom degrees by this method, due to the error accumulation of the transfer matrix, the calculation accuracy will decrease, thus the precision of higher frequencies computation is relatively low.

3.1.4 Energy method and amplification coefficient method

Both energy method and amplification coefficient method[2,23] belong to the resonance calculation method of forced torsional vibration, which are basic and the most important calculation methods of torsional vibration before electronic computer popularized. They are still widely used at present. The basic principle of energy method is that the input energy of the exciting moment within a system vibration period is completely consumed by system damping, namely, $W_M=W_C$. Amplification coefficient method was proposed by Tuplin in 1930's for resonance calculation, and then was further developed, becoming a guiding method that the Shipping Standard of British Lloyds Register recommends.

3.1.5 Modal analysis method

The basic thought of modal analysis method [24-25] is to decompose complex multi-freedom system into several sub systems. Firstly, when analyzing, compute the several lower modes of each sub system, then assemble each sub system into an integrated motion differential equation set according to displacement compatibilities or force balance relations between adjacent sub systems to derive comprehensive eigenvalue problem of shrinkage of Freedom

Degree., thus work out the inherent frequency A vibration mode and response of the system. Since modal analysis method reduces the freedom degree of the system, the time consumption and memory for calculation are significantly reduced compared with finite element method. If the sub systems are divided reasonably, its calculation precision is also satisfactory. In addition, modal analysis method can also combine with experimental research [26] to obtain system vibration modal parameters by measuring the transfer function of shaft vibration, e.g., natural frequency, vibration mode, damping, modal inertia and modal rigidity, etc.

3.1.6 Finite element method

Finite element method [1,2,7-10] is a numerical calculation method for solving mathematical physics equation based on variation principle. Its basic thought is to regard complex structure as finite set of discretized units. Each unit is connected into a unity through the common point of the neighboring units, namely, "joints". Take each unit as a continuous component and joint displacement as generalized coordinate. To establish torsional vibration mechanics model of the shafting of internal combustion engine, we need to define which units are selected as well as load positions and sizes, etc. Finite element method is currently accepted as with the highest calculation precision for torsional vibration calculation.

3.1.7 Substructure analysis method of the torsional vibration of systems with branch shafts [27]

In the torsional vibration analysis of shaft systems with branches, the main commonly used methods are transfer matrix method, matrix iterative method and system matrix method, etc [28-31]. But these methods are mainly used to analyze straight string structure or a particular branch. Their calculation efficiency is relatively low for the whole branching structure system. In general, substructure method has already formed systemic theory [32-33], whose basic idea is to divide large and complex structure system into several substructures and calculate the dynamic characteristics information of each subsystem by finite element method, analytical method and experimental method, and then integrate them into the dynamic characteristics of the whole structure system. But substructure method is used less in torsional vibration analysis of shaft systems with branches. Representative method is dynamic substructure matrix method. This method requires working out the compatible relation among all substructures in substructure integrating, which leads to the complicated and tedious modal synthesis process in case the amount of divided substructures is large. Thus this method has certain limit in solving complex shaft systems with branches.
Chu Hua [34] and Z P Mourelatos [35] combined substructure method with transfer matrix method in torsional vibration calculation, which became a new migration substructure method, and the new method was compared with finite element method.
According to the structure of shaft systems with branches, Li Shen and Zhao Shusen [27] put forward a method that divided substructures and integrate step by step based on gear meshing form to obtain the torsional vibration inherent characteristics of the whole system. This method was also used in analyzing the torsional vibration inherent characteristics of the structure of the main transmission system with branches of 650 rolling mill. The comparison with the results calculated by other methods shows the feasibility of substructure graded division and stepwise integration method. This principle has expanded the application scope of the substructure modal synthesis in solving the torsional vibration

of shaft systems with branches and effectively solved the problem of complex system, thus provided a good idea for solving the torsional vibration of shaft systems with branches.

3.1.8 New research methods for torsional vibration

In recent years, the number of scholars engaged in research of vibration has continuously increased and new algorithms kept on emerging continuously, such as elastic wave propagation method, eigenvector method and frequency analysis method, etc. According to the theory of torsional elastic wave, Bogacz [36] gave out a method to solve torsional vibration dynamic response by torsion wave method. Shu Gequn and Hao Zhiyong [11,37] also presented a new torsional vibration response calculation method based on the theory of torsional elastic wave, whose basic thought is: the torsional vibration of the shafting is caused by the torsional elastic wave propagation along the shaft; elastic wave propagates along the axis forward and back in traveling wave form; when one traveling wave meets with another after reflection or delay, , both waves will stack into standing wave causing torsional vibration if their phases are appropriate. The method can be used to analyze continuous parameter distribution boundary, transient response and steady-state response of the crankshaft axis with transient boundary conditions and other vibration characteristics. Since it only requires solving linear equations in calculation, its computational complexity is small, thus it is an accurate and fast vibration analysis method. According to electromechanical analog principle, Gu Yujiong [16] put forward a four-terminal network method for analyzing torsional vibration. State vector method, proposed by He Chengbing [38], was widely used in the analysis of torsional vibration. People are also exploring the calculation method of continuous mass model for torsional vibration response of forced vibration. Wang Ke she and Wang Zheng guang[39] used frequency analysis method in the calculation of torsional vibration to combine frequency change with the structural parameters of shafting, which was beneficial to visual analysis. It also worked out analysis mode, resonance frequency and resonant modes, etc.

3.2 Research direction of shafting solving methods

At present, the methods for solving torsional vibration are various and each has its own use. While in general, it shall be developed from the following aspects:

1. Improve the computation efficiency of current methods. For instance, calculation precision is high by finite element method, but its calculation is time consuming and resources occupying, so fewer and dimension-reduced units should be considered in model building when improving this method.
2. Combine the calculation method used in other industries with this direction.
   - 2.1 For example, Shen Tumiao [16] mentioned to apply electrical four-terminal network in the calculation of torsional vibration . This is an example of unified calculation method. In addition, integrating all calculation methods to construct a new method is also one of the research directions.
   - 2.2 Based on the torsional vibration numerical simulation study of the internal combustion engine shaft with the precise time integration method, Lin Sen [40] introduced and deduced the precise time integration based on Duhamel integral, described in detail the calculation characteristics of this method with example and comparison, simulated the torsional vibration of some type of internal combustion engine, compared the results with the calculation results

in literature, and analyzed their similarities and differences briefly, which, to a certain extent, solved the conflict between the accuracy and stability of calculation.

2.3 Along with the development of microcomputer technology, we can use professional software to analyze torsional vibration of internal combustion engine. Tong-Qun Han [41] introduced the functions and characteristics of engine simulation software EXCITE—designer developed by AVL company, analyzed torsional vibration and vibration reduction of the shaft system based on the software targeting at the problem of a car engine flywheel bolt fracture, and put forward correcting measures.

4. Experimental studies on engine crankshaft

4.1 Current torsional vibration measurement methods

Torsional vibration measurement is an important content in the study of crankshaft vibration. Compared with transverse vibration measurement, the extract and analysis of torsional vibration signals are both difficult. There are basically two kinds of torsional vibration measurements: contact measurement and non-contact measurement. The former installs sensor (such as strain gauge, accelerometer, etc) on the shaft, and the measured signal is transmitted to instrument by collector ring or radio signal. Non-contact measurement commonly uses "measuring gear method", which uses shaft encoder, gear, or other repeated structure to measure angular velocity in homogeneity to measure torsional vibration. If designing Doppler test method properly, laser can also be used to measure torsional vibration. The followings are introduction to various methods and analysis of their error sources and applications.

4.1.1 Mechanical measurement

Geiger torsional vibration analyzer is a typical mechanical torsional vibration measuring instrument [2, 23] and was used in torsional vibration study in the earliest stage. This instrument is designed dexterously, whose signal acquisition and signal record are both realized by mechanical devices, simple and practical. It is widely used in the study of torsional vibration. DVL torsional vibration instrument also belongs to this type of torsional vibration instrument. However, the torsional vibration of this method is transmitted to measuring head shelf by belt, the belt elastic vibration will cause distortion. The response bandwidth of mechanical measuring system is very limited, and also because disc springs cannot be too soft, so very low frequency torsional vibration cannot be measured. In addition, the measured signal cannot be analyzed directly by means of modern analytical instrument, thus it has gradually been eliminated.

4.1.2 Contact measurement

Contact measurement [42] is to install sensor (such as inductance, strain gauge, etc.) on crankshaft directly. The measured signals are transmitted to analytical instrument by collector ring or in radio frequency manner. To monitor the dynamic response of shaft or shaft parts (e.g., blade, etc.), the arrangement of strain gauge should eliminate the interference of transverse vibration, and can realize the automatic compensation of influence by temperature. Torsional vibration meters belonging to this measurement method include strain-gauge torsional vibration meter, piezoelectric torsional vibration meter and inductance-type torsional vibration meter, etc. Contact measurement, centered by sensing element, is widely used in the vibration test of internal combustion engine, thanks to its high sensitivity, wide frequency response range and convenience for measured signal record and analysis. But this kind of measuring device system itself has certain rotational inertia, which will inevitably impact on the system under test in measurement. In all kinds of contact measurement, measurement devices, such as sensors, are required being installed on the shaft, which sometimes has to destroy the original shaft structure. This is not allowed in many cases.

4.1.3 Non-contact measurement

The measurement device of non-contact torsional vibration measurement [43,44] is not installed directly on the crankshaft, but collects torsional vibration signals through photoelectric and magnetoelectricity conversion by code disc, gear or other indexing structure on the crankshaft. These kinds of method are based on the principle of "gear testing". When the shaft is rotating, the teeth structure installed on the shaft can induce bell shaped pulse leveling signal sequences on the sensor, whose amplitude and phase might carry the information on axial torsional vibration, which is demodulated by phase detectors into torsional vibration signals. Torsional vibration meters belonging to this measurement methods include TV - 1 torsional vibration meter of British Econocruise Company, VED - 233A torsional vibration meter of American Shaker Company and DTV - 88 torsional vibration meter[45] developed by Shanghai Institute of Electrical Equipment, etc. Non-contact measurement method does not need installing special devices on the shaft, but uses the existing shaft repeated structure, whose measurement preparations is less, and measurement process does not interfere with the normal operation of shaft. It's especially suitable for the long-term monitoring of torsional vibration. At present it has become a major means of torsional vibration measurement.

4.1.4 Laser measurement

Laser Doppler Torsional Vibration Measurement Technique [46-48] is put forward and developed from fluid velocity measurement. When laser beam irradiates on shaft surface, the linear velocity of shaft surface make scattered light produce Doppler frequency shift. The transient angular velocity of shaft represents the transient value of frequency shift volume of the instantaneous axis. Torsional vibration is obtained by removing dc component. 2523 torsional vibration meter launched out by the Denmark B&K Company was a typical representative of Doppler laser torsional vibration meter. In 1994, Ge Weijing [49] and others from Tianjin University applied laser Doppler velocimetry on the torsional vibration measurement of internal combustion engine shaft. Only a smooth section on the surface of the shaft is required, and measuring point is easy to be set up. This method can realize absolute measurement and measurement datum is not required to be specially established. However, since the transverse vibration of the shaft and the form and position errors of the shaft section directly affect measurement precision, it is rather difficult to improve its accuracy.

4.1.5 The latest torsional vibration measurement method

Wang Ting and Cheng Peng [50] introduced a kind of digital measurement system of crankshaft using PC computer. The measurement system consists of angle encoder, self-made count plate, PCL724 digital input/output card and PC computer, etc. installed on the tested crankshaft. Angle encoder is crankshaft angle sensor with high precision, and is connected with crankshaft by flange. The grating disc fixed in the angle encoder has two reticules, e.g., outer and inner rings. The outer ring is a uniform reticule, which can produce CDM signals, and the inner ring is a TRIG reticule for judging tdc signal. The light emitting components in the angle encoder are two infrared light emitting diodes, and there are two infrared light receptors respectively corresponded with CDM reticule and TRIG reticule. When angle encoder operates together with crankshaft, a TRIG signal will be outputted in each rotation, and a series of CDM square-wave pulse signal will be outputted. Thus, the crankshaft torsional vibration can be directly reflected on the time width of the CDM square-wave pulses outputted by angle encoder. Count each CDM pulse width with frequency division by high count circuit board. The counted data is inputted into PC by parallel data I/O card PCL - 724. Then crankshaft torsional angle can be obtained after program processing. The measurement system measures torsional angle directly, thus it's with convenient measurement, high precision and simple process.

4.2 Research direction of torsional vibration measurement of internal combustion engine

The focus and future development of torsional vibration measurement is to improve its accuracy and real-time performance to realize the torsional vibration monitoring of internal combustion engine in operation, especially the monitoring of severe torsional vibration caused by emergencies, such as the severe torsional vibration by transient large torque incentive resulted from cylinder flameout. At the same time, eliminating the interference of lateral vibration and establishing reliable measurement datum are still the problem requiring to be solved. Finally, the problem of system calibration should also be solved.

5. The latest research direction of torsional vibration of internal combustion engine

The traditional research methods of torsional vibration can not meet the needs on the precise study. In recent years, many scholars have continuously broadened research field and scope to further explore the various problems of shaft torsional vibration, making the research on torsional vibration closing to ideal level unceasingly. Some main research directions of torsional vibration in resent years are introduced as follows.

5.1 Nonlinear research

With the further research on torsional vibration of the shaft, many nonlinear vibration problems are met [51-55]. At the same time, crank shaft is a complex nonlinear system, thus it often needs to consider all sorts of complex nonlinear factors to construct a model that can reflect actual system. However, current relevant studies are mostly on single degree-of-freedom nonlinear vibration problems that considers single factor, which obviously cannot meet the need on the accurate calculation of crankshaft vibration. Therefore, it is necessary to further consider the crankshaft nonlinear vibration problems with multiple nonlinear factors, multi-degree-of-freedom and even continuous mass distribution. In the current calculation models, equivalent moment of inertia (constant) is usually adopted to consider the inertia of piston and connecting rod. But in fact, the inertia of engine crank module is:

$$I = I_0\left[1 - \varepsilon\cos(2\varphi)\right]$$

Where ε is variable inertia coefficient, a value below 1. We can see from the formula, the moment of inertia of crank component is a variable related with rotation angle. Sheng Gang [56] researched on the solution methods of some simplified models of single cylinder engine and established the equation of motion of crankshaft vibration of multi-cylinder engine under the condition of considering variable inertia. At the same time, in literature [1], the problems of variables caused by machining error and assembling error were also considered. In literature [57], forward and inverse Fourier transformation was applied to numerically solve nonlinear torsional vibration system. While Lin Ruilin [58] took the diesel engine shaft with a third-order rigidity component as research object, deduced the calculation formula and numerical calculation formula iterative procedure for solving periodic response of nonlinear torsional vibration by incremental harmonic balance method (IHB). This method is used to solve linear and nonlinear torsional vibration response of diesel engine shaft. Compared with the existing methods, it is more effective to solve strong nonlinear vibration response. What's more, it has virtues of less operation time and accurate calculated result.

The discussion about nonlinear components is mainly concentrated on the non-linear shock absorber, coupling and other components. Literature[59] of as early as 1987 analyzed the nonlinear problems of diesel engine shaft with piecewise linear components (cylindrical spring-loaded buffers), and calculated vibration response of shaft by step-by-step integration method. Gong Xiansheng[55] introduced theoretical and experimental research on the calculation method of steady state vibration response of marine propulsion shafting with hysteretic nonlinear coupling subjected to eccentric mass exciting force action. Farshidianfar [60] solved nonlinear problems of driving shaft by substructure modeling, and compared the results with the results of whole structure modeling.

The research on nonlinear torsional vibration has made many important achievements [50]. But so far, the majority of nonlinear torsional vibration problems are still analyzed by some approximate methods or by ignoring nonlinear factors, in most cases, the results obtained have greater errors compared with actual results. Therefore, there are still many problems waiting to be solved in further exploring the nonlinear problem, mainly including:

1. The modeling, system parameters identification method and test of complex nonlinear torsion vibration problems;
2. Accurate solving methods for multi-degree-of-freedom strong nonlinear torsional vibration problems;
3. Self-excited vibration of complex nonlinear torsional vibration system;
4. Decoupling, numerical calculation and optimization methods of complex nonlinear structure.

5.2 Coupling vibration analysis

Torsional vibration of shaft has huge harm on the system, so people paid attention to and researched on it at very early period. However, many phenomenon produced by vibration in practice need to take longitudinal/bending/tortional vibration together into account. The bending vibration caused by unbalanced mass has certain weight on the torsional direction and can couple to the torsional vibration; on the other hand, torsion also has certain weight on horizontal and vertical directions, thus couples to the bending vibration. In recent years, significant progress has been made in the aspects of theoretical calculation method and testing technology of longitudinal/bending/tortional coupling vibration.

Li Bozhong [62] discussed about the axial vibration problems caused by torsional vibration and established a relatively simple analysis model for this kind of model. In the following two literatures[63,64] of same series, the longitudinal twist coupling vibration was tested and further analyzed and the coupled vibration model was established, and the model calculation was compared and analyzed with actual measurement. In paper [65], the author put forward a kind of spring - mass model with non-linear rigidness being used in calculating torsional - vertical coupled vibration of engine shaft. It explained the doubled-frequency problem of the torsional - longitudinal coupling, and also revealed the presence of quadruple frequency and octuple frequency in the longitudinal - torsional coupled response. It is more reasonable than just simply giving an assumption doubled-frequency excitation torque in the right of the motion equation. Zhang Yong and Jiang Zikang[66,67,68] adopted distributed mass model in analyzing bending - torsional coupling vibration of shaft, which divided the actual unit shaft system structure into several sections with equal diameter according to orders in simplifing, treated each segment as continuous mass, and listed the vibration differential equation of each segment, then united them to solve. Finally, some results of the analysis for bending - torsional coupling vibration of shaft by numerical method were given. In literature [69], system matrix model was established for longitudinal twisting coupling vibration of shaft, whose general rule of coupling vibration was studied based on the calculation and analysis of the practical examples of longitudinal twisting coupling vibration of shaft. In this paper, the test equipment used for measuring coupling vibration is only eddy current sensor for non-contact measuring the condition of axis vertical vibration. In contrast, multi-dimensional measurement is relatively rare.

Okamura[70] and Shen Hongbin[71] used the longitudinal / bending / tortional vibration test device in all research processes of shaft vibration. This kind of measuring device can acquire three-dimensional vibration signals simultaneously. As shown in figure 5, an electromagnetic sensor (measuring torsional vibration signal) and three acceleration sensors (of which, one measuring longitudinal vibration signals and the other two measuring bending vibration signals) are mounted on its shell. It shows that testing technology has also been developed from single parameter measurement to multi-parameter measurement method.

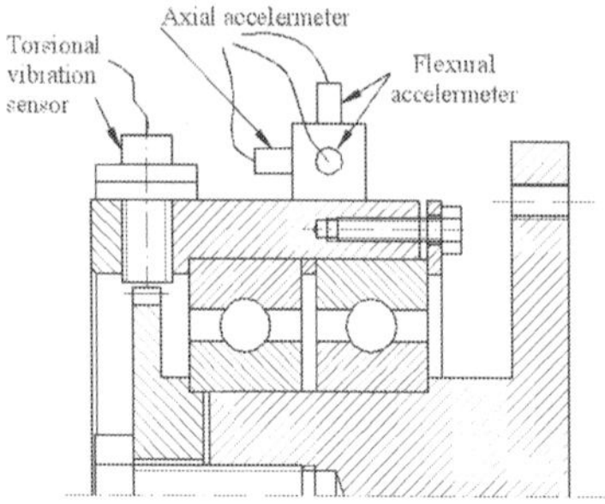

Fig. 5. Three-dimensional torsional vibration measurement schematic diagram

For the research of torsional vibration with coupling vibration and transversal vibration, the current research level is far insufficient. Especially in the study of theoretical models, traditional method can not unify the physical model and mathematical model of coupling vibration simultaneously. In view of this, the future research on coupling vibration can roughly focus on the following aspects:

1. Mechanism research on coupling vibration and relay relation of mutual excitation;
2. Model unification and solve using universal algorithm;
3. Precise treatment of measuring equipment, long distance measurement and the implementation of long-term test, etc.

5.3 The analysis of torsional vibration response based on multi-body dynamics of soft body crank shaft

The forces on engine are very complex. Traditional analysis method is to calculate rotation inertia and reciprocating inertia produced by each force based on the motion analysis of each component, then combine them with the maximum combustion pressure of gas to solve the force on the main body and excitation force of shaft vibration. This is a very complicated process[73]. By using mechanical system simulation software ADAMS, by establishing crankshaft multi-body dynamics model including pistons, connecting rod, crankshaft and flywheel, we can not only calculate the motion law and the force among each component, but can also further analyze balance and vibration. Due to the interaction between inertial load and transverse bending deformation of shaft and the coupling behavior with lubrication problem, the bearing load problem based on rigid body dynamics becomes complicated, and there exist errors in calculation precision. If transforming engine crankshaft into flexible body, the tiny deformation can guarantee the completely accurate dynamic equation to deformation generalized coordinates first-order items. In order to sufficiently study the effect of crankshaft flexible body on the calculation results of dynamics, based on the finite element analysis of crankshaft system and by establishing rigid-flexible coupling multi-body dynamics system model with multiple degrees, Liang Xingyu and Shu Gequn[72] analyzed the torsional vibration response of crankshaft system that constitutes main flexible body, and obtained the time history response of system dynamics, and then made assessment on the power quality and safety of the system. Then it measured the torsional vibration of the crankshaft free end of an inline four cylinder diesel engine with a newly developed test device. Through calculation and comparison between sub-harmonic analysis of test results, both reflected higher equality, and explained the correctness of rigid-flexible coupling multi-body dynamics system model.

5.4 The method of compensating divisional error in shafts torsional vibration measurement and program implementation [74]

Now non-contact measurement method are generally used for torsional vibration measurement, namely, by using repeated structure in the shaft, pulses are produced in non-contact sensor, and the interval dimension reflects the transient angular velocity dimension of the shaft. The shaft torsional vibration information can be obtained by processing interval data of the pulses. When measuring torsional vibration by this method, the indexing error of the shaft repeated structure directly influences the precision of measurement results. If indexing error is very great, the measurement results will have serious distortion. In the measurement of torsional vibration by non-contact measurement method in practice, selected sensors mainly include photoelectric encoder, hall sensor and photoelectric sensor. The three kinds of sensors in practical measurement have their advantages and disadvantages. Photoelectric encoder has large indexing number and small indexing error, so its measurement is more accurate. But its installation is inconvenient and it requires using shaft coupling to connect with measured shaft. If encoder shaft is eccentric with measured shaft after installation, transmission eccentric error will be introduced[75]. Hall sensor requires gearing disc with equal division to measure. Since the teeth number of equal division disc is usually less than the indexing number of encoder, and gear disc has certain indexing error in processing, its accuracy is lower than that of encoder. But its installation is convenient. It can be directly installed on the measured shaft, or can directly measure by gear disc on the shaft without additional modification on the shaft. The use of photoelectric sensor is of the most convenience. It needs only uniformly pasting a certain number of reflective strips on the component with circular surface of the shaft. If the shaft is very thick, those reflective strips can be directly pasted on the shaft. However, currently, the reflective strips can only be manually pasted. Great degree error will definitely occur leading to the distortion of measurement results. So in the actual torsional vibration test, if the indexing error of selected repeated structure can not be ignored, such as selected manually pasted turntable of reflective strips, test results should be dealt with to compensate for the effects of indexing error, then correct torsional vibration information can be calculated. Guo Wei-dong [74] described the compensation principle of indexing error in detail and listed the compensation program of indexing error compiled based on LabVIEW. So we could find out from test results that the data curves after the compensation of indexing error became smooth, and the effect that indexing error on measured results is obviously reduced and test result is more accurate.

6. Control technologies in torsional vibration of internal combustion engine

For the internal combustion engine with reciprocating motion, due to the property of periodical work, the torque on the shaft is a periodic compound harmonic torque, and then forms excitation source. When the frequency of the excitation source is equal to the inherent vibration frequency, resonance phenomenon will occur, and torsional vibration will be subjected to huge dynamic amplification effect, then the torsional stress on the shaft greatly increases, leading to various accidents on the shaft, and even fracture. These are the causes and consequences of torsional vibration.

To avoid the destructive accident of torsional vibration of internal combustion engine, it's not only required to conduct detailed calculation of torsional vibration in design phase, torsional vibration measurement is also required timely after manufacturing completion. This can not only check and modify the theoretical calculation results, but also detect and so as to solve the torsional vibration problems promptly.

Based on the above analysis, main vibration control technology includes two parts: study on the avoidance of vibration and on shock absorber.

6.1 Study on the avoidance of vibration

If great torsional vibration does exist on internal combustion engine according to the calculation of and actual test on torsion vibration, proper measures shall be taken to avoid or remove it.

There are a lot of preventive measures for avoiding torsional vibration [61], classified roughly into the following two methods.

6.1.1 Frequency adjustment method [2, 76]

According to torsional vibration characteristics, when the frequency of excitation torsional vibration is equal to some inherent frequency n_w of torsional vibration system, extremely severe dynamic amplification phenomenon will occur, namely resonance phenomenon, thus the possibility of $w = n_w$ shall be avoided, i.e., avoidance of the most severe conditions of dynamic amplification means the possibility of avoidance of all consequences caused by excessive torsional vibration. The basic concept of this method is that let w actively avoid n_w. The main measures of this kind of method include: inertia adjustment method and flexibility adjustment method, etc. By adjustment, let the natural vibration frequency of the system itself avoid excitation frequency. Reduce vibration stress to be within the instantaneous allowable stress range, thus avoid the damage on engine by bigger torsional vibration. This method is one of the most widely applied measures in torsional vibration prevention measures, not only because of it being a simple and feasible measure, but also because of it being effective and reliable when meeting the requirement of frequency modulation. But its disadvantage is small scale of frequency modulation, which restricts its practical application.

6.1.2 Vibration energy deducing method [23]

Incentive torque is the power source causing torsional vibration. Since the input system energy of incentive torque is the source of maintaining torsional vibration., if the vibration energy of input system can be reduced, the magnitude order of torsional vibration can also be reduced immediately. One way is to change the firing sequence of internal combustion engine. When the dangerous torsional vibration is deputy critical rotation speed within machine speed range, this method might be used to reduce the dangerous torsional vibration and reduce the risk degree. The second method is to change crank arrangement. Deliberately choosing unequal interval firing in multi-cylinder engine and appropriately choosing crank angle to change crank arrangement can let some simple harmonic torsional vibration in any main-subsidiary critical speed counteract mutually to avoid dangerous torsional vibration. The third method is to choose the best relative position between crank and power output device, make the disturbance torque between them counteract mutually, which can reduce the torsional vibration of the crankshaft.

6.1.3 Impedance coordination method

Considering the complexity of solving above problems by the conventional dynamics method, energy wave theory can be used to solve this problem. According to energy wave theory and by coordinating the impedance of various component loops, resonance can be avoided to realize the target of reducing vibration intensity. Impedance coordination method can modify the inferior design in design phase, or design directly correct transmission shaft system, to ensure the shaft working with sound dynamic characteristics without resonance and reducing dynamic load.

6.2 Study on shock absorber

As is known to all, engine installed on shock absorber can greatly reduce the vibration transmitted to the foundation. Likewise, torsional vibration can also be eliminated before it reaches the foundation. If vibration reducing device is installed on the front head of the crankshaft of the engine, then shock absorber will absorb the torsional vibration of rotating shaft generated by engine. It shows the important role of shock absorber in internal combustion engine system. The technical requirements on shock absorber are very high, mainly including: elastic material strength should be reliable in use and storage, the fixation with metal should be firm, rigid fluctuation range in installation stage should be small, and technical characteristics do not change with time.

Now, main shock absorbers include the following kinds: dynamic shock absorber, damping shock absorber and dynamic-damping shock absorber.

6.2.1 Dynamic shock absorber [2, 23, 77]

This kind of shock absorber is connected with crankshaft by spring or short shaft. By the dynamic effect of shock absorber at resonance, an inertia moment with the size and frequency same with excitation torque, but direction opposite to excitation torque is produced at the vibration reduction location to achieve the purpose of vibration reduction. This kind of shock absorber doesn't consume the energy of the shaft. They can be divided into two types: one type is constant fm dynamic shock absorber, namely undamped elastic shock absorber, shown in the schematic of figure 6, and the other type is variable fm dynamic shock absorber, such as undamped tilting shock absorber, drawing as shown in figure 7.

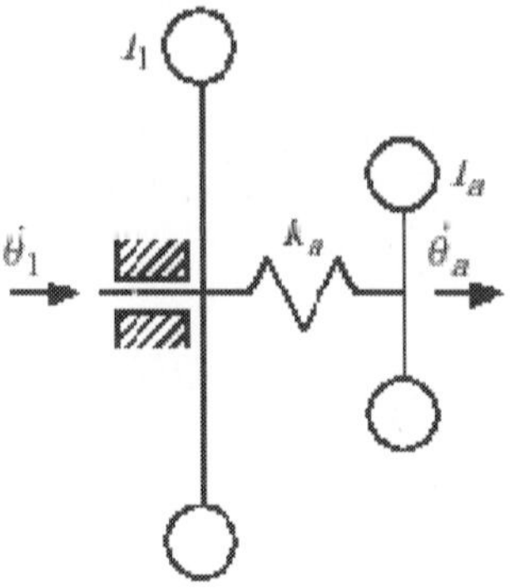

Fig. 6. Undamped elastic shock absorber schematic diagram

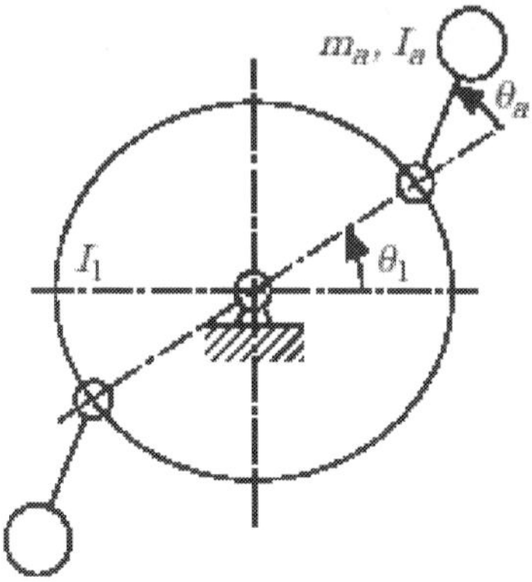

Fig. 7. Undamped tilting shock absorber schematic diagram

6.2.2 Damping shock absorber

Damping shock absorber achieves the purpose of vibration reduction by damping consuming excitation energy (shown in schematic diagram 8).The main type is silicone oil damper[78,79], whose shell is fixed to the crankshaft, high viscosity silicone oil is filled between ring and shell. When the shaft is under torsional vibration, the shell and the crankshaft vibrate together, and the ring moves relatively with the shell due to inertia effect. Silicone oil absorbs vibration energy by friction damper, thus reduce vibration for the torsional vibration system. This kind of shock absorber is widely used with simple structure, good effect of vibration isolation and reliable and durable performance.

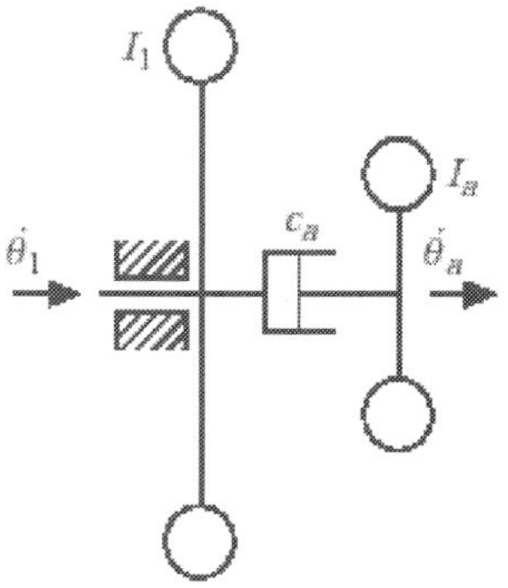

Fig. 8. Damping shock absorber schematic diagram

6.2.3 dynamic-damping shock absorber

Dynamic damping shock absorber features both of the above effects, such as rubber flexible shock absorber [80], rubber silicone oil shock absorber, silicone oil spring shock absorber [81], etc., shown in schematic diagram 9. Theoretically, the effect of dynamic damping shock absorber is the best, since it can not only produce dynamic effect by elastic, but also consume excitation energy by damping. But the elastic elements, such as springs and rubber, etc., that connect the shock absorber and crankshaft often work under great amplitude and high stress, thus the process is relatively complex and the cost is higher.

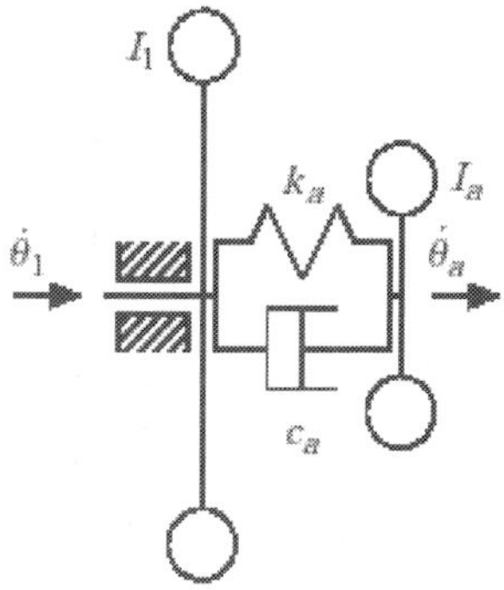

Fig. 9. Dynamic-damping shock absorber schematic diagram

6.2.4 Study on new shock absorber

Along with the deepening of the research on shock absorber, many new shock absorbers have appeared. Several typical kinds are listed as below:

Yan Jiabin [82] proposed an elastic metal shock absorber, which fixed two disks connected with elastic materials at the ends of the crankshaft. disks are tightened in the way that one disk rotates in the opposite direction of the other disk (see figure 10). If loosen both disks simultaneously, they will complete torsional vibration with low amplitude, till stop. At this moment, one section of the shaft rotates in one direction, and another section of the shaft rotates in the other direction. In this case, one end face of the crankshaft will produce displacement. Due to the effect of vibration absorption, the vibration will proceed with reduced amplitude but constant speed, which depends on the internal friction or delayed quantity of elastic material. There are three kinds of elastic shock structure of shock absorber: welding metal elastic elements, combination elastic metal components and welding-combination elastic metal elements. Welding-combination elastic metal elements consist of driving and inertia members that connect each other with elastic material. This kind of shock absorber is suitable for application with simple structure and convenient maintenance.

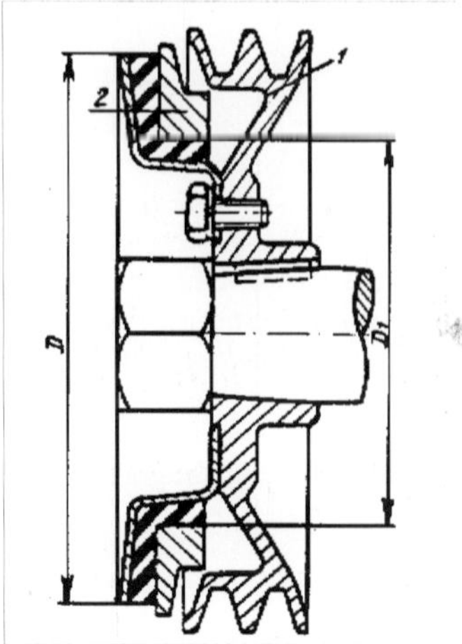

Fig. 10. Monolayer thin-type elastic metal shock absorber

Huo Quanzhong [83] and Hao Zhiyong[84] introduced the research on driving control shock absorber. Figure 2 is the diagram of the shock absorber. The shock absorber itself is similar to a dc motor, whose stator and the shell of the shock absorber compose as a whole entity, and rotor is connected with the shell by radial leaf spring, forming a dynamic shock absorber. The shell of the shock absorber is fixed on the main vibration body. According to the conditions of main vibration body, the regulation apparatus produces control signals with fixed size, phase and frequency, which, by power amplifying, make armature generate control torque (namely, electromagnetic torque). Active torsional vibration shock absorber is feasible both in theory and practice. What's more, its damping effect is better than that of dynamic shock absorber.

Liu Shengtian [85] proposed a double-mass flywheel torsional vibration shock absorber, which was a new type of torsional vibration shock absorber occurred in the middle of

1980's. The double-mass flywheel torsional vibration shock absorber at early period was to remove torsional vibration absorber from the clutch driven plate, place it among engine flywheels, thus double-mass flywheel torsional vibration shock absorber was formed. The basic structure of double-mass flywheel torsional vibration shock absorber has three major parts, i.e. the first mass, the second mass and the shock absorber between the two masses. Relative rotation can exist between the first and the second masses, which are connected with each other by shock absorber.

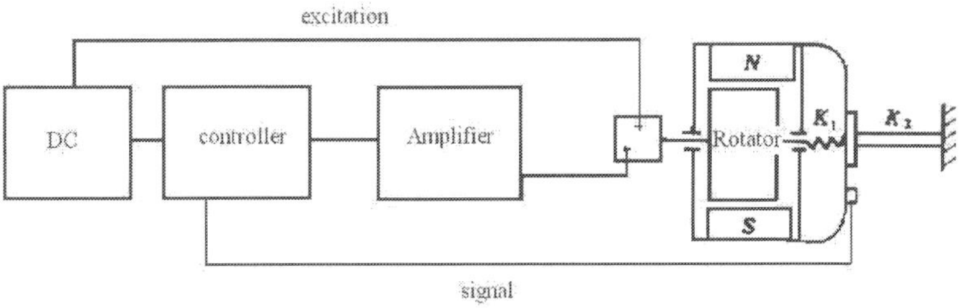

Fig. 11. Active control shock absorber functional diagram

Double mass flywheel torsional vibration shock absorber can very effectively control torsional vibration and noise of automobile power-transmission system. Compared with the traditional clutch disc torsional vibration shock absorber, its effect of damping and isolation of vibration is not only better within the common engine speed range, it can also realize effective control over idle noise. After developing the double-mass flywheel torsional vibration shock absorber, the author also introduced hydraulic pressure into shock absorber and developed hydraulic double-mass flywheel shock absorber [86], which is the latest structure style in the family of double-mass flywheel torsional vibration shock absorber. It lets the technology of car powertrain and noise control of torsional vibration step further into the direction of excellent performance and simple structure.

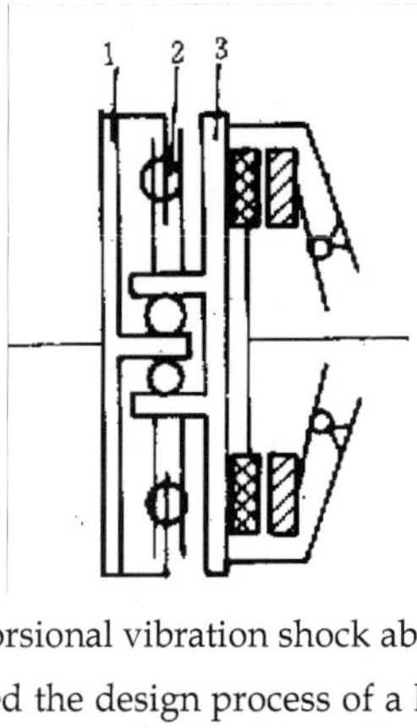

Fig. 12. Double-mass flywheel torsional vibration shock absorber

M Hosek, H Elmal [87] introduced the design process of a kind of FM tilting shock absorber, which was developed based on centrifugal tilting shock absorber, and was named as centrifugal delay type resonator by the author. Based on the study of centrifugal tilting shock absorber, the author installed a sliding globule between the end of pendulum and the rotary table, thus, when the rotation of the shaft fluctuates, the pendulum will delay duet to the effect of damper, while the sliding globule will coordinate actively with the changes. By this shock absorber, minor disturbance can be quickly completely eliminated; and broadband disturbance, especially the disturbance that obviously increases speed, also can be completely eliminated.

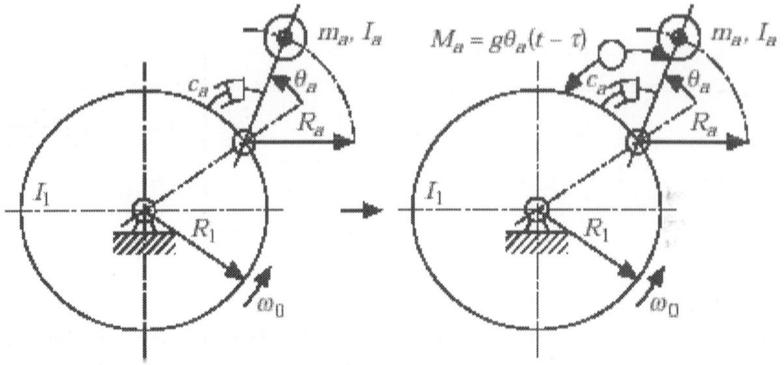

Fig. 13. Centrifugal delay resonator

Shu Gequn [88,89] presented a research approach of coupling shock absorber. Since torsion vibration is the most dangerous vibration mode in shaft vibration, torsional vibration shock absorber is the main damping device, and for coupling damping, bending shock absorber or lateral shock absorber will be installed on the basis of torsional vibration shock absorber. Through the experimental research, the author concluded that, compared with single torsion vibration damper, after installing bending vibration shock absorber, due to the effective damping act on shaft bending vibration, twist/bending shock absorber can control engine vibration and noise effectively. Normally, the parameters setting of bending vibration shock absorber depends on the bending vibration model of crankshaft, but due to its effect on torsion vibration reduction, the design of coupling shock absorber should consider its damping effect on torsional vibration and bending vibration of the shaft. Shu Gequn [90] investigated the effect of bending shock absorber on the performance of twist/bending shock absorber by theoretical analysis.

Through the above analysis, we can see that torsional vibration absorber is being developed towards the aspects of broadband, high efficiency, being timely, multi-function, etc. So research on torsional vibration shock absorber still has considerable prospect, worthy more efforts from scholars. The following aspects can be studied and explored.

1. Study on active control of torsional vibration of internal combustion engine;
2. Study on shock absorber with coupling between torsional vibration with longitudinal vibration and transverse vibration, etc;
3. Study on integrating torsion vibration absorber with clutch or other components of internal combustion engine;
4. Finite element optimization design of torsional vibration shock absorber [91].

7. Utilization of torsional vibration of internal combustion engine crankshaft

Restricted by various factors, we can only decrease the degree of torsional vibration, while the occurrence of torsion vibration is inevitable. Torsional vibration is directly related to the various incentive factors of internal combustion engine, such as the combustion sequence of various cylinders, the change of crankshaft inertia and the sudden change of loads, etc. So how to use torsional vibration signals to monitor the changes of these quantities is the main purpose of utilizing torsional vibration. The utilization of torsional vibration is mainly embodied in identifying faults by torsional vibration signals[92, 93].

7.1 The progress of torsional vibration utilization

The diagnosis of diesel engine faults by the change of torsional vibration parameters of the shaft is a new fault diagnosis technology. Torsional vibration signals of diesel engine shaft often have strong repeatability and regularity, and fault diagnosis by torsional vibration signals is used to diagnose cylinder flameout. Diagnosing cylinder flameout fault by torsional vibration signal of diesel engine has been developed in recent years. The work process fault of diesel engine cylinder directly affect the changes of torsional vibration characteristics, and such changes of torsional vibration parameters also reflect directly the work state of cylinders. Torsional vibration signals of diesel engine shaft can be used as the basis for fault diagnosis. Ying Qiguang explored this issue at the beginning of 1990's of the 20th century, who thought that diagnosis of the technical condition and fault of diesel engine by the response characteristics of the frequency, amplitude (and phase) and damping of shaft torsional vibration is a new and promising fault diagnosis technology. The author judged cylinder flameout fault by the comparison of amplitude size between normal torsional vibration and in the circumstance of cylinder flameout. This method is convenient and intuitive. However, it requires normally torsional vibration amplitude figure for comparison under the same conditions, so its application is limited [94]. In the application of fault diagnosis by torsional vibration, Lin Dayuan and Shu Gequn studied on the sensitivity of various torsional vibration modals and frequency response characteristics on crack by torsional vibration modal experiment, who recommended modal damping, damping attenuation factor, frequency response function modal and self-spectral modal as the optimum evaluation factors for the crack fault, and further discussed the change law between crack and the above evaluation factors, thus provided an effective method of intermittent diagnosis for the engine stops[95,96].

7.2 The development direction of torsional vibration utilization

Fault diagnosis by torsional vibration signal of internal combustion engine crankshaft is a new type of fault diagnosis theory. Developing this theory towards new application field is an inevitable trend in internal combustion engine industry. Thus, the development direction of torsional vibration is mainly oriented to broader fault diagnosis fields and continuously make this achievement become more mature and its application become more skilled.

7.3 Sub-conclusions

1. Damping technology becomes more mature.
2. Multi-function shock absorbers are innovated constantly.
3. Application fields of torsional vibration become more and more wide.
4. Modern design theory is used unceasingly in control field.

8. Conclusions

Research on torsional vibration of internal combustion engine will become more and more deepen with the development of science and technology. Corresponding new research methods will appear in modal building, solving, test and control of the shaft model, making research contents more wide, method more scientific, object more specific and application more direct.

9. Acknowledgements

Authors wish to express their sincere appreciation to the financial support of the NSFC(50906060) and State Key Laboratory of Engines.

10. References

Tan Daming. Vibration control of internal combustion engine [M]. Southwest Jiaotong university press, 1993.

Yu Qi. Torsional vibration of internal combustion engine [M].Beijing: National Defense Industry Press, 1985.

W J Hshen. on the Vibration Analysis of Multi-Branch Torsional System[J]. Journal of Sound and Vibration,1999,224(2):209-220.

Nestorides E J. A Handbook on Torsional Vibration[M]. London: Cambridge University Press, 1958.

Wilson W K. Practical So1ution of Torsional Vibration Problem[M]. London: Campman and Hall, 1963.

A Boysal and H Rahnejat. Torsional Vibration Analysis of a Multi-body Single Cylinder Internal Combustion Engine Model[J]. Appl. Math. Modeling,1997,21(8):481-493.

Wang Changmin, Zhu Dejun. Calculation of Torsional Stiffness of Engine Crankthrow by Finite Element Method [J]. Transactions of CSICE, 1991,9 (2) : 177-183.

Qu xiaoming, Gu yanhua, Wang xiaohu. Modeling and calculation of torsional vibration of internal combustion engine [J]. Internal combustion Engines, 2000,4:6-9.

Li Huizhen, Zhang Deping, Crankshaft Torsional Vibration Calculation by Finite Element Method [J]. Transactions of Csice, 1991,9 (2) : 157-162.

K Koser, F Pasin. Continuous Modeling of the Torsional Vibrations of the Drive Shaft of Mechanisms[J]. Journal of Sound and Vibration,1995,188(1):17-24.

Hao Zhiyong, Shu Gequn. An Investigation on the Torsional Elastic Wave Theory for the Calculation of Crankshaft [J]. Torsional Vibration Response in I. C. Engines. Transactions of Csice, 2000,18 (1) : 29-32.

J Q Pan, J Pan, R S Ming, T Lin. Three-Dimensional Response and Intensity of Torsional Vibration in a Stepped Shaft [J]. Journal of Sound and Vibration,2000,236(1):115-128.

Chen Yong, Lu Songyuan, Gu Fang, Wang Yumin. The Calculation and Analysis of Torsional Vibration Characteristic of Turbine Generator Rotor System. [J]. Turbine Technology, 1999,41(5):276-279.

Liang Xingyu, Shu Gequn, Li Donghai, Shen Yinggang. Torsional Vibration Analysis Based on Multi-body Dynamics of Flexible Crankshaft System [J]. Chinese Combustion Engine Engineering, in 2008, the fourth period.

Li Renxian. Calculation of Loads on Crankshafts of Internal Combustion Engines Based on Continuous Beam Model [J]. Journal of Southwest Jiaotong University, 1998,33(2): 164-169.

Gu Yujiong, Yang Kun, He Chengbing, ect. Simulations on Torsional Vibration Characteristics of Turbogenerators Based on Four-Terminal Network Method [J]. China Mechanical Engineering, 1999,10(5):540-542.

He Shanghong, Duan Jian. Network method for dynamic modeling of complex shafting torsional vibration system [J]. The Chinese Journal of Nonferrous Metals, 2002,12(2): 388-392.

Xiang Jianhua, Liao Ridong, Zhang Weizheng. Study and Application of Graphic Modeling Techniques of Shaft Torsional Vibration Based on System Matrix Method [J]. Acta Armamentarii, 2005-03-001

Yu Yinghui, Zhang Baohui. Development And Prospect of Research on Turbine-Generator Shaft Torsional Oscillation [J]. Automation Of Electric Power Systems, 1999,23(10):56-60.

He Chengbing, Gu Yujiong, Yang Kun. Summarization on torsional vibration model and algorithm of turbogenerator unit [J]. Journal of North China Electric Power University, 2003,30(2):56-60.

Liu Hui, Xiang Changle, Zheng Muqiao. Torsional Vibration on the Complex Shaft System of a Vehicular Powertrain [J]. Journal of Beijing Institute of Technology, 2002,22(6):699-703.

Honda Ym, Wakabayashi K, Matsuki K. Torsional Vibration Characteristics of a Diesel Engine Crankshaft with a Torsional Rubber Damper by Using Transition Matrix Method [J]. JSAE Review, 1996,17(1):89.

Li BoZhong, Chen ZhiYan, Ying QiGuang. The torsional vibration of internal combustion engine crankshaft system[M]. Beijing: National Defense Industry Press, 1984.

G D Jenning, R G Harley. Modal Parameters and Turbo-Generator Torsional Behavior[C]. IEEE 3rd AFRICON Conference,1992:517-520.

Wang Guozhi, Modal Analysis Method in Research on Torsional Vibration of Diesel Engine Shafting System [J]. Transactions of Csice, 1986,4(3):249-259.

Shen Tumiao. Dynamic Measurement and Calculation of Twisting Stress of Crankshaft and Flywheel System in Diesel Type TC 387 [J]. Transactions of Csice, 1993, 11(6):243-248.

Li Shen, Zhao Shusen. Study on Substructure Method for Torsional Vibration of a Branch Shafting System [J]. Journal of Vibration and Shock, 2007-10-033.

Qu Zhihao, Chai Shaokuan, Ye Qianyuan. Analysis of Dynamic Characteristics and Chatter Of A 1420 Cold Tandem Rolling Mill [J]. Journal of Vibration and Shock, 2006, 25(4): 25-29.

Xiang Jianhua, Liao Ridong, Zhang Weizheng. Study and Application of Graphic Modeling Techniques of Shaft Torsional Vibration Based on System Matrix Method [J]. Acta Armamentarii, 2005, 26(3): 294-298.

Wei Haijun. Modification of Some Formulas for Calculating Shaft Torsional Vibration [J]. Journal of Vibration and Shock, 2006, 25 (2): 166-168.

Ji Chen, Zeng Fanming. Study on the general model and system matrix method for torsion vibration calculation of complex shafting [J]. Ship & Ocean Engineering, 2006(4): 55-57.

Li Lin, Sheng Jun. Mixed Interface Direct Component Modal Synthesis Method [J]. Chinese Journal of Applied Mechanics, 2005 22 (2): 315-319.

Xiang Shuhong, Qiu Jibao, Wang Dajun. The Resent Progresses on Modal Analysis and Dynamic Sub-Structure Methods [J]. Advances In Mechanics, 2004, 34(3): 289-303.

Chu Hua, She Yinghe. A Substructure Transfer Matrix Method in Torsional Vibration Analysis of Rotor System [J]. Journal of Southeast University, 1989,19(3):87-95.

Z P Mourelatos. An Efficient Crankshaft Dynamic Analysis Using Sub structuring with RITZ VECTORS[J]. Journal of Sound and Vibration, 2000,238(3):495-527.

Bogacz, Szolc T, Irretier H. An Application of Torsional Wave Analysis to Turbo generator Rotor Shaft Response[J]. Trans. of ASME, Journal of Vibration and Acoustics, 1992, 114(2):149-153.

Shu Gequn, Hao Zhiyong. An Investigation on the Torsional Elastic Wave Theory for the Calculation of Multi-Stepped Shaft Torsional Vibration Response [J]. China Mechanical Engineering, 1999,10(11):1277-1279.

He Chengbing, Gu Yujiong, Yang Kun. The Analysis of the Shaft Torsional Vibrations Response of Turbine-generator Units Based on State-Space Method [C]. Asia-Pacific Vibration Conference, China, 2001.

Wang Ke she, Wang Zheng guang. Frequency analysis method of shafting crankle vibration [J]. Journal of Beijing Institute of Machinery, 2001,16(4):11-18.

Lin Sen, Zhang Hongtian, Geng Ruiguang, et al. Study on numerical emulator of shafts' torsion vibration of internal-combustion engine based upon time step integration method [J]. Journal of Heilongjiang Institute of Technology, 2008, 22(1):1-7.

Tong-Qun Han, Sheng-Hua Yao, Sheng-Jun Wu. Analysis of Torsional Vibration of Crankshaft System of a Vehicle Diesel Engine Based on EXCITE-designer Software. Journal of Hubei Automotive Industries Institute, 2005, 19(4): 5-8.

Yang Yanfu, Measurement and Application of Torsional Vibration in Engine[J]. Ccec Science & Technology,1997,4:1-6.

Huang Diangui. Experiment on the Characteristics of Torsional Vibration of Rotor-to-stator Rub in Turbo Machinery[J]. Tribology International,2000,33:75-79.

Xia Qingjie. 16V240ZJC Diesel Engine Torsional Vibration Measurement and Analysis[J]. Si Ji Science & Technology,1998,4:16-19.

Guo Li, Li Bo. , Assessment of Torsional Vibration Measurement Method[J], Grinder and Grinding. 2000,3:53-56.

N A Halliwell. The Laser Torsional Vibrometer: A Step Forward in Rotating Machinery Diagnostics[J]. Journal of Sound and Vibration, 1996,190(3):399-418.

T J Miless, M Lucast, N A Halliwell. Torsional and Bending Vibration Measurement on Rotors Using Laser Technology[J]. Journal of Sound and Vibration, 1999,226(3): 441-467.

Hua Jianwen, Liu Liren. Study of Measurement by Torsional Vibration Transducer Type MM0071 [J]. Laser & Infrared, 1996,26(3):193-195.

Ge Weijing, Wang Weisheng. Small Internal Combustion Engine, 1994,23(2):53-57.

Wang Ting, Cheng Peng. Development of Computer Testing System of Crankshaft Torsional Vibration [J]. Natural Science Journal of Jilin University of Technology, 201,21(3):74-77.

B O AL-Bedoor. Modeling the Coupled Torsional and Lateral Vibrations of Unbalanced Rotors[J]. Computer Methods in Applied Mechanics and Engineering, 2001,190: 5999-6008.

S F Asokanthan, P A Meehans. Non-linear Vibration of a Torsional System Driven by a Hook's Joint[J]. Journal of Sound and Vibration, 1999,226(3):441-467.

K Koser, F Pasin. Torsional Vibration of the Drive Shafts of Mechanisms[J]. Journal of Sound and Vibration, 1997,199(4):559-565.

E Brusa, C Delprete, G Genta. Torsional Vibration of Crankshafts Effects of Non Constant Moments of Inertia[J]. Journal of Sound and Vibration, 1997,205(2):135-150.

Gong Xiansheng Xie Zhijiang Tang Yike. Research on Steady Vibration Response of Shafting with Nonlinear Coupling [J]. Chinese Journal of Mechanical Engineering, 2001,37(6):19-23.

Sheng Gang, Chen Zhiyan, Li Liangfeng. Torsional Vibration of the Diesel Crankshaft System with Variable Inertia [J]. Transactions of Csice, 1991,9(2):143-149.

Zhu Menghua. Direct-Inverse Fourier Transformation Technique for Responses of Non-Linear Torsional Vibration of Diesel Shafting [J]. Transactions of Csice, 1992,10(1):47-52

Lin Ruilin, Huang Cihao, Tang Kaiyuan. The Incremental Harmonic Balance Method for CalculatingResponse of Non-Linear Torsional Vibration of a Diesel Shafting [J].Transactions of Csice, 2002,20(2):185-187.

Zhang Zhihua, Tang Mi. The Numerical Computation on Torsional Vibration of Diesel Nonlinear Shafting, Transactions of Csice, 1987,5(4): 353-361.

Farshidianfar A, Ebrahimi M, Bartlett H. Hybrid Modeling and Simulation of the Torsional Vibration of Vehicle Driveline Systems[J]. Journal of Automobile Engineering, 2001.215(4):217-229.

Wen Bangchun, Li Yinong, Han Qingkai, Analytical Approach and Application of Nonlinear Vibration Theory [M]. Shenyang: Northeast University Press, 2001.

Li Bozhong, Song Tianxiang, Song Xigeng. On Coupled Vibrations of Reciprocating Engine Shaft Systems(Part 1) Axial Vibration Caused by Torsional Vibration [J]. Transactions of Csice, 1989,7(1): 1-6.

Li Bozhong, Song Xigeng, Song Tianxiang. On Coupled Vibrations of Reciprocating Engine Shaft Systems(Part 2)- Progressive Torsional-Axial Continued Vibrations [J].Transactions of Csice, 1990,8(4):317-322.

Song Xigeng, Song Tianxiang, Li Bozhong. On Coupled Vibrations of Reciprocating Engine Shaft Systems (Part 3)- Calculation Method of Coupled at Same and Double Frequencies [J]. Transactions of Csice, 1994,12(2):115-120.

Du Hongbing, Chen Zhiyan, Jing Bo. The Mathematical Model for the Coupled Torsional-Axial Vibration of Internal Combustion Engine Shaft System[J]. Chinese Combustion Engine Engineering 1992,13(2):66-74.

Zhang Yong, Jiang Zikang. Numerical Analyses of the Coupling of the Flexural and Torsional Vibrations of Rotary Shaft System[J]. Turbine Technology,1999,41(5): 280-283.

Zhang Yong, Jiang Zikang. Analysis of the coupled flexural-torsional vibrations of rotary shaft systems [J]. Journal of Tsinghua University(Science And Technology), 2000,40(6):80-83.

Zhang Yong, Jiang Zikang. Mathematic model of coupled bending and torsional vibration of shaft systems [J]. Journal of Tsinghua University(Science And Technology), 1998,38(8):114-117.

Zhang Hongtian, Zhang Zhihua, LIU Zhigang. A Study on Coupled Axial and Torsional Vibration of Marine Propulsion Shafting [J]. Shipbuilding of China, 1995,129(2):68-76.

0kamura H et. Simple Modeling and Analysis for Crankshaft Three-Dimensional Vibration[J]. Journal of Vibration and Acoustics, 1995,117(1): 70-86.

Shen Hongbin Shu Gequn. An Investigation into Dynamic Measurement of Longitudinal/Bend/Tortional Vibration in Shaft System of Internal Combustion Engines [J]. Small Internal Combustion Engine, 2000, 29(6):9-10.

Liang Xingyu, Shu Gequn, Li Donghai, Shen Yinggang. Torsional Vibration Analysis Based on Multi-body Dynamics of Flexible Crankshaft System [J]. Chinese Combution Engine Engineering, In 2008, the fourth period.

Natarajan M, Frame E A, Naegeli D, et al. Oxygenates for ad vanced petroleum-based diesel fuels: part1. Screening and selection methodology for the oxygenates[C]. SAE 2001-01-3631.

Guo Wei-dong; Gu Wen-gang; ZhanG Xiao-ling. The Method of Compensating Division Error in Shafts Torsional Vibration Measurement and Implementation with Program [J]. Machine Design & Research, 2007, 23(2): 88-89.

Zhang Xiaoling, Tang Xikuan. Analysis of torsional vibration measuring errors and their correct method [J]. Journal Of Tsinghua University (Science And Technology), 1997, 37(8): 9-12.

Xu Xintian. Measure of the torsional vibration control [J]. Wuhan Shipbuilding, 1999, 1:17-18.

C T Lee, S W Shaw. The Non-linear Dynamic Response of Paired Centrifugal Pendulum Vibration Absorbers [J]. Journal of Sound and Vibration, 1997,203(5): 731-743.

Weng Wenhua. Silicone Oil Damper [J]. CCEC Science & Technology, 1997,1:19-23.

Luo Zhouquan. Characters of Shafting Torsional Vibration and Analysis on Damping Measures of 6180 Series Marine Engine [J]. Design and Manufacture of Diesel Engine, 2000, 92 (3):3-7.

Zhou Shengfu, Li Cuncai. Function and Behaviour of Torsional Damper of Crankshaft [J]. Vehicle Engine, 1993,4:23-28.

Chen Xi'en. Study on Nonlinear High-damping Dampers in calculation and measurement of torsional vibration[J], China Ship Survey, 1996, 1:26-32.

Yan Jiabin. Engine Crankshaft Torsional Vibration Damper, World Rubbre Industry, 1998, 25(4):26-30.

Huo Quanzhong, Liang Jie, Hao Zhiyong. Operation Principle and Experimental Study of an Active Torsional Vibration Absorber [J]. Transactions of Csice, 1992,10(4): 329-334.

Hao Zhiyong, Gao Wenzhi. A Study on Active Control and Simulation Test for Torsional Vibration of Turbogenrator Shaft System [J]. Power Engineering, 1999,19(5): 338-341.

Liu Shengtian. Double Mass Flywheel Type Torsional Vibration Damper [J]. Automobile Technology, 1997,1: 23-27.

Liu Shengtian, Hydraulic dual-mass flywheel torsional vibration damper, [J]. Motor Transport,1998,9: 17-19.

M Hosek, H Elmal, N Olgac. A Tunable Torsional Vibration Absorber: the Centrifugal Delayed Resonator[J]. Journal of Sound and Vibration, 1997,205(2):151-165.

Shu Gequn, Hao Zhiyong. Experiments on the Crankshaft Torsional/Bending Vibration Damper in Internal Combustion Engine [J]. Journal of Tianjin University, 1997,30(6): 806-811.

Shu Gequn, Lü Xingcai, QIin De, Su Yanling. Experimental Study on Hybrid Damper of HighSpeed Diesel Engine Crankshaft [J]. Transactions of Csice, 2000,20(6): 537-540.

Shu Gequn. Effect of Bending Vibration Damper on Properties of Torsional [J]. Transactions of Csice, 1998,16(3): 348-353.

Zhao Qian, Hao Zhiyong. Optimal Design of Torsional Damper of Diesel Engine Crankshaft System [J]. Transactions of the Chinese Society of Agricultural Machinery, 2000,31 (5): 91-93.

Song Jingbo. Internal combustion Engines, 1993, (3):38-40.

Ying Qiguang, Bao Defu and Others. A New Fault-diagnosing Technique of Diesel Engine based on Shafting Vibration [J]. Ship Engineering, 1995, (4):33-35.

Ying Qiguang, Li Bozhong. A New Effective Method to Diagnose the Troubles in the Technical Condition of the Diesel Engine [J]. Journal of Shanghai Maritime University, 1990, (3):20-30.

Lin Dayuan, Wang Xu, Song Xueren. Diagnosis of Crankshaft Cracks via Measurement of Crankshaft—Flywheel System Torsional Vibration Mode [J]. Transactions of Csice, 1989, 7(3):215-222.

Shu Gequn, Lü Xingcai. Calculation Method for Continuous Distribution Model of Torsional Vibration of a Crankshaft [J]. Transactions of The Chinese Society of Agricultural Machinery, 2004,35 (4) 36-39.

Time-Frequency Analysis for Rotor-Rubbing Diagnosis

Eduardo Rubio and Juan C. Jáuregui
CIATEQ A.C., Centro de Tecnología Avanzada
Mexico

1. Introduction

Predictive maintenance by condition monitoring is used to diagnose machinery health. Early detection of potential failures can be accomplished by periodic monitoring and analysis of vibrations. This can be used to avoid production losses or a catastrophic machinery breakdown. Predictive maintenance can monitor equipments during operation. Predictions are based on a vibration signature generated by a healthy machine. Vibrations are measured periodically and any increment in their reference levels indicates the possibility of a failure.

There are several approaches to analyze the vibrations information for machinery diagnosis. Conventional time-domain methods are based on the overall level measurement, which is a simple technique for which reference charts are available to indicate the acceptable levels of vibrations. Processing algorithms have been developed to extract some extra features in the vibrations signature of the machinery. Among these is the Fast Fourier Transforms (FFT) that offers a frequency-domain representation of a signal where the analyst can identify abnormal operation of the machinery through the peaks of the frequency spectra. Since FFT cannot detect transient signals that occur in non-stationary signals, more complex analysis methods have been developed such as the wavelet transform. These methods can detect mechanical phenomena that are transient in nature, such as a rotor rubbing the casing of a motor in the machine. This approach converts a time-domain signal into a time-frequency representation where frequency components and structured signals can be localized. Fast and efficient computational algorithms to process the information are available for these new techniques.

A number of papers can be found in the literature which report wavelets as a vibration processing technique. Wavelets are multiresolution analysis tools that are helpful in identifying defects in mechanical parts and potential failures in machinery. Multiresolution has been used to extract features of signals to be used in classifications algorithms for automated diagnosis of machine elements such as rolling bearings (Castejón et al., 2010; Xinsheng & Kenneth, 2004). These elements produce clear localized frequencies in the vibration spectrum when defects are developing. However, a more complex phenomena occurs when the rotor rubs a stationary element. The impacts produce vibrations at the fundamental rotational frequency and its harmonics, and additionally yield some high frequency components, that increase as the severity of the impacts increases (Peng et al., 2005).

Rotor dynamics may present light and severe rubbing, and both are characterized by a different induced vibration response. It is known that conditions that cause high vibration levels are accompanied by significant dynamic nonlinearity (Adams, 2010). The resonance frequency is modified because of the stiffening effect of the rubbing on the rotor (Abuzaid et al., 2009). These systems are strongly nonlinear and techniques have been applied for parameter identification. These techniques have developed models that explain the jump phenomenon typical of partial rub (Choi, 2001; Choi, 2004).

The analysis of rubbing is accomplished with the aid of the Jeffcott rotor model for lateral shaft vibrations. This model states the idealized equations of rotor dynamics (Jeffcott, 1919). Research has been done to extend this model to include the nonlinear behavior of the rotor system for rubbing identification. It has been shown that time-frequency maps can be used to analyze multi-non-linear factors in rotors. They also reveal many complex characteristics that cannot be discovered with FFT spectra (Wang et al., 2004). Other approaches use analytical methods for calculating the nonlinear dynamic response of rotor systems. Second-order differential equations which are linear for non-contact and strongly nonlinear for contact scenarios have been used (Karpenko et al., 2002). Rub-related forces for a rotor touching an obstacle can be modeled by means of a periodic step-function that neglects the transient process (Muszynska, 2005).

In this chapter the phenomenon of rotor rubbing is analyzed by means of a vibrations analysis technique that transforms the time-domain signal into the time-frequency domain. The approach is proposed as a technique to identify rubbing from the time-frequency spectra generated for diagnostic purposes. Nonlinear systems with rotating elements are revised and a nonlinear model which includes terms for the stiffness variation is presented. The analysis of the signal is made through the wavelet transform where it is demonstrated that location and scale of transient phenomena can be identified in the time-frequency maps. The method is proposed as a fast diagnostic technique for rapid on-line identification of severe rubbing, since algorithms can be implemented in modern embedded systems with a very high computational efficiency.

2. Nonlinear rotor system with rubbing elements

Linear models have intrinsic limitations describing physical systems that show large vibration amplitudes. Particularly, they are unable to describe systems with variable stiffness. To reduce the complexity of nonlinear problems, models incorporate simplified assumptions, consistent with the physical situation, that reduce their complexity and allow representing them by linear expressions. Although linearized models capture the essence of the problem and give the main characteristics of the dynamics of the system, they are unable to identify instability and sudden changes. These problems are found in nonlinear systems and the linear vibration theory offers limited tools to explain the complexity of their unpredictable behavior. Therefore, nonlinear vibration theories have been developed for such systems.

The steady state response of the nonlinear vibration solution exhibits strong differences with respect to the linear approach. One of the most powerful models for the analysis of nonlinear mechanical systems is the Duffing equation. Consider the harmonically forced Duffing equation with external excitation:

$$\ddot{x} + c\dot{x} + \alpha x + \beta x^3 = a cos(\omega t + \phi) \qquad (1)$$

Curves of response amplitude versus exciting frequency are often employed to represent this vibration behavior as shown in Fig. 1. The solid line in this figure shows the response curve for a linear system. The vertical line at ω/ω_n=1 corresponds to the resonance. At this point vibration amplitude increases dramatically and it is limited only by the amount of damping in the system. It is important to ensure that the system operates outside of this frequency to avoid excessive vibration that can result in damage to the mechanical parts. In linear systems amplitude of vibrations grows following a straight line as excitation force increases.

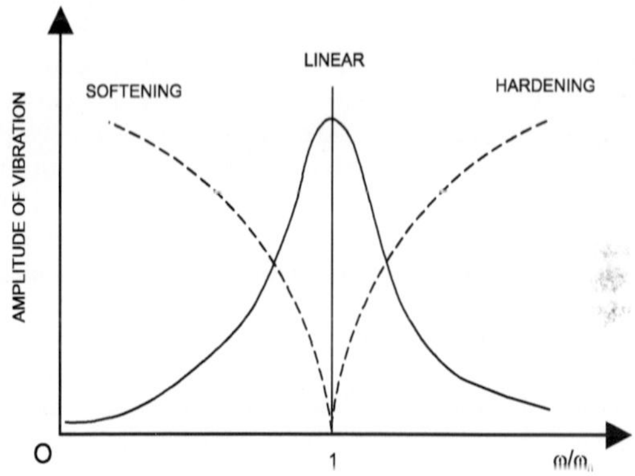

Fig. 1. Resonant frequency dependency in nonlinear systems

In nonlinear systems the motion follows a trend that is dependent upon the amplitude of the vibrations and the initial conditions. The resonance frequency is a function of the excitation force and the response curve does not follow a straight line. When the excitation force increases, the peak amplitude "bends" to the right or left, depending on whether the stiffness of the system hardens or softens. For larger amplitudes, the resonance frequency decreases with amplitude for softening systems and increases with amplitude for hardening systems. The dashed lines in Fig. 1 show this effect.

When the excitation force is such that large vibration amplitudes are present, an additional "jump" phenomenon associated with this bending arises. This is observed in Fig. 2. Jump phenomenon occurs in many mechanical systems. In those systems, if the speed is increased the amplitude will continue increasing up to values above $1.6\omega_n$.

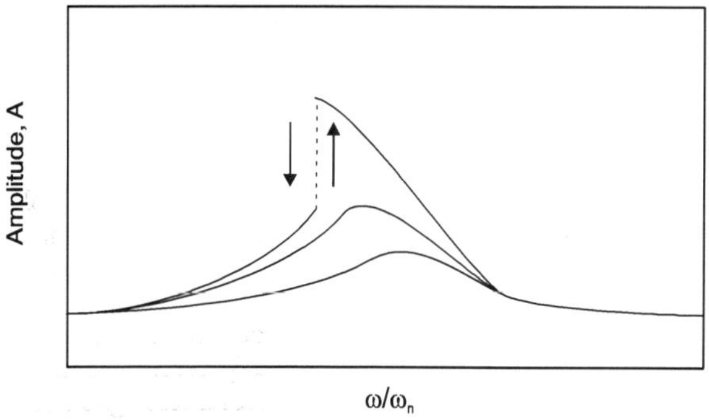

Fig. 2. Jump phenomenon typical of nonlinear systems

When the excitation force imposes low vibration amplitudes, or there is a relative strong damping, the response curve is not very different from the linear case as it can be observed in the two lower traces. However, for large vibration amplitudes the bending effect gets stronger and a "jump" phenomenon near the resonance frequency is observed. This phenomenon may be observed by gradually changing the exciting frequency ω while keeping the other parameters fixed. Starting from a small ω and gradually increasing the frequency, the amplitude of the vibrations will increase and follow a continuous trend. When frequency is near resonance, vibrations are so large that the system suddenly exhibits a jump in amplitude to follow the upper path, as denoted with a dashed line in Fig. 2. When reducing the excitation frequency the system will exhibit a sudden jump from the upper to the lower path. This unusual performance takes place at the point of vertical tangency of the response curve, and it requires a few cycles of vibration to establish the new steady-state conditions.

There is a region of instability in the family of response curves of a nonlinear system where such amplitudes of vibration cannot be established. This is shown in Fig. 3. It is not possible to obtain a particular amplitude in this region by forcing the exciting frequency. Even with small variations the system is unable to restore the stable conditions. Therefore, from the three regions depicted in this figure, only the upper and lower amplitudes of vibration exist. The same applies for a hardening system but with the peaks of amplitude of vibrations bending to the right.

A rotor system with rub impact is complex and behaves in a strong nonlinearity. A complicated vibration phenomenon is observed and the response of the system may be characterized by the jump phenomena at some frequencies. Impacts are associated with stiffening effects; therefore, modeling of rotor rub usually includes the nonlinear term of stiffness.

When the rotor hits a stationary element, it involves several physical phenomena, such as stiffness variation, friction, and thermal effects. This contact produces a behavior that worsens the operation of the machine. Rubbing is a secondary transient phenomenon that arises as a result of strong rotor vibrations. The transient and chaotic behavior of the rotor impacts generate a wide frequency bandwidth in the vibrational response.

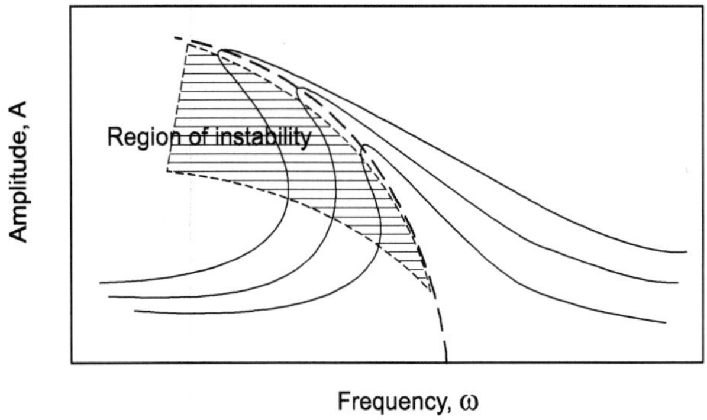

Fig. 3. Region of instability

Dynamics of the rotor rubbing can be studied with the Jeffcott rotor model (Jeffcott, 1919). This model was developed to analyze lateral vibrations of rotors and consists of a centrally mounted disk on a flexible shaft. Rigid bearings support the ends of the shaft as shown in Fig. 4. The model is more representative of real rotor dynamics for the inclusion of a damping force proportional to the velocity of the lateral motion. The purpose of this model was to analyze the effect of unbalance at speeds near the natural frequency, since the vibration amplitude increase considerably in this region.

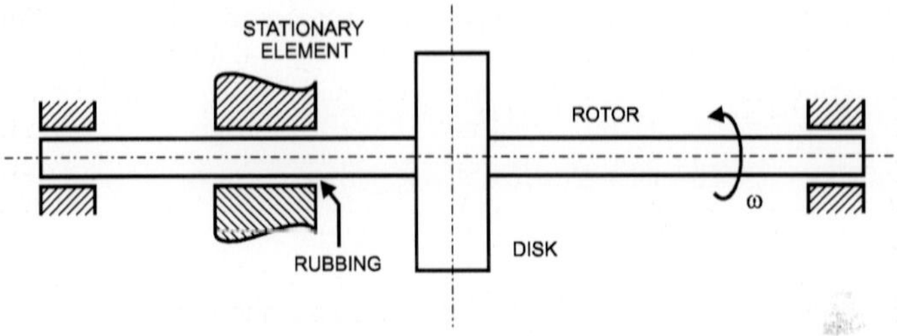

Fig. 4. Diagram of a rotor rubbing with a stationary element

Modifying the Jeffcott´s model, the rubbing phenomenon can be studied. A stationary element can be added to the model to take rubbing into consideration. A diagram of the forces that are involved during the rub-impact phenomenon is shown in Fig. 5.

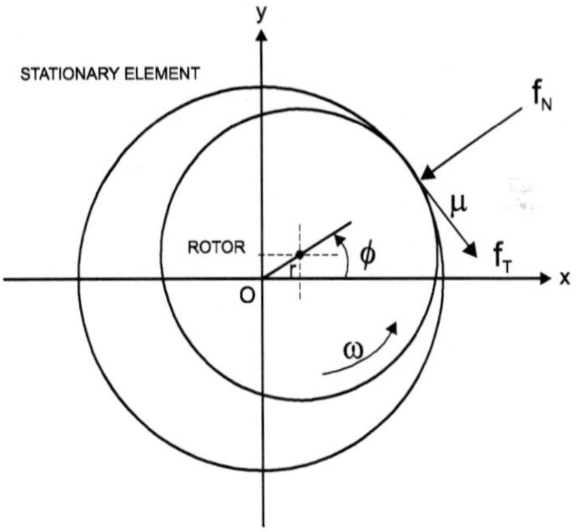

Fig. 5. A Jeffcott rotor model with rubbing

At the contact point, normal and tangential forces are described by the following expressions:

$$f_N = K_R(R - c) \tag{2}$$

$$f_T = \mu f_T \tag{3}$$

Where K_R is the combined stiffness of the shaft and the contact stiffness.
This is valid for

$$R \geq c \tag{4}$$

otherwise

$$f_N = f_T = 0 \tag{5}$$

For a Cartesian coordinate system forces are represented as:

$$f_x = -f_N cos\ \phi + f_T\ sin\ \phi \tag{6}$$

$$f_y = -f_N sin\ \phi - f_T\ cos\ \phi \tag{7}$$

And the motion equations of the rotor system are described by:

$$M\ddot{x} + C\dot{x} + K(\omega)x = F_x + mp\omega^2 cos\ (\varphi_0 + \omega t) \tag{8}$$

$$M\ddot{y} + C\dot{y} + K(\omega)y = F_y + MP\omega^2 sin\ (\varphi_0 + \omega t) \tag{9}$$

Where K is the stiffness of the system and C is the damping coefficient of the system.
The contact of the rotor with the stationary element creates a coupling of the system that causes a variation in the stiffness because of the non-continuous f_N and the model becomes nonlinear. The rotor rubs the element only a fraction of the circumferential movement and the stiffness value varies with respect to the rotor angular position.
The nonlinear behavior can be related to the stiffness variation. As shown in Fig. 6, the system´s stiffness can be related to the shaft stiffness K_S, and it increases to K_R during contact. This increment can be estimated using the Hertz theory of contact between two elastic bodies placed in mutual contact.

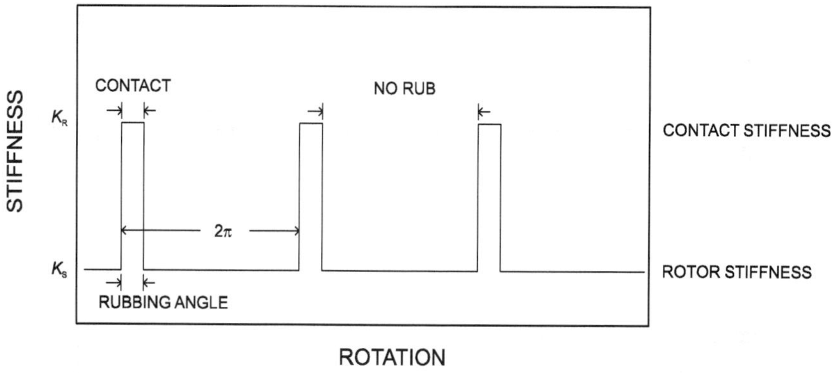

Fig. 6. Stiffness increase during contact

Assuming that the system´s stiffness can be represented as a rectangular function, then the stiffness variation can be approximated as a Taylor series such that

$$K(\omega) = \alpha + \beta^3\omega \tag{10}$$

3. Vibrations analysis with data-domain transformations

The vibrational motion produced by a rotating machine is complicated and may be analyzed by transforming data from the time-domain to the frequency-domain by means of the

Fourier Transform. This transform gives to the operator additional information from the behavior of the machine that a signal in time-domain cannot offer.
Fourier developed a theory in which any periodic function *f(t)*, with period T, can be expressed as an infinite series of sine and cosine functions of the form:

$$f(t) = \frac{a_0}{2} + \sum_{n=1}^{\infty} (a_n cos\, n\omega_T t + b_n\, sin\, \omega_T t) \tag{11}$$

Where ω denotes the fundamental frequency and 2ω, 3ω, etc., its harmonics. This series is known as the Fourier series expansion and a_n and b_n are called the Fourier coefficients. By this way, a periodic waveform can be expanded into individual terms that represent the various frequency components that make up the signal. These frequency components are integer multiples of ω.
The following identity can be used to extend the Fourier series to complex functions:

$$e^{ix} = cos(x) + isin(x) \tag{12}$$

$$f(t) = \sum_{-\infty}^{\infty} c_n e^{int} \tag{13}$$

Where c_n can be obtained by the following integration:

$$c_n = \frac{1}{2\pi} \int_{-\pi}^{\pi} f(t) e^{-int} dt \tag{14}$$

This applies to periodic functions on a 2π interval.
Fourier series can be extended to functions with any period T with angular frequency ω=2π/T. Sine and cosine functions have frequencies that are multiples of ω as in Eq. (11).
For non-periodic functions, with period T, discrete frequencies $n\omega$ separated by Δω=2π/T, and taking the limit as T→∞, $n\Delta\omega$ becomes continuous and the summation can be expressed as an integral. As a result, the continuous Fourier transform for frequency domain is defined as:

$$F(\omega) = \int_{-\infty}^{+\infty} f(t) e^{-i\omega t}\, dt \tag{15}$$

While for time domain the inverse Fourier transform is defined as:

$$f(t) = \frac{1}{2\pi} \int_{-\infty}^{+\infty} F(\omega) e^{i\omega t}\, d\omega \tag{16}$$

And *f=f(t)* satisfying the condition

$$\int_{-\infty}^{+\infty} |f(t)|^2 dt < \infty \tag{17}$$

Since computers can't work with continuous signals, the Discrete Fourier Transform (DFT) was developed and implemented through the Fast Fourier Transform (FFT). The FFT is a fast algorithm for computing the DFT that requires the size of the input data to be a power of 2. FFT is a helpful engineering tool to obtain the frequency components from stationary signals. However, non-stationary phenomena can be present in signals obtained from real engineering applications, and are characterized by features that vary with time.

A difficulty that has been observed with FFT is that the complex exponentials used as the basis functions have infinite extent. Therefore, localized information is spread out over the whole spectrum of the signal. A different approach is required for this type of signals. Time-frequency methods are used for their analysis and one of the most used methods is the Short Time Fourier Transform (STFT). This was the first time-frequency technique developed. The solution approach introduces windowed complex sinusoids as the basis functions.

The STFT is a technique that cuts out a signal in short time intervals, which can be assumed to be locally stationary, and performs the conventional Fourier Transform to each interval. In this approach a signal $S_t(\omega)$ is multiplied by a window function $h(\tau)$, centered at τ, to obtain a modified signal that emphasises the signal characteristics around τ:

$$S_t(\omega) = \frac{1}{2\pi}\int_{-\infty}^{\infty} e^{-j\omega t} s(\tau) h(\tau - t) d\tau \tag{18}$$

Frequency distribution will be reflected around τ after applying the Fourier Transform to this window. The spectral density of the energy at time τ can be written as follows:

$$P(t,\omega) = |S_t(\omega)|^2 = \left|\frac{1}{2\pi}\int_{-\infty}^{\infty} e^{-j\omega\tau} s(\tau) h(t - \tau) d\tau\right|^2 \tag{19}$$

As expected, a different spectrum will be obtained for each time. A Spectrogram, which is the time-frequency distribution, can be constructed with the resulting spectra. Resolution in time and frequency depends on the width of the windows $h(\tau)$. Large windows will provide a good resolution in the frequency domain, but poor resolution in time domain. Small windows will provide good resolution in time domain, but poor resolution in frequency domain. The major disadvantage of this approach is that resolution in STFT is fixed for the entire time-frequency map. This means that a single window is used for all the frequency analysis. Therefore, only the signals that are well correlated in the time interval and frequency interval chosen will be localized by the procedure. It may be thought of as a technique to map a time-domain signal into a fixed resolution time-frequency domain.

This drawback can be surpassed with basis functions that are short enough to localize high frequency discontinuities in the signal, while long ones are used to obtain low frequency information. A new transform called wavelet transform achieves this with a single prototype function that is translated and dilated to get the required basis functions.

The wavelet transform is a time-frequency representation technique with flexible time and frequency resolution. Conversely to the STFT where the length of the windows function remains constant during the analysis, in the wavelet approach a function called the mother wavelet is operated by translation and dilation to build a family of window functions of variable length:

$$\psi_{s\tau}(t) = \frac{1}{\sqrt{s}}\psi\left(\frac{t - \tau}{s}\right) \tag{20}$$

Where $\psi(t)$ is the mother wavelet function, s the scale parameter, and τ the time shift or dilation parameter. Based on the mother wavelet function, the wavelet transform is defined as:

$$W_{\psi}x(s,\tau) = \int_{-\infty}^{\infty} x(t)\,\overline{\psi_{s\tau}}(t)dt \tag{21}$$

And $W_{\psi}x(s,\tau)$ are the wavelet coefficients.
The wavelet transform is different from other techniques in that it is a multiresolution signal analysis technique that decomposes a signal in multiple frequency bands. By operating over s and τ the wavelets permit to detect singularities, which makes it an important technique for nonstationary signal analysis.
Due to this characteristic, the wavelet transform is the analysis technique that we found more suitable for the analysis of the rubbing phenomenon.

4. Experimental methodology

An experimental test rig was implemented to get a deeper understanding of the main characteristics of the rubbing phenomenon, and to apply the wavelet analysis technique in the processing and identification of the vibrations produced by the rub-impact of the system. Elements were included to run experiments under controlled conditions. Fig. 7 shows the experimental set-up.

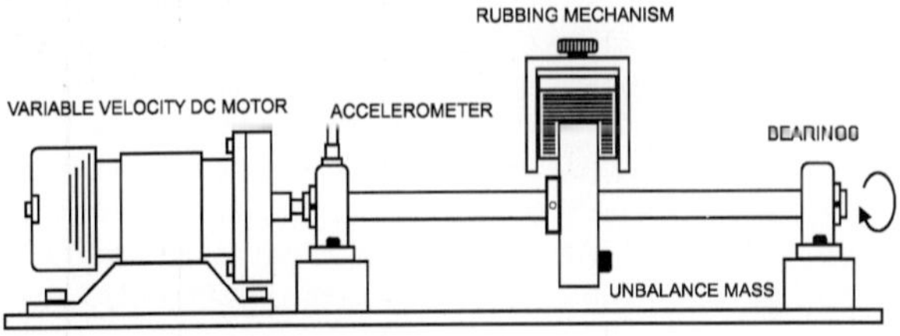

Fig. 7. Test rig for the rubbing experiments

The experimental system is composed of a shaft supported by ball bearings and coupled to an electrical motor with variable rotational speed. The velocity of the motor was controlled with an electronic circuit. A disk was installed in the middle of the shaft, which was drilled to be able to mount bolts of different masses to simulate unbalance forces. An adjustable mechanism was designed in order to simulate the effect of a rotor rubbing a stationary element. The position of the device, acting as the stationary element, was adjusted with a threaded bolt that slides a surface to set the clearance between the rotating disk and the rubbing surface. The shaft and disk were made of steel, and the rubbing device of aluminium alloy. Light and severe rubbing were simulated by controlling the speed of the rotor. Low velocities caused light rubbing while high velocities generated severe impact-like rubbing vibrations. Both types of rubbing were analyzed with the proposed methodology.
An accelerometer was used to measure the vibrations amplitude. Output of the accelerometer was connected to a data acquisition system to convert analog signals to digital data with a sampling rate of 10 kHz. An antialias filter stage was included to get a band limited input signal.
Experimental runs were carried out for fixed and variable rotor velocities. Fixed velocities were tested for values between 350 rpm and 1900 rpm. Continuous variable velocity experiments were also carried out to simulate a rotor system under ramp-up and ramp-down conditions, to verify the preservation of the scale and temporal information with the processing technique used.

Daubechies 4 wavelet transform was implemented to convert the signal from the time-domain to the time-frequency domain. Scaling and wavelet functions for this transform are shown in Fig. 8.

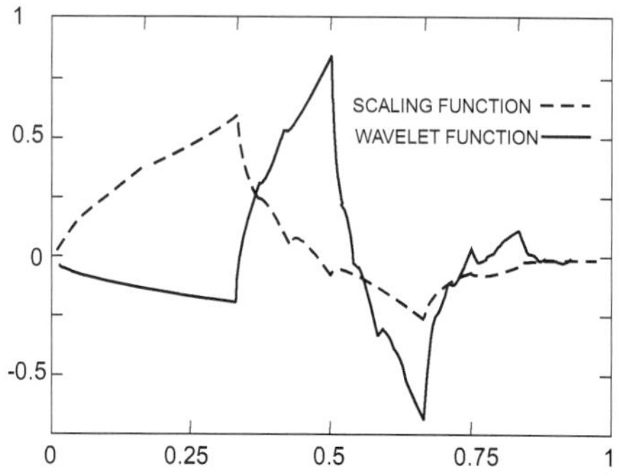

Fig. 8. Daubechies D4 scaling and wavelet functions

Implementation of the continuous wavelet transform is impractical, especially for on-line detection devices for process monitoring purposes. This implementation consumes a significant amount of time and resources. An algorithm for the discrete wavelet transform (DWT) is used to overcome this situation. It is based on a sub-band coding which can be programmed with a high computational efficiency. Subband coding is a multiresolution signal processing technique that decomposes the signal into independent frequency subbands.

With this approach, the DWT applies successive low-pass and high-pass filters to the discrete time-domain signal as shown in Fig. 9. This procedure is known as the Mallat algorithm.

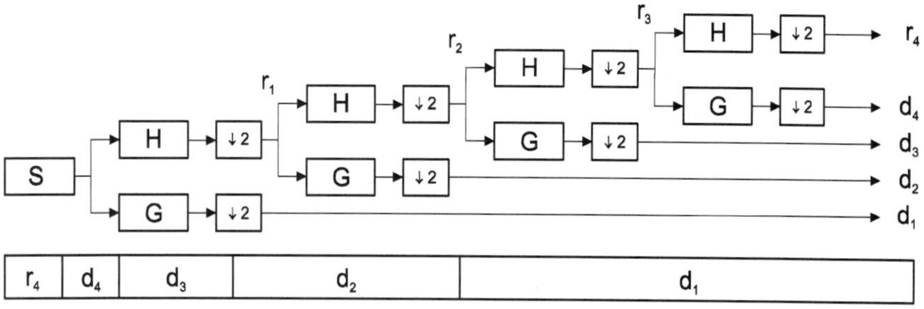

Fig. 9. Algorithm for the sub-band decomposition of the input signal

The algorithm uses a cascade of filters to decompose the signal. Each resolution has its own pair of filters. A low-pass filter is associated with the scaling function, giving the overall picture of the signal or low frequency content, and the high-pass filter is associated with the wavelet function, extracting the high frequency components or details. In Fig. 9 the low-pass filter is denoted by H and the high-pass filter is denoted by G. Each end raw is a level of decomposition. A sub-sampling stage is added to modify the resolution by two at each step of the procedure. As a result of this process, time resolution is good at high frequencies, while frequency resolution is good at low frequencies.

For each transform iteration the scale function to the input data is applied through the low-pass and high-pass filters. If the input array has N cells, after applying the scale function, the down-sampling by two, which follows the filtering, halves the resolution and an array with N/2 values will be obtained. With the low-pass branch, coarse approximations are obtained. The high-pass filtered signal will reflect the fluctuations or details content of the signal. By iterating recursively a signal is decomposed into sequences. The successive sequences are lower resolution versions of the original data.

The implemented form of the Daubechies 4 wavelet transform has a wavelet function with four coefficients and a scale function with four coefficients. The scale function is:

$$A_i = H_0 S_{2i} + H_1 S_{2i+1} + H_2 S_{2i+2} + H_3 S_{2i+3} \tag{22}$$

Where scaling coefficients are defined as

$$H_0 = \frac{1+\sqrt{3}}{4\sqrt{2}} \tag{23}$$

$$H_1 = \frac{3+\sqrt{3}}{4\sqrt{2}} \tag{24}$$

$$H_2 = \frac{3-\sqrt{3}}{4\sqrt{2}} \tag{25}$$

$$H_3 = \frac{1-\sqrt{3}}{4\sqrt{2}} \tag{26}$$

The wavelet function is:

$$C_i = G_0 S_{2i} + G_1 S_{2i+1} + G_2 S_{2i+2} + G_3 S_{2i+3} \tag{27}$$

Where wavelet coefficients are defined as

$$G_0 = H_3 \tag{28}$$

$$G_1 = -H_2 \tag{29}$$

$$G_2 = H_1 \tag{30}$$

$$G_3 = -H_0 \tag{31}$$

Each wavelet and function value is calculated by taking the product of the coefficients with four data values of the input data array. The process is iterated until desired results are reached.

5. Experimental results and discussion

The methodology described in the previous section was applied and experimental runs were carried out with the aid of the test rig to obtain a deeper comprehension of the rubbing phenomenon. Fig. 10 shows results for time and frequency domains for the rotor rubbing and no-rub experimental runs. The upper row corresponds to time-domain signals, while the lower row shows the frequency-domain signals.

With no-rub (upper left), the signal in time-domain is characterized by a uniform trace with a small dispersion of data produced by the low-level noise of the measuring system. However, when rub occurs (upper right), as acceleration is the measurement engineering unit, even for low level rubbing the energy content of the signal is high, and spikes appear at the location of each rub-contact.

The spectral distribution of the signal obtained when the rotor is rubbing shows the wide frequency bandwidth in the vibrational response, produced by the chaotic behavior of the rubbing phenomenon.

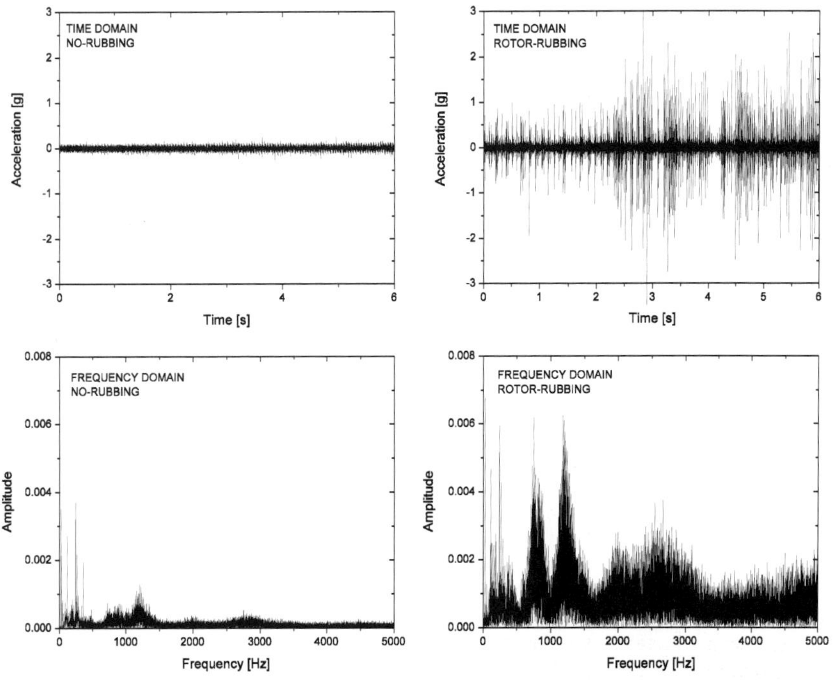

Fig. 10. Spectral distribution of the vibrations for the rotor rubbing and no-rubbing

When rubbing is present, the response is dependent on the angular frequency of the rotor. For low rotor velocities rub generates low vibration amplitudes as shown in the acceleration values in the upper graph of Fig. 11. This can be considered a light-rubbing, but when the rotor velocity is high, the time-domain response of the vibrations produced is quite different and get closer to an impact response characterized by spikes with high acceleration values. This response can be seen in the lower graph of Fig. 11. The amplitudes of vibrations for light rubbing are within ±0.1 g, while for severe rubbing peak values reach ±1 g, about ten times higher.

Processing results of the signals for the rotor with rubbing and without rubbing to obtain the spectral distribution are shown in Fig. 12. The graph localizes the natural frequency of the test rig for both cases. As explained in the introductory section, the nonlinear nature of the system produced a slight different vibrations pattern when rubbing is present. Solid line shows the natural frequency for the rub-free experiments. However, the dotted line corresponds to the experiments with rubbing induced to the rotor, and as expected there is a shift in the natural frequency.

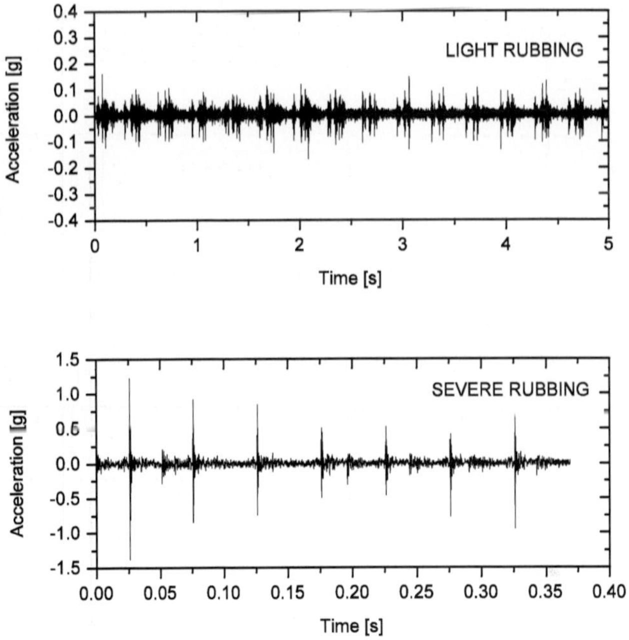

Fig. 11. Vibrations amplitude for light rubbing and severe rubbing

Frequency shift occurs to the right, with the trend to move to the high frequency side of the spectrum, which means that the system is hardening as a result of the stiffness increase produced by the contact of the rotor with the stationary element. The amplitude of the natural frequency also increases as a product of the higher energy content of the rub-impacts.

The signal of the vibrations was processed to transform the data from the time-domain to the time-frequency domain. Wavelet transform Daubechies 4 was used for the transformation and results are shown in Fig. 13.

As stated before, a vector is obtained with this procedure which is the same size as the original vector. Recalling the subband coding, upper half of the vector contains the high frequency content of the information (subband 1). From the remaining data, upper half contains the next subband with mid-frequency content (subband 2), and so on. This way, the low frequency content of the information is coded and located in the lower part of the vector while the high frequency content is coded into the higher indexes of the vector. Indexes represent the transformed values in the resulting vector which amplitude is a function of the correlation between the input signal and the mother wavelet. A higher value for the index means a stronger correlation and therefore a major content of that frequency corresponding to a particular value of scale and translation.

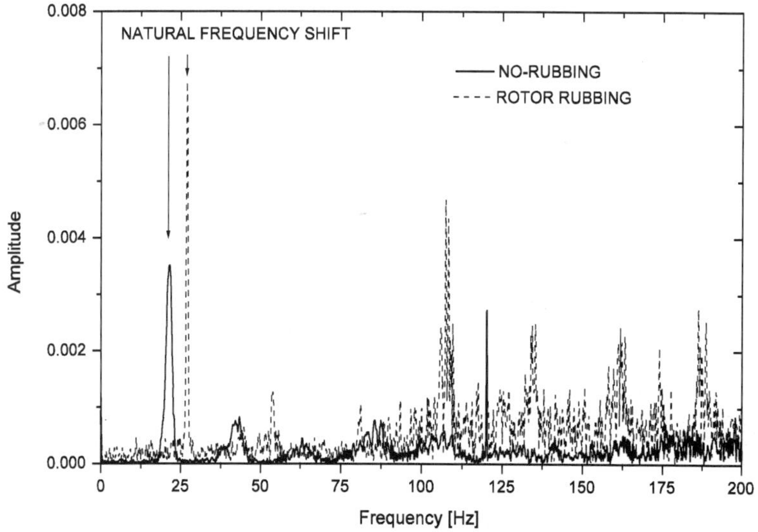

Fig. 12. Resonance frequency dependence observed for a rotor with rubbing

Taking this into consideration, it can be observed that for light rubbing the correlation gets stronger for mid-value indexes, which means that light rubbing is characterized by frequencies that fall in the lower middle of the frequency spectrum.

On the other side, rub-impacts produced by the contact of the rotor with the stationary element at high rotational frequencies, are characterized by spikes with a high frequency content. The wavelet transformation enhances this type of rubbing as can be observed in the upper half of the vector for severe rubbing shown in Fig. 13 (subband 1), although some rubbing information can be found in the next subband. As both light and severe rubbing may be present in a rotor, the sum of the frequency content produced by the phenomena reflects again a wide spectral distribution in the vibrational response.

To test the wavelet approach as a rubbing detection technique, especially for severe rubbing where it is desirable to assess alert signals before a catastrophic failure occurs as it can happens under some rubbing conditions, a vibration signal which presents rub-impacts was chosen. The test data are shown in Fig. 14.

There are two spikes in the graph produced by the rotor rubbing at high velocity rotation. These spikes can be treated as singularities of transient nature whose occurrence cannot be predicted. A technique like wavelets that analyzes a signal by comparison of a basis wavelet that is scaled and translated to extract frequency and location information is ideal for this situation. The procedure enhances these singularities and makes it easier their detection as it is demonstrated next.

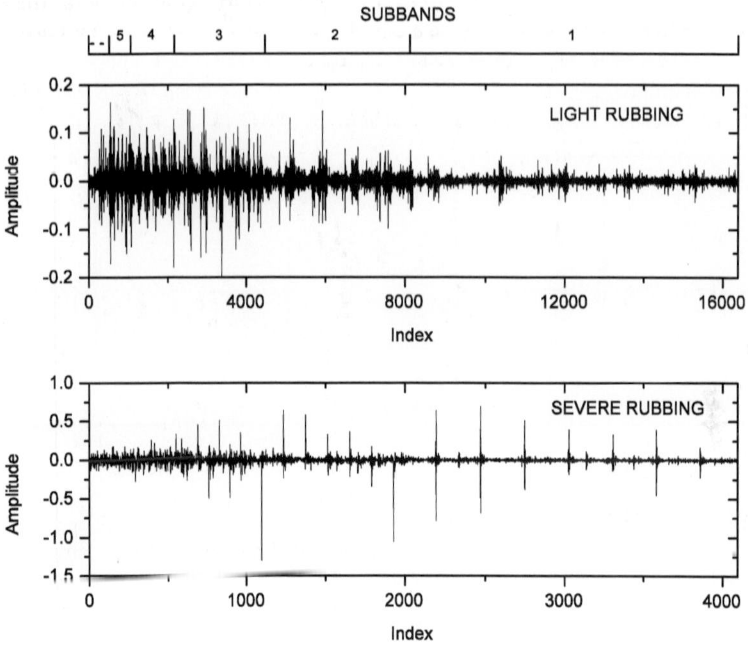

Fig. 13. Rotor rubbing signal transformed to time-frequency domain with wavelets

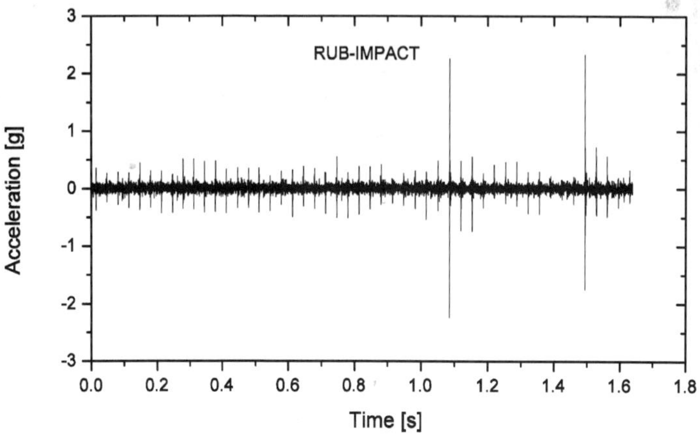

Fig. 14. Time-domain vibrations with rub-impact

Graphs showing the details of the impacts are shown in Fig. 15. The signal is characterized by a sudden excitation that generates a mechanical oscillation that grows to peak amplitude and decays as the impact energy dissipates. Each impact is characterized by only a few cycles that analyzed with the traditional FFT would not have enough energy to obtain a clear spectral definition.

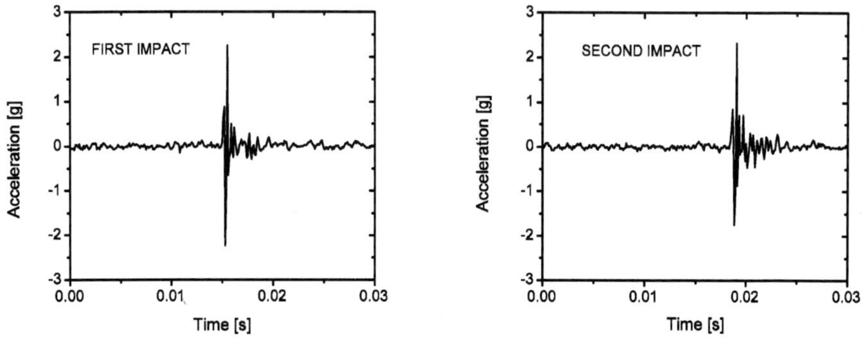

Fig. 15. Details of the transient impacts analyzed

A wavelet decomposition of this signal was made and the main subbands are shown in Fig. 16. Subbands are associated with their corresponding frequency range according to the sampling rate established during the experiments. The graph shows the frequency content between 78 Hz and 5 kHz.

From this graph it is observed that the subband with the higher frequency content encompasses the information of the transient signals. The correlation technique enhances the spikes giving amplitude values higher than the coefficients where impacts are not present. This makes it easier to establish a discrimination criterion and an estimation of their values to determine the presence and severity of the rubbing for diagnostic purposes. Additionally, as the transient signals produced are in the first subband, only the first level of decomposition in the wavelet transformation is necessary reducing the computing time that it takes to make the analysis and optimizing the detection process.

The experimental results of the vibrations presented in the previous discussion were analyzed through one of the approaches that wavelets offer to the vibration analyst. This is a time-frequency representation of the data from which it can be extracted the information of interest to apply the necessary processes and criteria for the rubbing detection. This approach permits the characterization of the signal from which it can be obtained the necessary information for the implementation of the technique for the design of testing equipment with automatic detection and recognition of the rubbing phenomena.

Another type of representation of the information that wavelets offer are the time-frequency maps. These are contour plots where the wavelet coefficient values are plotted against the time and scale parameter, that is, translation and frequency. One axis represents time, the other axis frequency, and the amplitude of the vibrations is color-coded. The contour maps permit to visualize the whole picture of the frequencies present in the signal as well as their occurrence or location in time.

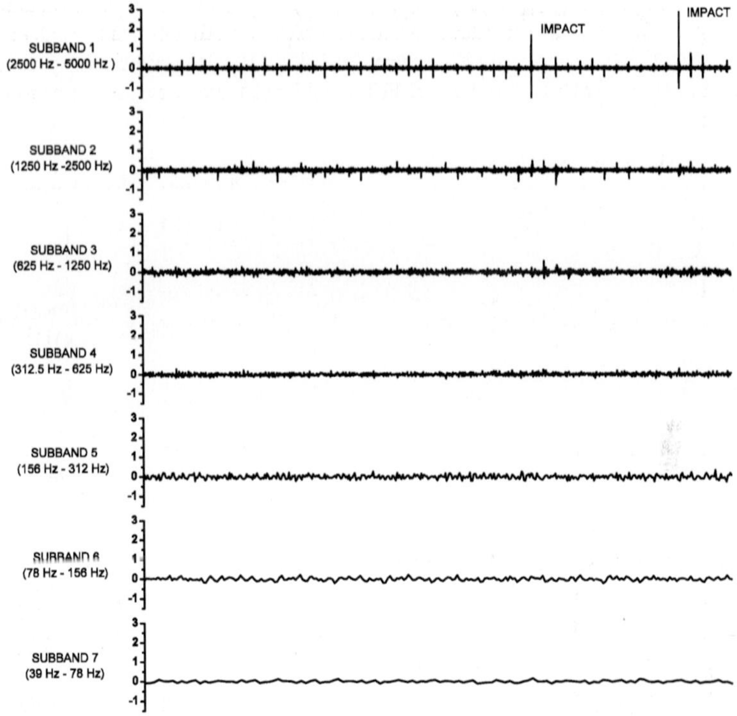

Fig. 16. Subband coding with wavelets of the vibrational response with impact-rubbing

The light rubbing data was processed with commercially available software by means of the Morlet continuous wavelet transform and results are shown in Fig. 17. In this graph, the color coding is red for low amplitude vibrations throughout blue for high amplitude vibrations. It can be observed the intermittent nature of the rubbing and, in concordance with Fig. 13, that the main vibrations are limited to frequencies below 1 250 Hz.

An analogous process was applied to severe rubbing data and results are shown in Fig. 18. The image shows that mid-range frequencies get stronger while high frequencies appear as a result of the increase in the vibrations amplitude as in the lower graph of Fig. 13. Upper spots in the time-frequency map (rub-impacts) appear elongated and lower spots stretched due to the compromise between the time and frequency resolution of the technique as stated in the introduction.

Additionally, an experimental run was carried out varying continuously the rotating conditions to obtain a sweep from a low to a high velocity and then decreasing the velocity until a minimum value. Results are shown in Fig. 19.

It can be seen that as time runs throughout the experiment, velocity increases and higher frequency components appear. These components get their peak value near the middle of the time axis where the maximum velocity is reached, and then begin to fade showing the trend of the higher frequencies to disappear as velocity decreases. This graph confirms the wide spectral bandwidth of the rubbing phenomena.

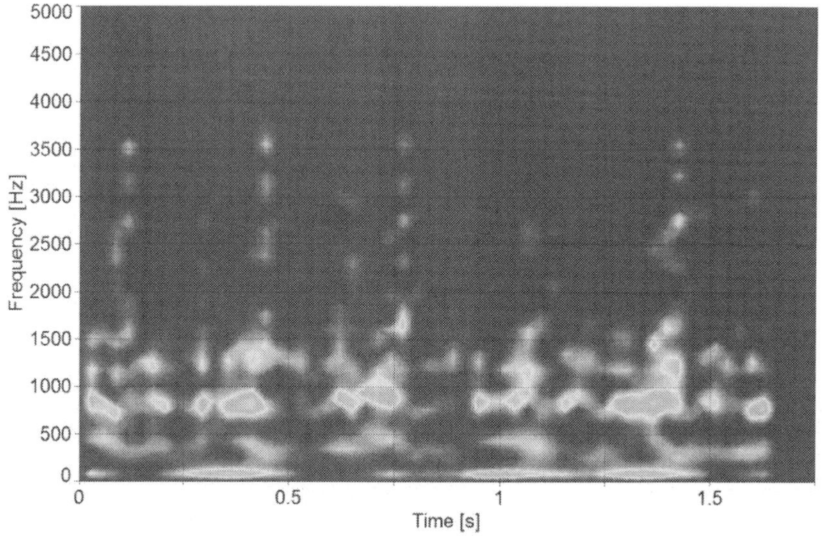

Fig. 17. Time-Frequency map for light rubbing

It is important to notice the evolution of the frequencies as time passes by. There is an unsteady variation of frequencies, and in Fig. 19 it can be seen how they have an unsymmetrical pattern even with speed variations.

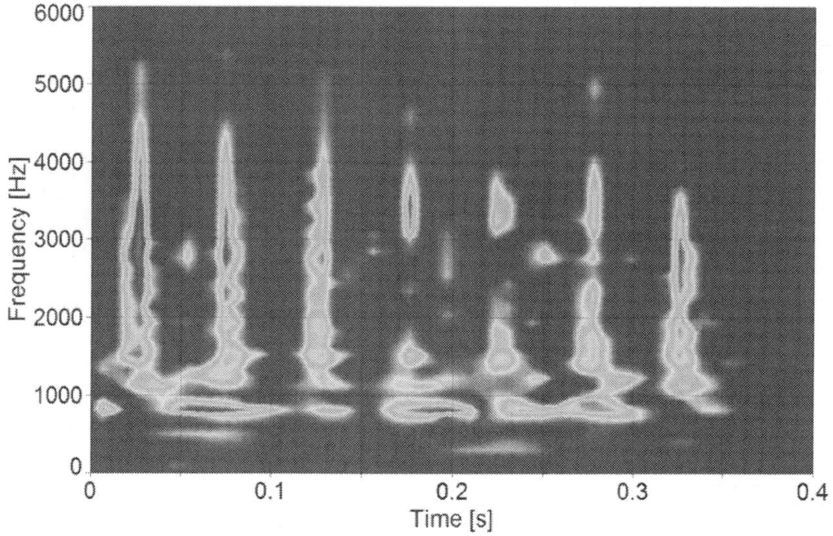

Fig. 18. Time-frequency map for severe rubbing

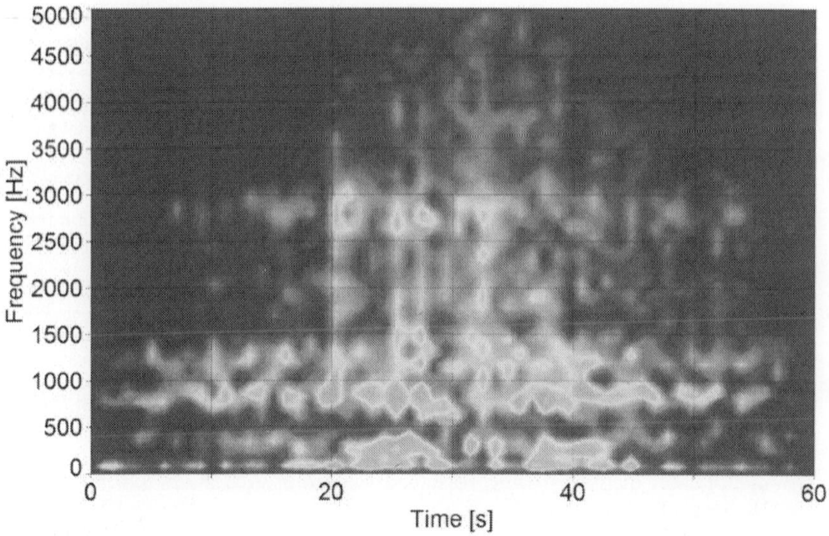

Fig. 19. Time-frequency map for a run-up and run-down rotor velocity sweep

6. Conclusions

Rotor rubbing has been analyzed with a methodology that processes the vibrations signal in such a way that time and scale information is preserved. It was demonstrated that with this approach vibrations of transient nature can be studied through a controlled subband coding scheme and time-frequency spectra. The technique revealed additional information that traditional processing techniques cannot, such as FFT. Experimental results showed that light rubbing presents a vibrational response characterized by a rich frequency content spectrum, and that severe rubbing is more adequately described as an impact-like transient behavior. Accordingly, impacts could be identified and localized with wavelets in the upper-frequency subbands which resulted after the algorithm was applied.

Since rotor-to-stator contact changes the effective stiffness of the coupled elements, a frequency shift was identified that shows the nonlinear response of the system. Time-frequency maps evidenced again the wide spectral response and differences between light and severe rotor rubbing, and location in time of the rub-impacts. The processing algorithm can be implemented with a high computational efficiency for on-line inspection systems for continuous machinery condition monitoring.

7. References

Abuzaid, M.A. ; Eleshaky, M.E. & Zedan, M.G. (2009). Effect of partial rotor-to-stator rub on shaft vibration. *Journal of Mechanical Science and Technology*, Vol. 23, No. 1, 170-182.

Adams, M.L. (2010). *Rotating Machinery Vibration*, 84, CRC Press, Taylor & Francis Group, ISBN 978-1-4398-0717-0, Boca Raton, FL, USA.

Castejón, B; Lara, O. & García-Prada, J.C. (2010). Automated diagnosis of rolling bearings using MRA and neural networks. *Mechanical Systems and Signal Processing,* Vol. 24, No. 1, 289-299.

Choi, Y.-S. (2001). On the contact of partial rotor rub with experimental observations. *KSME International Journal,* Vol. 15, No. 12, 1630-1638.

Choi, Y.-S. (2004). Nonlinear parameter identification of partial rotor rub based on experiment. *KSME International Journal,* Vol. 18, No. 11, 1969-1977.

Jeffcott, H.H. (1919). The lateral vibration of loaded shafts in the neighborhood of a whirling speed-the effects of want of balance. *Philosophical Magazine, Series 6,* Vol. 37, 304-314.

Karpenko, E.V. ; Wiercigroch, M. ; Pavlovskaia, E.E. & Cartmell, M.P. (2002). Piecewise approximate analytical solutios for a Jeffcott rotor with a snubber ring. *International Journal of Machine Sciences,* Vol. 44, No. 3, 475-488.

Muszynska, A. (2005). *Rotordynamics,* 646-648, CRC Press, Taylor & Francis Group, ISBN 978-0-8247-2399-6, Boca Raton, FL, USA.

Peng, Z.K. ; Chu, F.L. & Tse, P.W. (2005). Detection of rubbing-caused impacts for rotor-stator fault diagnosis using reasigned scalogram. *Mechanical Systems and Signal Processing,* Vol. 19, No. 2, 391-409.

Xinsheng, L. & Kenneth, A.L. (2004). Bearing fault diagnosis based on wavelet transform and fuzzy inference. *Mechanical Systems and Signal Processing,* Vol. 18, No. 5, 1077-1095.

Wang, F. ; Xu, Q. & Li, S. (2004). Vibration analysis of cracked rotor sliding bearing systems with rotor-stator rubbing by harmonic wavelet transform. *Journal of Sound and Vibration,* Vol. 271, No. 3-5, 507-518.

ADVANCES IN VIBRATION ENGINEERING AND STRUCTURAL DYNAMICS

Edited by **Francisco Beltran-Carbajal**

Rotors on Active Magnetic Bearings: Modeling and Control Techniques

Andrea Tonoli, Angelo Bonfitto, Mario Silvagni and
Lester D. Suarez

Additional information is available at the end of the chapter

http://dx.doi.org/10.5772/51298

1. Introduction

In the last decades the deeper and more detailed understanding of rotating machinery dynamic behavior facilitated the study and the design of several devices aiming at friction reduction, vibration damping and control, rotational speed increase and mechanical design optimization. Among these devices a promising technology is represented by magnetic actuators used as bearings which found a great spread in rotordynamics and in high precision applications. A first classification of magnetic bearings according to the physical working principle allows to pick out two main families: a) Active Magnetic Bearings [1], [2], making use of an electronic control unit to regulate the current flowing in the coils of the actuators. They need external source of energy. b) Passive Magnetic Bearings [3], [4], [5]: they do not need any electronic equipment. The control of the mechanical structure is achieved without the introduction of any external energy source. They exploit the reluctance force or the Lorentz force due to the generation of eddy currents developed in a conductor in a relative motion in a magnetic field.Active Magnetic Bearings require sensors an electronic equipment but, although more expensive respect to classical ball bearings, they offer several technological advantages:

- The absence of all fatigue and tribology issues due to contact: it allows the use of such bearings in vacuum systems, in clean and sterile rooms, or for the transport of aggressive or very pure media, and at high temperatures;
- No lubrication needed;
- No contamination by the dust created by friction between the rotor and the stator;
- Low bearing losses: at high operating speeds are 5 to 20 times less than in conventional ball or journal bearings, result in lower operating costs;

- Viscous friction can be avoided if the rotor is confined in high vacuum;
- Low vibration level;
- Dynamics adaptable to the desired application by tuning of the control loop;
- Precise positioning of the rotor due to the control loop: this is mainly determined by the quality of the measurement signal within the control loop. Conventional inductive sensors, for example, have a measurement resolution of about 1 ÷ 1000μm of a millimeter;
- Achievable fast positioning and/or high rotational speed of the rotor;
- The small sensitivity to the operating conditions;
- The predictability of the behavior.
- Further statements about the technology of realization can be done:
- The gap between rotor and bearing amounts typically to a few tenths of a millimeter, but for specific applications it can be as large as 20 mm. In that case, of course, the bearing becomes much larger;
- The rotor can be allowed to rotate at high speeds. The high circumferential speed in the bearing, only limited by the strength of material of the rotor, offers the possibilities of designing new machines with higher power density and of realizing novel constructions. Actually, about 350 m/s are achievable, for example by using amorphous metals which can sustain high stresses and at the same time have very good soft-magnetic properties, or by binding the rotor laminations with carbon fibers. Design advantages result from the absence of lubrication seals and from the possibility of having a higher shaft diameter at the bearing site. This makes the shaft stiffer and less sensitive to vibrations;
- The specific load capacity of the bearing depends on the type of ferromagnetic material and the design of the bearing electromagnet. It will be about 20 N/cm2 and can be as high as 40 N/cm2. The reference area is the cross sectional area of the bearing. Thus the maximum bearing load is mainly a function of the bearing size;
- The bearing and the rotor can be integrated on the same shaft by realizing bearingless configurations which allow to reduce the size of the system and to perform a cost saving solution.
- Retainer bearings are additional ball or journal bearings, which in normal operation are not in contact with the rotor. In case of overload or malfunction of the AMB they have to operate for a very short time: they keep the spinning rotor from touching the housing until the rotor comes to rest or until the AMB regains control of the rotor. The design of such retainer bearings depends on the specific application and despite a variety of good solutions still needs special attention;
- The unbalance compensation and the force-free rotation are control features where the vibrations due to residual unbalance are measured and identified by the AMB. The signal is used to either generate counteracting and compensating bearing forces or to shift the rotor axis in such a way that the rotor is rotating force-free;

- Diagnostics are readily performed, as the states of the rotor are measured for the operation of the AMB anyway, and this information can be used to check operating conditions and performance. Even active diagnostics are feasible, by using the AMB as actuators for generating well defined test signals simultaneously with their bearing function;
- The lower maintenance costs and higher life time of an AMB have been demonstrated under severe conditions. Essentially, they are due to the lack of mechanical wear. Currently, this is the main reason for the increasing number of applications in turbomachinery;
- The cost structure of an AMB is that of a typical mechatronics product. The costs for developing a prototype, mainly because of the demanding software, can be rather high. On the other side, a series production will lower the costs considerably because of the portability of that software.

Active Magnetic Bearings can be classified as a typical mechatronic product due to its nature which involves mechanical, electrical and control aspects, merging them in a single system. Rotordynamic field offer several examples of application areas [1], [6] : (a) Turbomachinery, (b) Vibration isolation, (c) Machine tools and electric drives, (d) Energy storing flywheels, (e) Instruments in space and physics, (f) Non-contacting suspensions for micro-techniques, (g) Identification and test equipment in rotordynamics, (f) Microapplications such as gyroscopic sensors [7], [8].The attractive potential of active magnetic suspensions motivated a considerable research effort for the past decade focused mostly on electrical actuation subsystem and control strategies [3], [9], [10], [11], [12], [13], [14].

This chapter illustrates the design, the modeling, the experimental tests and validation of all subsystems of a rotor on a five-axes active magnetic suspension. The mechanical, electrical, electronic and control strategies aspects are explained with a mechatronic approach evaluating all the interactions between them. The main goals of the manuscript are: a) Illustrate the design and the modeling phases of a five-axes active magnetic suspension; b) Discuss the design steps and the practical implementation of a standard suspension control strategy; c) Introduce an off-line technique of electrical centering of the actuators. The experimental test rig is a shaft (Weight: 5.3 kg. Length: 0.5 m) supported by two radial and one axial cylindrical active magnetic bearings and powered by an asynchronous high frequency electric motor.The chapter starts on an overview of the most common technologies used to support rotors with a deep analysis of their advantages and drawbacks with respect to active magnetic bearings. Furthermore a discussion on magnetic suspensions state of the art is carried out highlighting the research efforts directions and the goals reached in the last years.In the central sections, a detailed description of each subsystem is performed along with the modeling steps. In particular the rotor is modeled with a FE code while the actuators are considered in a linearized model.

The last sections of the chapter are focused on the control strategies design and the experimental tests.An off-line technique of actuators electrical centering is explained and its advantages are described in the control design context. This strategy can be summarized as follows. Knowing that: a) each actuation axis is composed by two electromagnets; b) each electromagnet needs a current closed-loop control; c) the bandwidth of this control is depending on the mechanical Airgap,then the technique allows obtaining the same value of the closed-loop bandwidth of the current control of both the electromagnets on the same actuation axis. This approach improves performance and gives more steadiness to the control behavior.The decentralized approach of the control strategy allowing the full suspensions on five axes is illustrated from the design steps to the practical implementation on the control unit.Finally, the experimental tests are carried out on the rotor to validate the suspension control and the off-line electrical centering. The numerical and experimental results are superimposed and compared to prove the effectiveness of the modeling approach.

2. System Architecture

The rig used for the modeling, the design and the experimental tests is an electrical spindle (picture reported in Figure 1) consisting of a shaft supported by two radial and one axial active magnetic bearings with cylindrical geometry and powered by an asynchronous high frequency electric motor. Two mechanical ball bearings, with radial and axial airgaps equal to half of the levitation ones, are positioned at the ends of the shaft to guarantee a safely touch-down of the shaft for anomalous working conditions with excessive whirling amplitude. The rotation axis is horizontal and the weight has the direction of each bearing.

Figure 1. Picture of the rotor.

Parameter	Symbol	Value	Unit
Rotor mass	m	5.31	kg
Rotor transversal Inertia	$J_x=J_y$	1.153•10-1	Kg/m2
Rotor polar Inertia	J_z	1.826•10-3	Kg/m2
Bearing rad. 1 location	a	214.5	mm A/V
Bearing rad. 2 location	b	212.6	mm A/V
Axial/Radial Airgap	g	0.75e-3	mm
Isotropic support stiffness	f / x	2.5•10-5	N/m

Table 1. Rotor mechanical and geometric parameters.

Table 1 reports the main parameters of the rotor. Figure 2 illustrates the section view of the system showing the layout of sensors, actuators, motor and rotor.

Active Magnetic Bearings applied to rotating machines can be considered as a typical mechatronic application, since it involves the control of mechanical system (the rotor) by means of an electronic control unit which elaborates the commands to feed electrical power drivers regulating the electromechanical actuators. The information to perform closed loop control architecture is given by displacement and current sensors.

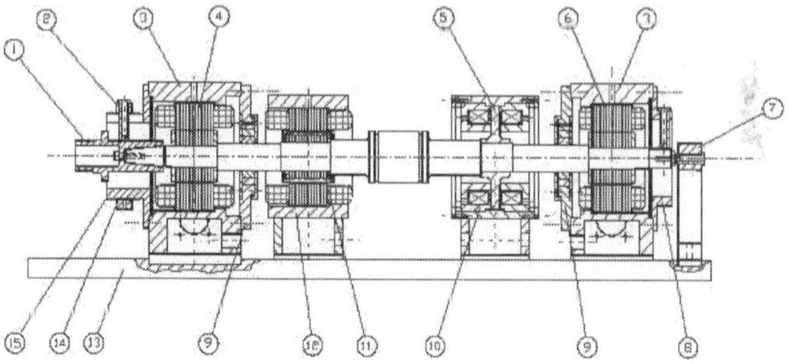

Figure 2. System section view: 1) tang, 2) radial sensor, 3) radial AMB support, 4) radial AMB laminated stator, 5) axial AMB disc, 6) radial AMB laminated stator, 7) axial sensor, 8) sensor cap, 9) bushing cap, 10) axial AMB electromagnet, 11) electric motor, 12) motor support, 13) foundation, 14) threaded ring, 15) sensor cap.

The interactions and the main functions of these subsystems are highlighted in Figure 3.The reported scheme is a standard representation of the system. However each block can be of different nature depending on the application. A short summary of the technologies typically used for each subsystem and the technologies used in the rig described in this chapter are reported in the following sections.

2.1. Control

Two main families of control architecture can be listed for active magnetic bearings:

- Decentralized SISO control: the action of each actuator is independent from the others and exploits a dedicated control law and sensors information;

- Centralized MIMO control: actuators are coupled as well as sensors information. A single control action is devoted to feed power drivers.

Several control strategies have been implemented and tested on rotors equipped with active magnetic bearings:

- Gain scheduled control [15];

- Adaptive control [16], [17];

- Robust H∞ control [18];
- Robust sliding mode control [19];
- Robust control via eigenstructure assignment dynamical compensation [20];
- Optimal control [21];
- Dynamic programming control [22];
- Genetic algorithm control [23];
- Fuzzy logic control [24];
- Feedback linearization control [25];
- Time-delay control [26];
- Control by transfer function approach [27];
- μ-synthesis control [28].

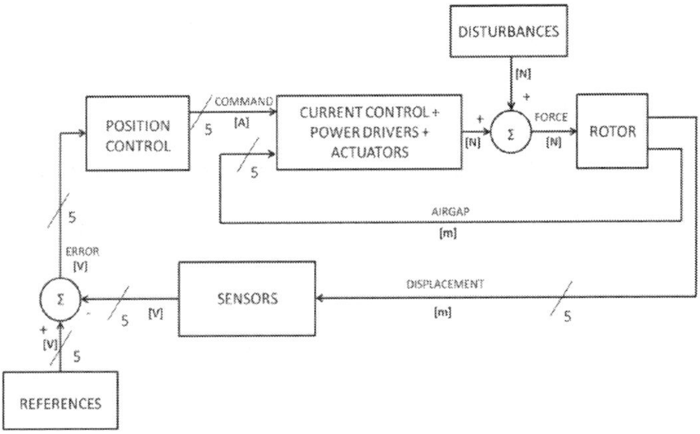

Figure 3. Overall system architecture.

In this chapter a decentralized PID strategy is implemented on a control module equipped with a DSP/FPGA–based digital control unit (EKU2.1). This digital platform allows the rapid reconfigurations of the overall system throw up to 108 (from FPGA) and 46 (from DSP) configurable digital I/O lines for input/output, event, PWM, capture/generation and user functions.Both DSP and FPGA have a dedicated Hard Real-Time Operating System (HRTOS) based on a non-pre-emptive scheduler (DSP side), involving ISR time or event triggered. EKU2.1 uses a single-master (DSP) multi-slave (FPGA) point-to-point communication protocol, based on Wishbone format; a system bus manages data exchange between the two cores. Software code is developed using the target-dependent tools Texas Instruments® Code Composer Studio.

2.2. Power drivers

The electronic circuits of power amplification stage to convert low power controller output signal to a high power stator input signal is chosen according to the kind of application. Basically, three main families of electronic circuits can be identified:

1. Linear analogue amplifiers have push–pull transistors at the output stage. They allow to enhance the current capability and to integrate a high-gain linear amplifier, such as a power operational amplifier. The linear amplifiers have the advantage of precise current and voltage regulation as well as low noise and they have a current rating of less than 10 A. Operation at the rated current is available only with effective cooling with heat sinks. Therefore the amplifier dimensions are large, resulting in high cost. The efficiency is low because of high losses in the push–pull transistor.

2. Switched-mode amplifiers enhance efficiency. Since the losses in the power devices are reduced, the heat sinks are much smaller and, as a result, switched mode amplifiers are compact in dimension so that the cost is low. Switched-mode operation of power devices is widely used in industry, e.g., for general-purpose inverters in ac drives and computer power supplies. This category of amplifiers is dominant in magnetic bearing drivers.

Hybrid amplifiers take advantage of linear and switched-mode amplifiers. At low current the push–pull transistors operate as a linear amplifier but at high current they operate in switched mode. To take advantage of a hybrid amplifier, it is quite important to modify the winding structure in magnetic bearings.

The rig object of study in this chapter is equipped with an H-bridge switching amplifier for each actuator.The power stage consists of an Embedded Isolated Power Module Board with four fully independent MOSFET/IGBT legs that supports up to 25 amperes with 100 volt of DC Bus. Also a maximum PWM switching frequency is 80kHz making this module suitable for high performance driving applications where control loop bandwidth and current ripple are important factors.The scheme used to feed power drivers is reported in Figure 4.

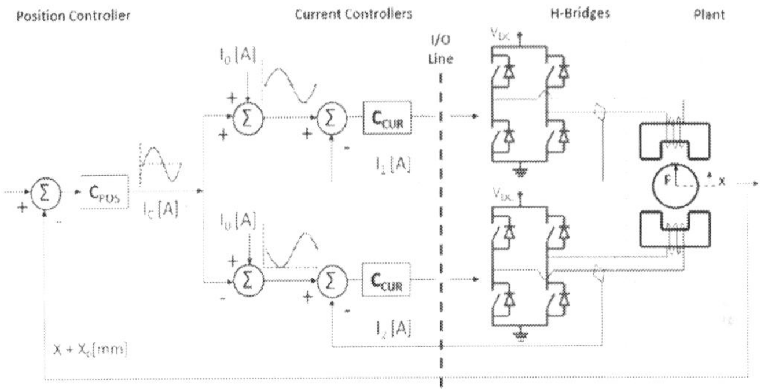

Figure 4. Power driver scheme.

Standard AMBs equipment for rotors suspensions are realized with five couples of cylindrical shape electromagnets to perform five dof active control. Conical shape of magnetic bearings exerting forces both in axial and radial direction simultaneously allow to save one couple of electromagnets and hence to reduce the size although the bearing design results more complex than standard cylindrical solution. This geometry permits to reachhigher rotation speed, limited in cylindrical solution by the strains growing in axial bearing disc.

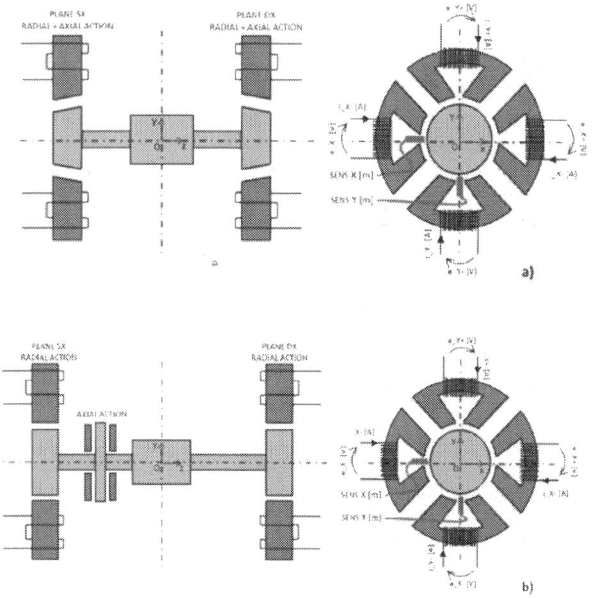

Figure 5. Geometry of actuation stage. a) Conical profile. b) Cylindrical profile.

2.3. Actuators

Actuators geometry and configuration depends on the electromagnets profile and on the number of actuators per actuation stage.

In this work classical configuration with four cylindrical actuators per actuation stage is dealt with. The disposition of the ten electromagnets is illustrated in Figure 6.

The main electrical and geometrical actuators parameters are listed in Table 2.

Parameter	Symbol	Value	Unit
Vacuum permeability	μ_0	1.26e-006	H/m
Voltage supply	V_{DC}	50	V

Parameter	Symbol	Value	Unit
	AXIAL Actuator		
Number of turns	N_{AX}	120	-
Circuitation length	l_{AX}	48e-3	m
Active section on airgap	S_{AX}	1210e-6	m^2
Nominal airgap	$g0_{AX}$	0.75e-3	m
Resistance	R_{AX}	0.5	Ω
Nominal inductance	$L\ 0_{AX}$	0.0146	H
	RADIAL actuator		
Number of turns	N_{RAD}	110	-
Circuitation length	l_{RAD}	135.2e-3	m
Active section on airgap	S_{RAD}	480e-6	m^2
Nominalairgap	$g0_{RAD}$	0.75e-3	M
Resistance	R_{RAD}	0.5	Ω
Nominal Inductance	$L\ 0_{RAD}$	0.0049	H

Table 2. Actuators parameters.

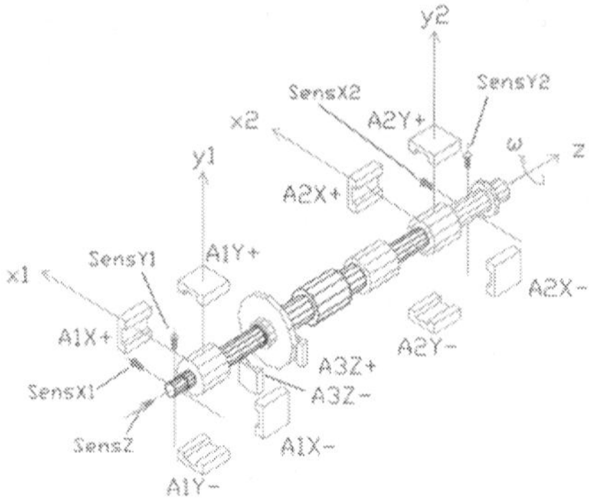

Figure 6. Actuators configuration.

2.4. Sensors

An important part of the performance of a magnetic bearing depends on the characteristics of the displacement sensors used. In order to measure the position of a moving rotor, contact-free sensors must be used which, moreover, must be able to measure on a rotating surface. Consequently, the geometry of the rotor, i.e. its surface quality, and the homogeneity of the material at the sensor will also influence the measuring results. A bad surface will thus produce noise disturbances, and geometry errors may cause disturbances with the rotational frequency or with multiples thereof.

In addition, depending on the application, speeds, currents, flux densities and temperatures are to be measured in magnetic bearing systems.

When selecting the displacement sensors, depending on the application of the magnetic bearing, measuring range, linearity, sensitivity, resolution, and frequency range are to be taken into account as well as:

- Temperature range, temperature drift of the zero point and sensitivity;
- Noise immunity against other sensors, magnetic alternating fields of the electromagnets, electromagnetic disturbances from switched amplifiers;
- Environmental factors such as dust, aggressive media, vacuum, or radiation;
- Mechanical factors such as shock and vibration;
- Electrical factors such as grounding issues associated with capacitive sensors.

The most important displacement sensors technologies are:

- Inductive sensors;
- Eddy-current sensors;
- Eddy Current Radial Displacement Sensor on a PCB (Transverse Flux Sensor)
- Capacitive sensors;
- Magnetic sensors.

The rig described in this chapter is equipped with five eddy current displacement sensors: high-frequency alternating current runs through the air-coil embedded in a housing. The electromagnetic coil section induces eddy currents in the conductive object whose position is to be measured, thus absorbing energy from the oscillating circuit. Depending on the clearance, the inductance of the coil varies, and external electronic circuitry converts this variation into an output signal. The usual modulation frequencies lie in a range of 1 - 2 MHz, resulting in useful measuring frequency ranges of 0 Hz up to approximately 20 kHz.

3. Modeling

The dynamic behavior of the rotor can be described by means of different models. This section describes the modeling techniques and the adopted assumptions, how acting forces and displacements are ordered and selected, the equation of motion used for the dynamic description and the variable used to describe the models.

Many modeling techniques can be adopted; an analytical rigid approach (based on the 4 d.o.f. rotor modeling) is here presented beside the most common Finite Element (FE) approach. The discretization for FEM software is reported in Figure 7.b.

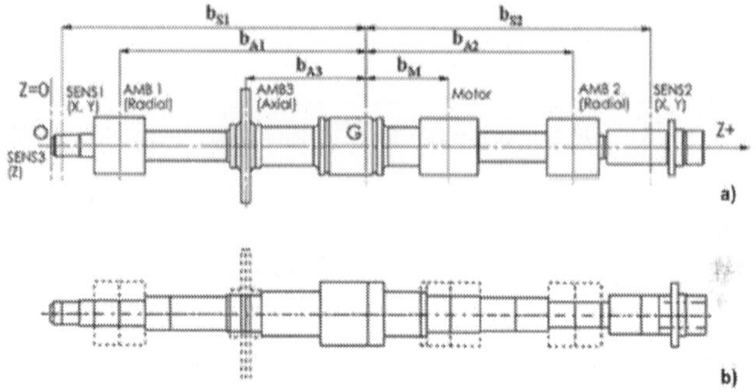

Figure 7. Rotor section view. a) View with dimensions. b) Discretization for FEM modeling.

The main hypothesis here adopted is to consider a constant spin speed. In this case the rotor behavior on the X-Y plane (known as flexural) is not coupled with the behavior on the Z-direction (axial). Other important assumption is that any rotation (except for spinning rotation) should be small.

3.1. Model Block Diagram

A simple description of the rotor model block diagram is presented in Figure 8. Forces (due to AMBs, motor and external) acting on the rotor are un-grouped for the X-Y and for the Z behavior, these signals are fed to the block describing the dynamic behaviors. Outputs of these blocks are the states (displacements and velocities) of the systems, reported as displacements and speeds to sensors, AMBs and motor. The constant spin speed Ω is used in the X-Y model for the gyroscopic behavior and reported as output.Spin speed and displacement on the sensors are physical entities measured by specific sensors, and signal are reported to the Sensor block; the other displacements and relative velocities (to AMBs and motor) should be used for intrinsic feedback such as back electromotive force in the motor or magnetic bearings.

3.1.1. Model inputs / outputs

Model inputs are the forces acting on the rotor, while outputs are typically displacements either on sensors or on AMBs and motor (Figure 9). The rotor is suspended by two radial magnetic bearing (AMB1 and AMB2) which generate four forces oriented as the reference plant reference frame and acting in the center of the relative AMB; these forces act the behavior on X-Y plane. A further magnetic bearing (AMB3) is used to constrain displacements along Z axis (axial). The five forces due to AMBs are collected in vector fAMB. The electric motor, used to generate rotation torque (not included in this kind of model where the spin speed is assumed to be constant), can also act two forces, in radial direction, while is not capable to generate a force in the Z direction. According to this, vector fMOT has two components acting in the center of the motor.

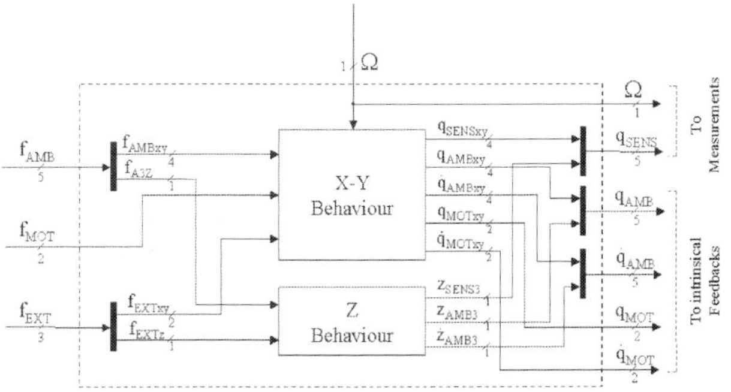

Figure 8. Rotor model block diagram.

In order to simplify the description of the system the generic external force are supposed to act directly to the center of mass of the rotor; these three forces are oriented as the plant reference frame. These components are the resultant of any external force, such as impact forces. While the X-Y behaviour is uncoupled from the Z behavior, acting forces due to AMBs and external can be rewritten dividing the forces acting in X-Y plane from forces acting on Z axis. Table 3 reports acting forces (inputs) on the rotor.

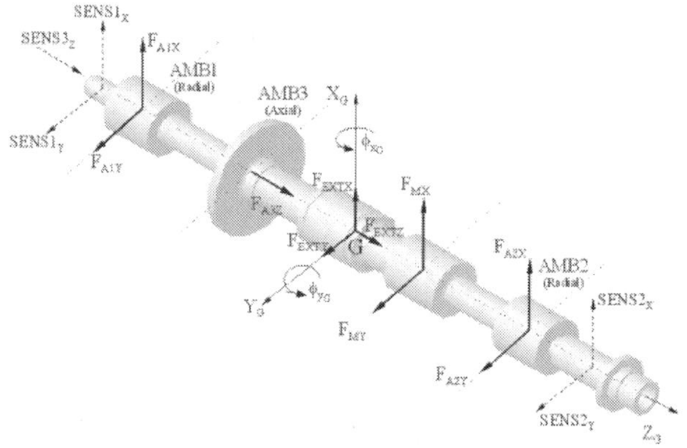

Figure 9. Actuation forces and sensors position.

AMB	Motor	External	
$\mathbf{f}_{AMB} = \begin{Bmatrix} F_{A1X} \\ F_{A1Y} \\ F_{A2X} \\ F_{A2Y} \\ F_{A3Z} \end{Bmatrix} = \begin{Bmatrix} \mathbf{f}_{AMBxy} \\ F_{A3Z} \end{Bmatrix}$	$\mathbf{f}_{MOT} = \begin{Bmatrix} F_{MX} \\ F_{MY} \end{Bmatrix}$	$\mathbf{f}_{EXT} = \begin{Bmatrix} F_{EXTx} \\ F_{EXTy} \\ F_{EXTz} \end{Bmatrix} = \begin{Bmatrix} \mathbf{f}_{EXTxy} \\ F_{EXTz} \end{Bmatrix}$	(1)

Table 3. Rotor Inputs.

Referring to Figure 8, a set of outputs is used for measurements (the spin speed Ω and the displacements on the sensor qSENS) and another set is used for intrinsical feedback (displacements and velocities on AMBs qAMB and $\dot{q}$ AMB, and on the motor qMOT and $\dot{q}$ MOT).

Table 4 reports displacements and velocities (outputs) on the rotor.

Sensors	AMB	Motor	
$\mathbf{q}_{SENS} = \begin{Bmatrix} x_{SENS1} \\ y_{SENS1} \\ x_{SENS2} \\ y_{SENS2} \\ z_{SENS3} \end{Bmatrix} = \begin{Bmatrix} \mathbf{q}_{SENSxy} \\ z_{SENS3} \end{Bmatrix}$	$\mathbf{q}_{AMB} = \begin{Bmatrix} x_{AMB1} \\ y_{AMB1} \\ x_{AMB2} \\ y_{AMB2} \\ z_{AMB3} \end{Bmatrix} = \begin{Bmatrix} \mathbf{q}_{AMBxy} \\ z_{AMB3} \end{Bmatrix}$	$\mathbf{q}_{MOT} = \begin{Bmatrix} x_{MOT} \\ y_{MOT} \\ z_{MOT} \end{Bmatrix}$	(2)

Table 4. Rotor outputs.

Input		Output		
X-Y Behaviour	Z Behaviour	X-Y Behaviour	Z Behaviour	
$\mathbf{f}_{xy} = \begin{Bmatrix} F_{A1X} \\ F_{EXTx} \\ F_{MX} \\ F_{A2X} \\ F_{A1Y} \\ F_{EXTx} \\ F_{MY} \\ F_{A2Y} \end{Bmatrix}$	$\mathbf{f}_z = \begin{Bmatrix} F_{A3Z} \\ F_{EXTz} \end{Bmatrix}$	$\mathbf{y}_{xy} = \begin{Bmatrix} x_{SENS1} \\ x_{AMB1} \\ x_{MOT} \\ x_{AMB2} \\ x_{SENS2} \\ y_{SENS1} \\ y_{AMB1} \\ y_{MOT} \\ y_{AMB2} \\ y_{SENS2} \end{Bmatrix}$	$\mathbf{y}_z = \begin{Bmatrix} z_{SENS3} \\ z_{AMB3} \end{Bmatrix}$	(3)

Table 5. Inputs / Outputs vector orders.

Input/Output vector reported in Table 5. and ordered to be compliant with blocks that generate such forces. In order to address the modeling technique (especially FE based), input/output vector should be reordered in the way reported in Table 5.

The dynamic behavior of the rotor can be described by mean of the equations of motion (EoM). In the following section the equations used for the model is described. The typical equations are reported for a generic rotor model and then applied to a rigid analytical model and to the FE model.While the spin speed is constant the X-Y behavior is uncoupled from Z behavior in the same way the equations can be separated.

X-Y Behavior

Equation of Motion:

$$\begin{aligned}&\mathbf{M}_{xy}\ddot{\mathbf{q}}_{xy}(t)+\left(\mathbf{L}_{xy}+\Omega\mathbf{G}_{xy}\right)\dot{\mathbf{q}}_{xy}(t)+\left(\mathbf{K}_{\Omega 0xy}+\Omega^2\mathbf{K}_{\Omega 2xy}+\Omega\mathbf{H}_{xy}\right)\mathbf{q}_{xy}(t)=\\&=\mathbf{f}_{sxy}+\Omega^2\mathbf{f}_{umb}\begin{Bmatrix}\sin(\Omega t)\\ \cos(\Omega t)\end{Bmatrix}+\mathbf{S}_{ixy}\mathbf{f}_{xy}(t)\end{aligned} \tag{1}$$

Measure Equation:

$$\begin{Bmatrix}\mathbf{y}_{xy}(t)\\ \dot{\mathbf{y}}_{xy}(t)\end{Bmatrix}=\mathbf{S}_{oxy}\begin{Bmatrix}\mathbf{q}_{xy}(t)\\ \dot{\mathbf{q}}_{xy}(t)\end{Bmatrix} \tag{2}$$

Z Behavior

Equation of Motion:

$$\mathbf{M}_z\ddot{\mathbf{q}}_z(t)+\mathbf{L}_z\dot{\mathbf{q}}_z(t)+\left(\mathbf{K}_{\Omega 0z}+\Omega^2\mathbf{K}_{\Omega 2z}\right)\mathbf{q}_z(t)=\mathbf{f}_{sz}+\mathbf{S}_{iz}\mathbf{f}_z(t) \tag{3}$$

Measure Equation:

$$\begin{Bmatrix}\mathbf{y}_z(t)\\ \dot{\mathbf{y}}_z(t)\end{Bmatrix}=\mathbf{S}_{oz}\begin{Bmatrix}\mathbf{q}_z(t)\\ \dot{\mathbf{q}}_z(t)\end{Bmatrix} \tag{4}$$

Refer to Table 6 for matrices naming and description.

Name	Description
Ω	Spin Speed
$\boldsymbol{q}(t)$	Generalized displacements

Name	Description
$\boldsymbol{M}$	Mass (symmetric) matrix
$\boldsymbol{L}$	Damping (symmetric) matrix
$\boldsymbol{G}$	Gyroscopic (skew-symmetric) matrix
$\boldsymbol{K}_{\Omega 0}$	Stiffness (symmetric) matrix: spin speed independent
$\boldsymbol{K}_{\Omega 2}$	Stiffness (symmetric) matrix: spin speed dependent
$\boldsymbol{H}$	Circulatory (skew-symmetric) matrix
$\boldsymbol{f}_s$	Static forces
$\boldsymbol{R}$	Rotation Matrix
$\boldsymbol{f}_{umb}$	Unbalance forces
$\boldsymbol{f}(t)$	External forces
$\boldsymbol{S}_i$	Input selection matrix
$\boldsymbol{y}(t)$	Output displacements
$\boldsymbol{S}_o$	Output selection matrix
Φ	Selected eigenvector for modal (MK) reduction

Table 6. Matrices names and description.

3.2. 4dof model

Generic Eom for rotors previously described, can be applied to a rigid analytical model based on the 4 d.o.f theory (for X-Y behavior), with an additional d.o.f. for the Z behavior. In this model the equation of motion are develop in a center of mass coordinate system as reported in Figure 10.

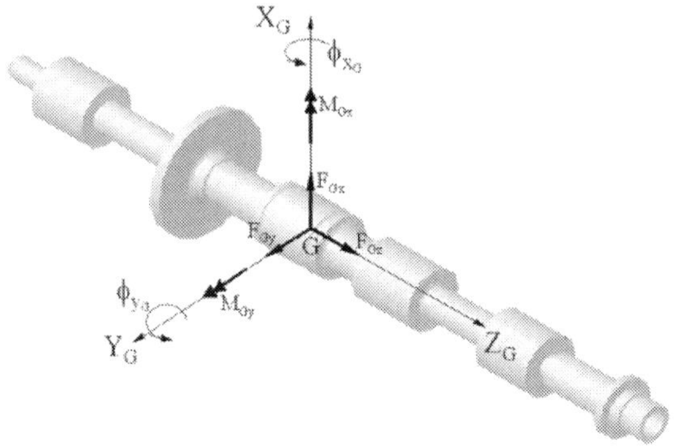

Figure 10. d.o.f. model with generalized displacements and forces.

The physical properties used in the model are:

Name	Description
m	Mass of the rotor [kg]
J_t	Transversal moment of inertia about any axis in the rotation plane [kgm2]
J_p	Transversal moment of inertia about any axis in the rotation plane [kgm2]
g_x, g_y	Gravity along x and y direction [m/s2]
ε_x, ε_y	Static unbalance eccentricity along x and y direction [m]
χ	Torque unbalance: angular error [rad]
b_{A1}	AMB1 distance from center of mass [m]
b_{A2}	AMB2 distance from center of mass [m]
b_M	Motor distance from center of mass [m]
b_{S1}	Sensor1 distance from center of mass [m]
b_{S2}	Sensor2 distance from center of mass [m]

Table 7. Physical properties of rigid analytical model.

X-Y Behavior

The generalized displacements vector is composed by the center of mass coordinates:

$$\mathbf{q}_{xy}(t) = \begin{Bmatrix} x_G \\ y_G \\ \varphi_{x_G} \\ \varphi_{y_G} \end{Bmatrix} \tag{5}$$

If the model has no damping the EoM (4) becomes:

$$\begin{bmatrix} m & 0 & 0 & 0 \\ 0 & m & 0 & 0 \\ 0 & 0 & J_t & 0 \\ 0 & 0 & 0 & J_t \end{bmatrix} \ddot{\mathbf{q}}_{xy}(t) + \Omega \begin{bmatrix} 0 & 0 & 0 & 0 \\ 0 & 0 & 0 & 0 \\ 0 & 0 & 0 & J_p \\ 0 & 0 & -J_p & 0 \end{bmatrix} \dot{\mathbf{q}}_{xy}(t) = \\ = \begin{Bmatrix} mg_x \\ mg_y \\ 0 \\ 0 \end{Bmatrix} + \Omega^2 \begin{bmatrix} -m\varepsilon_y & m\varepsilon_x \\ m\varepsilon_x & m\varepsilon_y \\ -\chi\left(J_t - J_p\right) & 0 \\ 0 & \chi\left(J_t - J_p\right) \end{bmatrix} \begin{Bmatrix} \sin(\Omega t) \\ \cos(\Omega t) \end{Bmatrix} + \mathbf{S}_{ixy} \mathbf{f}_{xy}(t) \tag{6}$$

The measure equation has the same structure reported in (5).

Z Behavior

Under the same assumptions equation (6) becomes:

$$m\ddot{z}(t) = \mathbf{f}_{sz} + \mathbf{S}_{iz}\mathbf{f}_z(t) \tag{7}$$

The measure equation has the same structure reported in (7).

In order to be compliant to input/output vector described in Table 8 the selection matrices of equation (9) and (10) with their relative measure equation should be:

X-Y Behaviour	Z Behaviour
$\mathbf{S}_{ixy} = \begin{bmatrix} 1 & 1 & 1 & 1 & 0 & 0 & 0 & 0 \\ 0 & 0 & 0 & 0 & 1 & 1 & 1 & 1 \\ 0 & 0 & 0 & 0 & b_{A1} & 0 & -b_M & -b_{A2} \\ -b_{A1} & 0 & b_M & b_{A2} & 0 & 0 & 0 & 0 \end{bmatrix}_{4x8}$	$\mathbf{S}_{iz} = \begin{bmatrix} 1 & 1 \end{bmatrix}_{1x2}$ (11)
$\mathbf{S}_{oxy} = \begin{bmatrix} \mathbf{S}'_{oxy} & \mathbf{0} \\ \mathbf{0} & \mathbf{S}'_{oxy} \end{bmatrix}_{20x8}, \mathbf{S}'_{oxy} = \begin{bmatrix} 1 & 0 & 0 & -b_{S1} \\ 1 & 0 & 0 & -b_{A1} \\ 1 & 0 & 0 & b_M \\ 1 & 0 & 0 & b_{A2} \\ 1 & 0 & 0 & b_{S2} \\ 0 & 1 & b_{S1} & 0 \\ 0 & 1 & b_{A1} & 0 \\ 0 & 1 & -b_M & 0 \\ 0 & 1 & -b_{A2} & 0 \\ 0 & 1 & -b_{S2} & 0 \end{bmatrix}$	$\mathbf{S}_{oz} = \begin{bmatrix} 1 & 0 \\ 1 & 0 \\ 0 & 1 \\ 0 & 1 \end{bmatrix}_{4x2}$

Table 8. Input/Output selection matrices for rigid analytical model.

3.3. Flexible Rotor Model (FE)

A flexible model is here described. This model is generated by using a finite element code especially designed for rotating machines (Dynrot). The outputs of this code are the matrices reported in equation (4) to (7).

3.3.1. *Full Model*

In the case all the nodes displacements are used the full dynamic behavior is described. The displacements vector q collects the translation displacement of the model nodes in the following order:

3.3.1. Reduced Model

Usually a reduced model is used. A typical reduction method is the modal (MK) reduction where only some modes of vibration are selected.

The displacements to modal transformation are reported in Table 10:

X-Y Behaviour	Z Behaviour	
$\mathbf{q}_{xy}(t)=\begin{Bmatrix} x_1 \\ x_2 \\ \ldots \\ x_n \\ y_1 \\ y_2 \\ \ldots \\ y_n \end{Bmatrix}$	$\mathbf{q}_z(t)=\begin{Bmatrix} z_1 \\ z_2 \\ \ldots \\ z_n \end{Bmatrix}$	(12)

Table 9. Generalized coordinates for full FE model.

X-Y Behaviour	Z Behaviour	
$\boldsymbol{\xi}_{xy}(t)=\Phi_{xy}\boldsymbol{q}_{xy}(t)$	$\xi_z(t)=\Phi_z\boldsymbol{q}_z(t)$	(13)

Table 10. Nodal to Modal transformation.

The equations of motion for the reduced model are formerly the same reported in equations (4) to (7), where the nodal displacements is substituted by the modal displacements and matrices and vector are reported in their modal forms.

Input/output selection matrices should be transformed, starting from FEM matrices, as indicated in:

X-Y Behaviour	Z Behaviour	
$\boldsymbol{S}_{\xi ixy}=\Phi_{xy}^T\boldsymbol{S}_{ixy}$	$\boldsymbol{S}_{\xi i}=\Phi_z^T\boldsymbol{S}_{iz}$	(14)
$\boldsymbol{S}_{\xi oxy}=\boldsymbol{S}_{oxy}\Phi_{xy}$	$\boldsymbol{S}_{\xi oz}=\boldsymbol{S}_{oz}\Phi_z$	

Table 11. Input / Output matrices conversion from Nodal to Modal.

3.4. State Space representationofrotordynamicequations

Dynamic equations (4) to (7) can be reported, with explicit spin speed, in the state space representation in the following way, either for X-Y or Z behavior:

$$\begin{aligned} \dot{\mathbf{x}}(t) &= \left(\mathbf{A}_{\Omega 0} + \Omega \mathbf{A}_{\Omega 1} + \Omega^2 \mathbf{A}_{\Omega 2}\right)\mathbf{x}(t)\mathbf{u}(t) + \mathbf{B}\mathbf{u}(t) \\ \mathbf{y}(t) &= \mathbf{C}\mathbf{x}(t) + \mathbf{D}\mathbf{u}(t) \end{aligned} \qquad (8)$$

Where:

Name	X-Y Behaviour	Z Behaviour	
$x(t)$	$\begin{Bmatrix} q_{xy}(t) \\ \dot{q}_{xy}(t) \end{Bmatrix}$	$\begin{Bmatrix} q_z(t) \\ \dot{q}_z(t) \end{Bmatrix}$	(16)
$A_{\Omega 0}$	$\begin{bmatrix} 0 & I \\ -M_{xy}^{-1} K_{\Omega 0xy} & -M_{xy}^{-1} L_{xy} \end{bmatrix}$	$\begin{bmatrix} 0 & I \\ -M_z^{-1} K_{\Omega 0z} & -M_z^{-1} L_z \end{bmatrix}$	
$A_{\Omega 1}$	$\begin{bmatrix} 0 & 0 \\ -M_{xy}^{-1} H_{xy} & -M_{xy}^{-1} G_{xy} \end{bmatrix}$	$[0]$	
$A_{\Omega 2}$	$\begin{bmatrix} 0 & 0 \\ -M_{xy}^{-1} K_{\Omega 2xy} & 0 \end{bmatrix}$	$\begin{bmatrix} 0 & 0 \\ -M_z^{-1} K_{\Omega 2z} & 0 \end{bmatrix}$	
B	$\begin{bmatrix} 0 & 0 & 0 \\ I & I & S_{ixy} \end{bmatrix}$	$\begin{bmatrix} 0 & 0 \\ I & S_{iz} \end{bmatrix}$	
$u(t)$	$\begin{Bmatrix} f_{sxy} \\ f_{umb} \\ f_{xy} \end{Bmatrix}$	$\begin{Bmatrix} f_{sz} \\ f_z \end{Bmatrix}$	
$y(t)$	$\begin{Bmatrix} y_{xy}(t) \\ \dot{y}_{xy}(t) \end{Bmatrix}$	$\begin{Bmatrix} y_z(t) \\ \dot{y}_z(t) \end{Bmatrix}$	
C	S_{oxy}	S_{oz}	
D	$[0]$	$[0]$	

Table 12. State Space Representation variables.

4. Control design and results

The aim of this section is to explain the steps followed to perform the suspension control design. A conventional decentralized control strategy is illustrated with two nested loops, the inner for current control and the outer for the position.The detailed description of this strategy is followed by the exposition of an off-line electrical centering strategy which is used equalize electrical parameters of the electromagnets on each actuation stage and allows to get a steady and balanced control action.

4.1. ActiveMagnetic Suspension Control

Figure 4 shows the classical control strategy for one axis AMB. The system is characterized with a nested control structure, where the inner loops describe the current loops used to achieve a direct actuator's effort drive (force) and the outer loop is used to compensate the rotor position error from the nominal air-gap. Generally, the same strategy is applied for each axis; so they are managed independently from each other and control is called decentralized. The driving of one axis is performed with two separate H-bridges. To exert a positive force on the rotor, current in the upper coil is increased by the control current while the current in the lower coil is decreased by control current and vice versa for negative forces. Also, to linearize the current to force characteristic of an electromagnet a constant bias current is applied to both coils respectively.Position control is performed by using five decentralized PID. The design and tuning of control laws parameters are well described in [1] and [2]. Here Bode diagram transfer function of position control law is reported (Figure11).

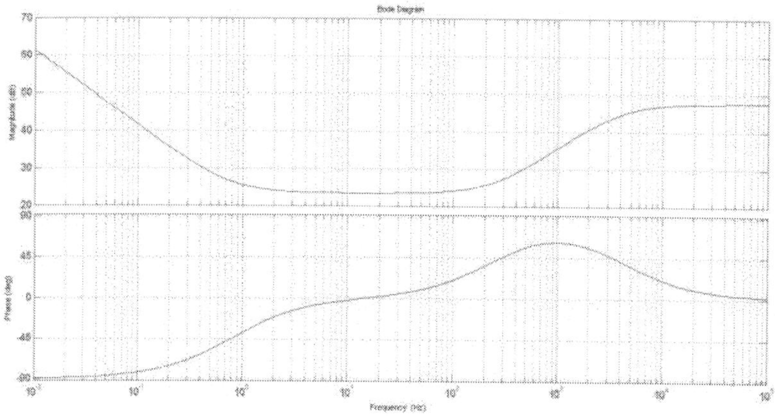

Figure 11. Control position Bode diagram.

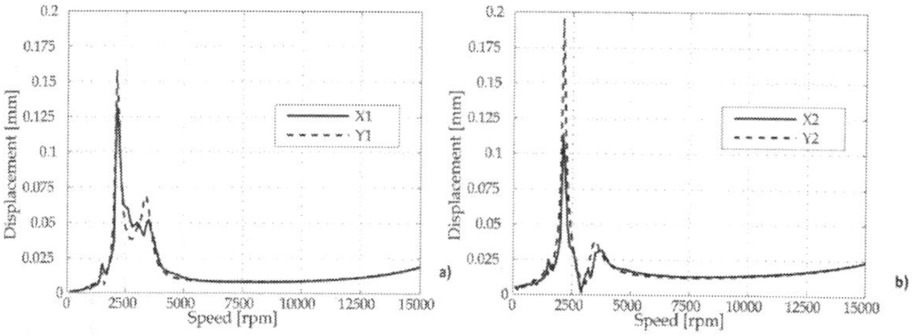

Figure 12. Unbalance response. a) Left actuation stage. b) Right actuation stage.

The experimental characterization has been performed by using classical tools of rotordynamic analysis: Unbalance responses (Figure 12), Waterfalls (Figure 13), orbital tube and orbital view (Figure 14, Figure 15). Theoretical notes on rotodynamic analysis can be found in [29].

4.2. Off-lineElectricalCentering

Active Magnetic Bearings offer several technological advantages that make their use mandatory in some particular applications, typically when clean environment is required or maintenance is expensive or difficult to manage. On the other hand, the main drawbacks are mostly the costs, higher than classical ball bearings due to the introduction of sensors and electronic equipment, and the complexity in the design phases of electrical and electronics subsystems and control strategies.

One of the aspects where this difficulty is more evident is the centering of the rotating part respect to the stator and in particular to the sensors. Sensors are indeed designed to detect microns of displacements and little inaccuracies in measurements lead to bad working or to system instability in the worst cases.

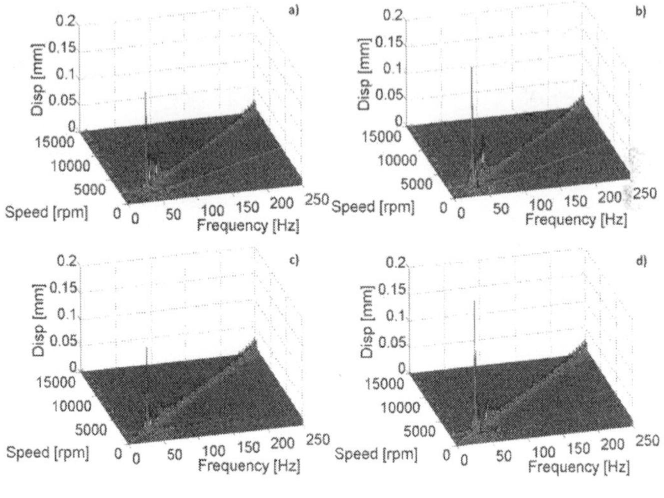

Figure 13. Waterfall plot. a) X1; b) Y1; c) X2; d) Y2.

The designer can choice to perform either a geometrical or an electrical centering, depending on the priorities of the application requirements. The first (Figure 16.a) consists in putting the rotor at the mechanical center neglecting the electrical differences in electromagnets coils parameters, inductance above all. The latter (Figure 16.b), on the contrary, leads to an equalization of electrical parameters, even if the rotor is not spinning around the geometrical center of the actuators.

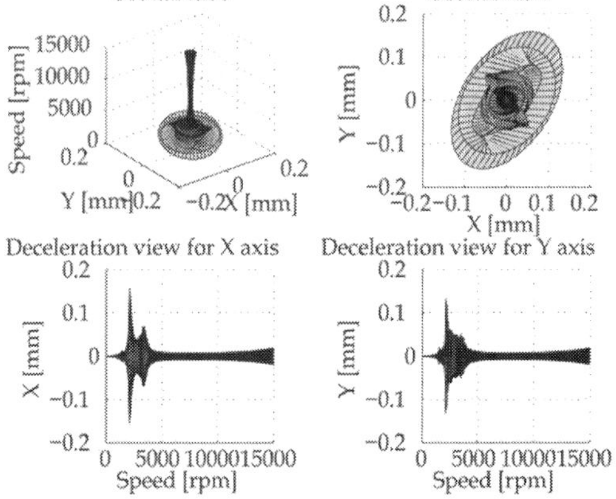

Figure 14. Orbital tube and orbital view representation on X1Y1 - plane. a) three dimensional view of the tube; b) Projection on the xy-plane; c) and d) projection on the Ωx and Ωy –planes.

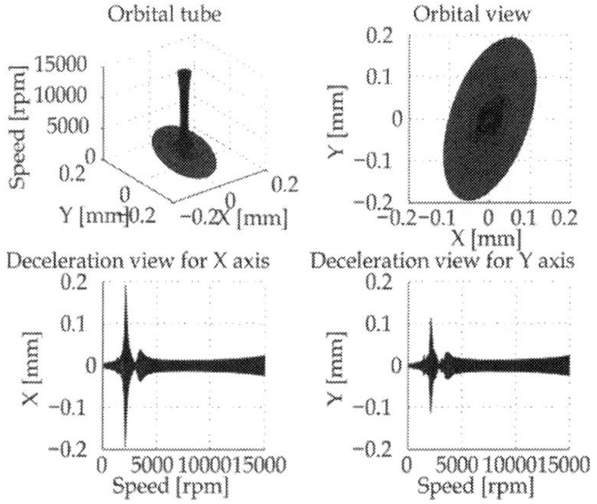

Figure 15. Orbital tube and orbital view on X2Y2 - plane. a) three dimensional view of the tube; b) Projection on the xy-plane; c) and d) projection on the Ωx and Ωy –planes.

In this section an off-line electrical centering technique is exposed. The goal is to make the rotor spins around a point which is not compulsorily coincident with the geometrical center of the actuators but grants the symmetry of the electrical parameters of them. It is well known that the inductance value of an electromagnet is depending on the distance between the ferromagnetic target (the rotor in this case) and the electromagnet itself.

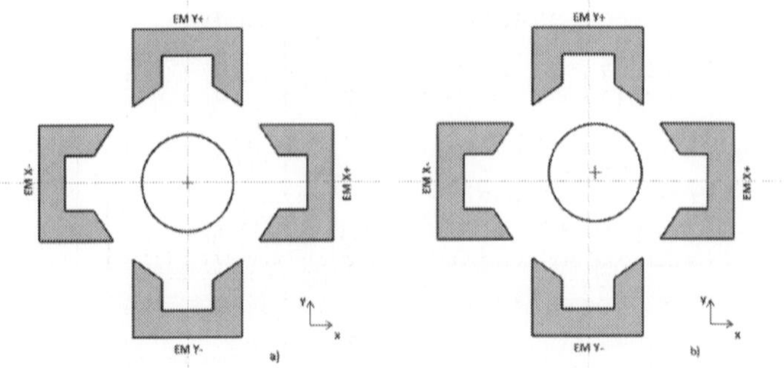

Figure 16. Mechanical centering (a) vs. Electrical centering (b).

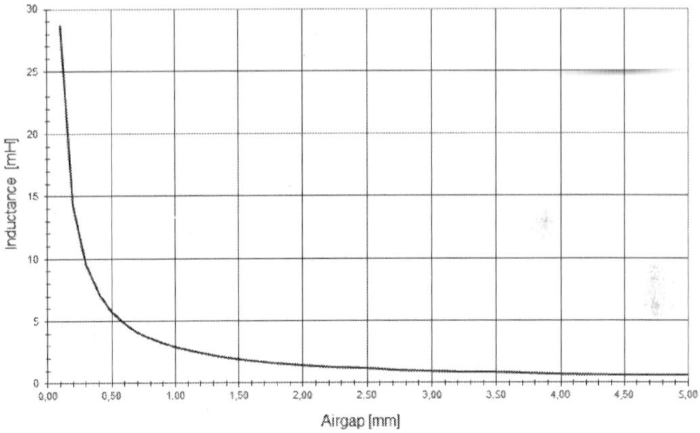

Figure 17. Inductance vs. Position.

This behavior is nonlinear as illustrated in Figure 17 and little variation of position and hence on inductance lead to big variation of actuator electrical pole and current control dynamics.Since the electrical dynamics of an electromagnet is depending on supply voltage, resistance and inductance values (Table 2) as in Eq. 17:

$$E = \frac{VDC}{R + Ls} \tag{9}$$

the electrical pole of the electromagnet is strictly dependent on the distance between the target and the actuator as reported in Figure 18. This issue has as consequence a different behavior of closed loop current control as illustrated in Figure 19 (a and b). It can be noticed that the same current control applied to the two opposite electromagnets of the same actuation axis without electrical centering generates two different closed loop responses. Few microns of Airgap generates differences of hundreds of Hertz on current control bandwidth.

Considering that this behavior is generated by a difference of inductance value of the two electromagnets, by acting on the position reference with offset corrections of the outer position control, the rotor can be set to spin around a point where electrical parameters are equalized and current loop bandwidths of both the electromagnets are the same (Figure 19 (c and d)). Further studies and research are being conducted on this strategy since this process can be performed in an on-line automatic routine with an adaptive technique, able to change the control parameters of the inner current loop while the Airgap is changing, i.e. when the rotor is oscillating.

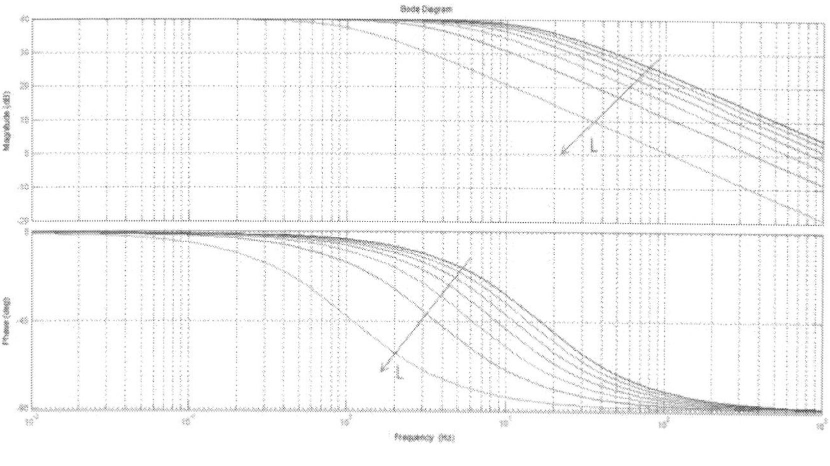

Figure 18. Electrical pole trend at varying of inductance value.

5. Conclusions

In this chapter the modeling, the design and the experimental tests phases of a rotor equipped with active magnetic bearings have been described. The description deals with rotordynamic aspects as well as electrical, electronic and control strategies subsystem. The control design of a standard decentralized SISO strategy and the details of an innovative off-line electrical centering technique have been exposed.Experimental results have been exposed highlighting mainly rotordynamics and control aspects.

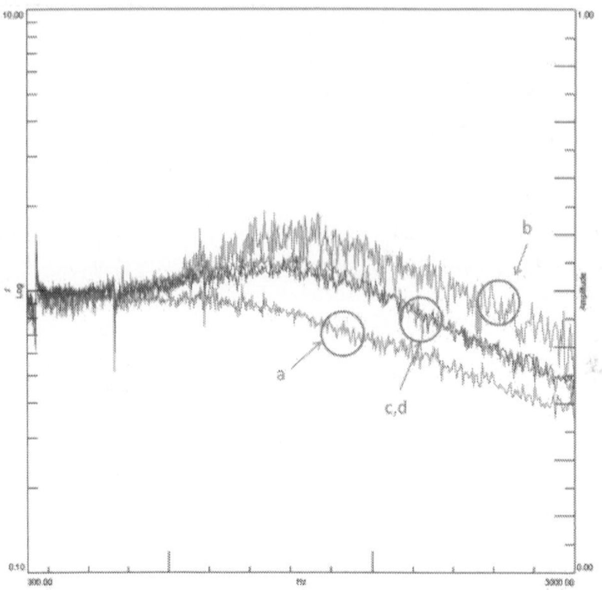

Figure 19. Electrical centering results. a) X- current loop Bode diagram before centering; b) X+ current loop Bode diagram before centering; c, d) X-/X+ current loop Bode diagrams after centering.

Author details

Andrea Tonoli, Angelo Bonfitto*, Mario Silvagni and Lester D. Suarez

*Address all correspondence to: angelo.bonfitto@polito.it

Mechanics Department, Mechatronics Laboratory – Politecnico di Torino, Italy

References

[1] Bleuler, H., Cole, M., Keogh, P., Larsonneur, R., Maslen, E., Nordmann, R., Okada, Y., Schweitzer, G., & Traxler, A. (2009). *Magnetic Bearings Theory, Design, and Application to Rotating Machinery*, Springer, Berlin Heidelberg.

[2] Chiba, A., Fukao, T., Ichikawa, O., Oshima, M., Takemoto, M., & Dorrell, D. G. (2005). *Magnetic Bearings and Bearingless drives*, Elsevier, Oxfor.

[3] Tonoli, A., Bonfitto, A., De Lépine, X., & Silvagni, M. (2009). Self-sensing Active Magnetic Dampers for vibration control. *Journal of Dynamic Systems, Measurement and Control,* 131.

[4] Tonoli, A., Amati, N., Impinna, F., Detoni, J. G., Bleuler, H., & Sandtner, J. (2010). Dynamic Modeling and Experimental Validation of Axial Electrodynamic Bearings. Wuhan, China. *Proceedings of the 12th International Symposium on Magnetic Bearings,* 978-7-90050-306-0.

[5] Genta, G., De Lépine, X., Impinna, F., Detoni, J. G., Amati, N., & Tonoli, A. (2009). Sensitivity Analysis of the Design Parameters of Electrodynamic Bearings. New Delhi, India. *IUTAM Symposium on Emerging Trends in Rotor Dynamics,* 287-296, DOI: 10.1007/978-94-007-0020-8, 1875-3507.

[6] Bonfitto, A. (2012). *Active Magnetic Control of Rotors,* LAP LAMBERT Academic publishing AG & Co. KG, 978-3-84842-104-6.

[7] Bosgiraud, T. (2008). Two Degrees of Freedom Miniaturized Gyroscope based on Active Magnetic Bearings. *PhD Thesis EPFL,* Lausanne, Switzerland.

[8] Bleuler, H., Bonfitto, A., Tonoli, A., & Amati, N. (2010). Miniaturized gyroscope with bearingless configuration. *In: Twelfth International Symposium on Magnetic Bearings (ISMB12),* Wuhan, China, August 22-25.

[9] Schammass, A., Herzog, R., Buhler, P., & Bleuler, H. (2005). New results for self-sensing active magnetic bearings using modulation approach. *IEEE Trans. Contr. Syst. Tech.,* 13(4), 509-516.

[10] Araki, Mizuno T., , K., & Bleuler, H. (1996). Stability analysis of self-sensing magnetic bearing controllers. *IEEE Trans. Contr. Syst. Tech.,* 4(5), 572-579.

[11] Maslen, E. H., Montie, D. T., & Iwasaki, T. (2006). Robustness limitations in self-sensing magnetic bearings. *Journal of Dynamic Systems, Measurement and Control,* 128, 197-203.

[12] Vischer, D., & Bleuler, H. (1990, 12-14 July). A new approach to sensorless and voltage controlled AMBs based on network theory concepts. Tokyo, Japan. *Second International Symposium on Magnetic bearings,* 301-306.

[13] Vischer, D., & Bleuler, H. (1993). Self-sensing magnetic levitation. *IEEE Trans. on Magn.,* 29(2), 1276-1281.

[14] Noh, M. D., & Maslen, E. H. (1997). Self-sensing magnetic bearings using parameter estimation. *IEEE Trans. Instr. Meas.,* 46(1), 45-50.

[15] Betschon, F., & Knospe, C. R. (2001). Reducing magnetic bearing currents via gain scheduled adaptive control. *IEEE/ASME Transactions on Mechatronics,* 6(4), 437-443.

[16] Lum, K. Y., Coppola, V. T., & Bernstein, D. S. (1996). Adaptive autocentering control for an active magnetic bearing supporting a rotor with unknown mass imbalance. *IEEE Trans. Contr. Syst. Technol.,* 4(5), 587-597.

[17] Betschon, F., & Schob, R. (1998). On-line adapted vibration control. Cambridge, MA. *in Proc. 6th Int. Symp. Magnetic Bearings*, Mass. Inst. Technol., 362-371.

[18] Hirata, M., Nonami, K., & Ohno, T. (1998). Robust control of a magnetic bearing system using constantly scaled H-infinity control. Cambridge, MA. *in Proc. 6th Int. Symp. Magnetic Bearings*, Mass. Inst. Technol., 713-722.

[19] Kanemitsu, Y., Ohsawa, M., & Marui, E. (1994). Comparison of control laws for magnetic levitation. Zuerich. *in Fourth International Symposium on Magnetic Bearings, ETH*, 13-18.

[20] Duan, G. R., & Howe, D. (2003). Robust magnetic bearing control via eigenstructure assignment dynamical compensation. *IEEE Trans. Contr. Syst. Technol.*, 11(2), 204-215.

[21] Zhuravlyov, Y. N. (2002). On LQ-control of magnetic bearing. *IEEE Trans. Contr. Syst. Technol.*, 8, 344-355.

[22] Steffani, H. F., Hofmann, W., & Cebulski, B. (1998). A controller for a magnetic bearing using the dynamic programming method of Bellman. Cambridge, MA. *in Proc. 6th Int. Symp. Magnetic Bearings*, Mass. Inst. Technol., 569-576.

[23] Li, L. (1998). On-line tuning of AMB controllers using genetic algorithms. Cambridge, MA. *in Proc. 6th Int. Symp. Magnetic Bearings*, Mass. Inst. Technol., 372-382.

[24] Hung, J. Y. (1995). Magnetic bearing control using fuzzy logic. *IEEE Trans. Ind. Applicat.*, 31, 1492-1497.

[25] Lindlau, J. D., & Knospe, C. R. (2002). Feedback linearization of an active magnetic bearing with voltage control. *IEEE Trans. Contr. Syst. Technol.*, 10, 21-30.

[26] Youcef-Toumi, K., & Reddy, S. (1992). Dynamic analysis and control of high speed and high precision active magnetic bearings. *ASME J. Dyn. Syst., Meas., Contr.*, 114, 623-633.

[27] Mizuno, T. (2001). Analysis on the fundamental properties of active magnetic bearing control systems by a transfer function approach. *JSME Int. J. Series C. Mech. Syst. Mach. Elements Manufact.*, 44, 367-373.

[28] Losch, F., Gahler, C., & Herzog, R. (1998, Aug. 5-7). synthesis controller design for a 3 MW pump running in AMBs. Cambridge, MA. *in Proc. 6th Int. Symp. Magnetic Bearings*, Mass. Inst. Technol., 415-428.

[29] Genta, G. (2005). *Dynamics of Rotating Systems*, Springer, New York, 1-658, 978-0-38720-936-4.

Vibration of Satellite Solar Array Paddle Caused by Thermal Shock When a Satellite Goes Through the Eclipse

Mitsushige Oda, Akihiko Honda, Satoshi Suzuki and Yusuke Hagiwara

Additional information is available at the end of the chapter

http://dx.doi.org/10.5772/52626

1. Introduction

Remote sensing satellites take images of the earth's surface in observing various activities by humans or nature. In order to obtain precise and high-resolution images from a satellite in Low Earth Orbit (LEO) at an attitude of 500 to 900 kilometers, the satellite's attitude must be stable when the onboard camera sensors take images of surface activities on the earth. Should the attitude stability of the satellite be disturbed while such images are being taken, poor image quality would probably result.

The fact that images taken when a satellite goes into or out from an eclipse do not provide good accuracy—due to degraded altitude stability at such timings—has been known for years. Such phenomena has long been attributed to the deformation and vibration of the satellite's solar array paddle that occurs when the satellite go into or out from an eclipse, along with the instantaneous change in solar energy received by the satellite. Several trials were conducted in the past to identify the phenomena, but all failed to achieve reasonable results.

2. Measurement of solar array paddle motion

The reason why past trials failed to observe the phenomena described above might be that the motion of the solar array paddle is too small or slow to be observed by such onboard sensors as an accelerometer. Therefore, JAXA decided to measure the phenomena by using an onboard camera that was originally mounted on the satellite to monitor solar array paddle deployment (Fig. 2). (For GOSAT, the solar array paddle consists of three solar panels, a yoke, wires that control deploying speed, and hinge/latch mechanisms that connect the panels and yoke.) A small CCD/CMOS camera (Fig. 4) is mounted on most JAXA satellites to monitor solar array paddle deployment, as failure to deploy the solar array paddle would become critical failure of the satellite itself. We thus developed a system that uses this CMOS camera to measure the distortion and vibration of the solar array paddle, with said vibration being measured as follows:

1. Attach small reflective target markers at the end of the paddle (as shown in Fig. 3).
2. Take images from this camera and transmit them to the on-ground station. Fig. 2 is an image taken by the camera.
3. Identify the locations of the target markers in the camera view with image processing.
4. Calculate displacement from the predicted target marker locations.

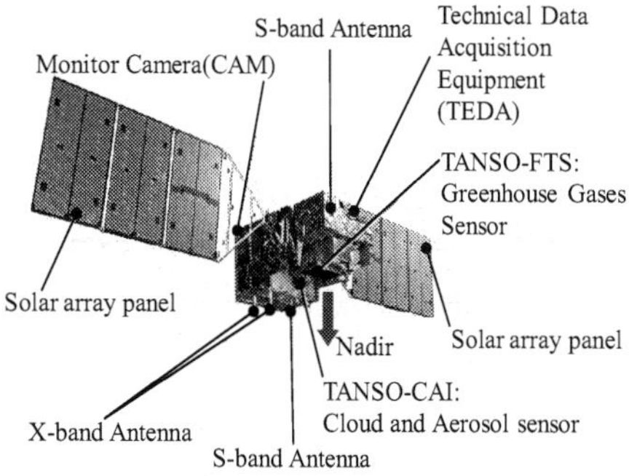

Figure 1. JAXA's earth observation satellite "GOSAT"

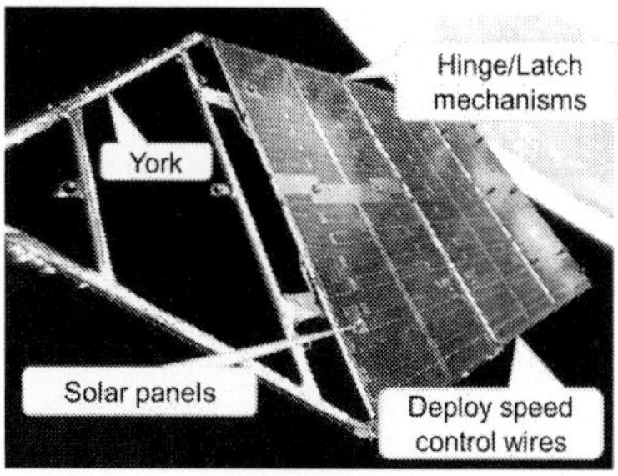

Figure 2. Image taken by camera

For GOSAT, the solar array paddle consists of three solar panels, a yoke, wires that control deploying speed, and hinge/latch mechanisms that connect the panels and yoke.

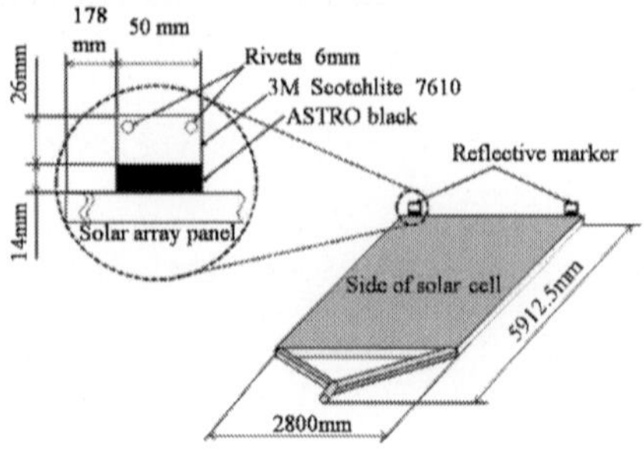

Figure 3. Target markers on the solar array paddle

Figure 4. CMOS camera and LED lights

3. Measurement algorithm

As the onboard camera used for this experiment was originally designed to monitor the deployment of a folded solar array paddle, its field of view is thus as wide as 90 [deg] (Fig. 5), while the camera's number of pixels is limited to SXGA (1280 X 1024 pixels). The size of the markers is also limited to 50 [mm] X 26 [mm], while distance from the camera to the target markers is as far as 6 [m]. These constraints mean that one pixel of the camera is equivalent to 7 [mm] at the target marker's position. Therefore, a technique for processing sub-pixel level image data is required to identify deformation of the solar array paddle.

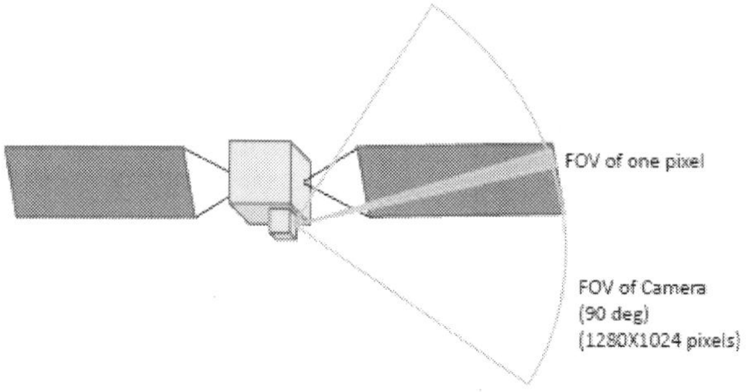

Figure 5. Constraints on image data processing

4. Previous results

The initial results of these measurements were reported in another paper in this series (Oda et al., 2011). Fig. 6 shows typical results. It shows an offset of a few millimeters appeared when solar array paddle illuminated by the Sun and when in an eclipse. However, these results pose certain difficulties in explaining the phenomena, as the tendency of the solar array paddle to bend does not agree with the observation results and conventional research.

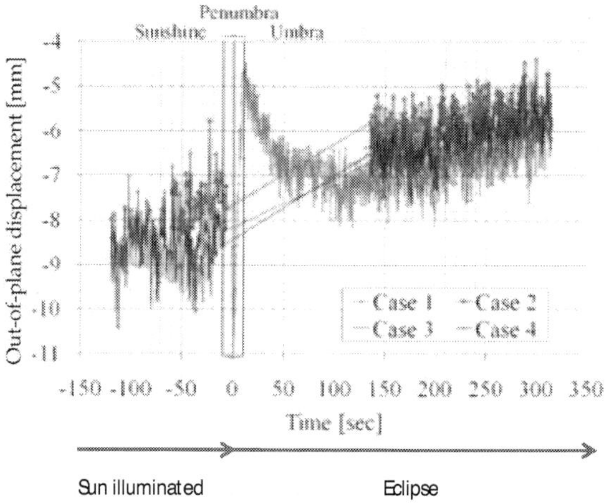

Figure 6. Motion of the solar array paddle's tip position as estimated from onboard camera images

Since solar energy from the Sun is the primary energy provided on the satellite and its solar array paddle, deformation of the solar array paddle should occur so that its structure on the Sun side is extended and causes bending of the solar panels toward the rear side of the solar array paddle. Previous results have indicated that the solar array paddle will bend toward its solar cell side, however. When the initial results of these measurements were recorded, we were unable to identify the source of the errors.

5. Revised results

Fig. 6 shows a sudden offset in the motion of the solar array paddle. When the solar array paddle is straight, the value of displacement may indicate -12 or -11 [mm] in Fig.6. Although we attempted to identify cause of this offset, we could not imagine a proper mechanism that would produce such a direction. We therefore assumed that the previous image data processing contained unidentified data processing errors, and consequently modified the algorithm that identifies the target markers. The earlier version of the algorithm used to identify locations of the target markers assumed such highly illuminated areas as those of the target markers. This algorithm works well when the markers are brightly illuminated. When the target markers are weakly illuminated, however, we found that this algorithm produces some data processing errors.

Fig. 7 and Fig. 8 illustrate the difference described above. Fig. 7 is based on the previous algorithm. The areas enclosed by a yellow line are pixels that are brighter than the threshold level and thus can be assumed to be the target markers as based on sub-pixel level image data processing. The red cross indicates the center position of the marker.

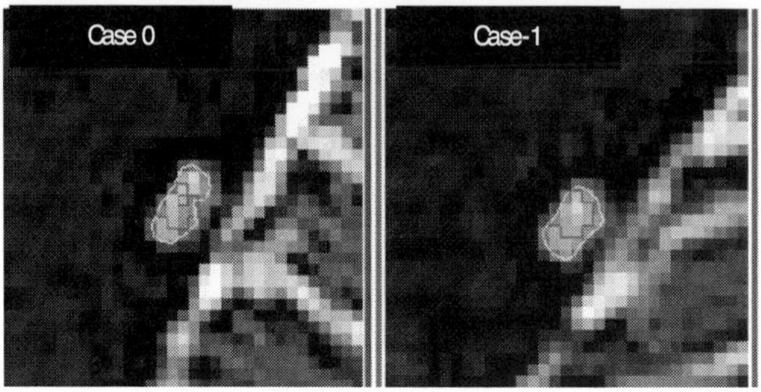

Figure 7. Target markers estimated by the previous algorithm under weak illumination. The areas of target markers assumed by the previous algorithm are smaller than the actual target marker size.

Fig. 7 shows that the sizes of the estimated target markers are smaller than the actual target markers. We therefore modified the algorithm used to estimate the area of a target marker so that the size of the predicted target marker is similar to that calculated from the actual target marker. This modification worked well to identify the target marker and its displacement. Fig. 8 is a result based on the revised algorithm. In this revised algorithm, the areas to be considered target makers are decided based on the brightness level of each pixel and also on the size of the areas considered to be target markers. When an assumed target marker is too small, then the threshold level of brightness is automatically adjusted to meet the possible size of the target markers.

Finally, Fig. 9 shows the motion of a target marker as based on the revised algorithm. We can see that the estimated motion of the solar array paddle has less dispersion.

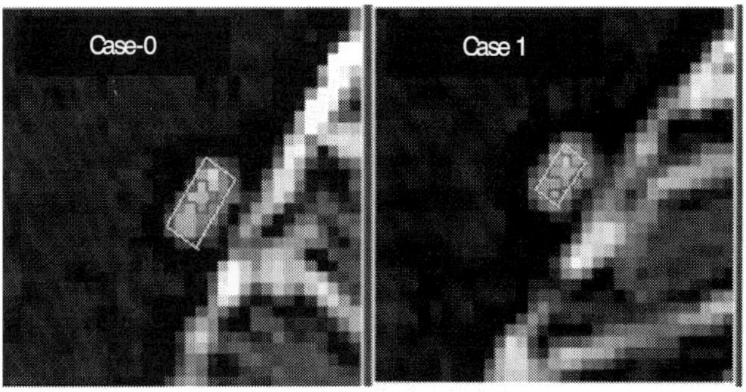

Figure 8. Target markers estimated by the revised algorithm

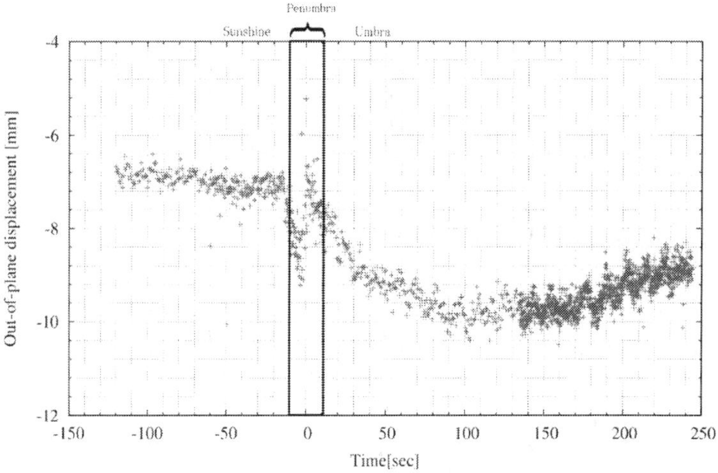

Figure 9. Displacement of GOSAT's solar array paddle when going into eclipse

From these data, we can conclude that the temperature of the solar array paddle changes steadily but not drastically, depending on its thermal capacity. And we also assume like as following.

1. When the solar array paddle is illuminated by the Sun, its temperature is governed by solar energy from the Sun (as well as that reflected from Earth).

2. When in an eclipse, the solar array paddle shows lower temperature depending on its thermal capacity. When the satellite's LST (Local Sun Time) is around noon, the solar cell side of the panels faces the Earth and receives solar radiation from Earth. In the evening and the early morning of satellite LST, however, both sides of the solar panels face toward space, resulting in a rapid drop in temperature.

6. Numerical simulation

In order to verify whether the observation results described above are correct, and to understand the features of thermal snap on the solar array paddle, we conducted numerical simulation of the solar array paddle. There are many studies analyzing the thermal snap (Thornton, 1996; Boley, 1972; Johnston, 1998; Lin, 2004; Xue2007), we develop the method using a thermal model and a structural model and revising these models by applying the observed data. This section presents the analytical system that we constructed, the result of a thermal-structural analysis and its problems. To solve the problems, we have developed a new model in considering the effects of hinge/latch mechanisms and friction. The result of using new model is also introduced.

6.1. Thermal snap analysis procedure

This section describes the thermal snap analysis procedure. Fig. 12 shows a flowchart of thermal snap analysis. The analysis can be broken down into to three parts: construction of the structural model, calculation of temperature distribution, and thermal snap analysis for the penumbra.

In the first part, a structural model of the solar array paddle is constructed for thermal snap analysis. To verify the structural model, we conducted modal analysis to obtain the natural frequency and vibrational mode of the solar array paddle model. These results will be compared with an on-orbit preliminary experiment, and if necessary, we will then revise the model (see Section 6.2 for details). In the second part, a thermal model is developed for the solar array paddle. Thermal analysis for the whole orbit is then conducted to verify the thermal model. The thermal analysis results will be applied to GOSAT's trajectory information, and also compared with data obtained by GOSAT, in order to verify accuracy. After accuracy is verified, thermal analysis will focus on GOSAT during its integration and testing. In order to improve the accuracy of thermal analysis, a profile of thermal input was determined based on the brightness of the solar paddles (see Section 6.3 for details).

In the third part, the thermal snap analysis is conducted using the structural FEM model of GOSAT.

Fig. 10 and Fig. 11 below show the structural model and the thermal model, respectively.

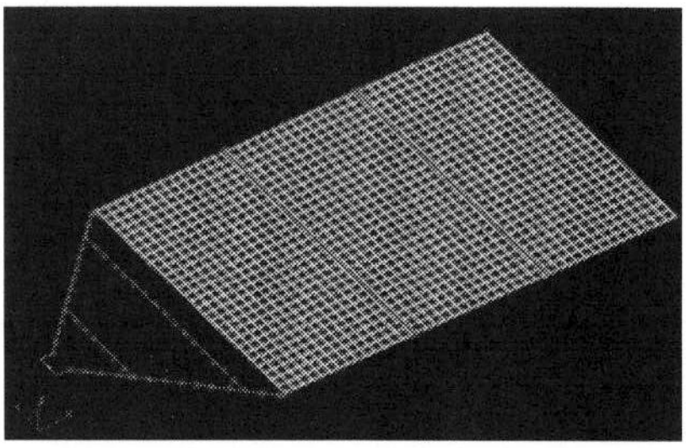

Figure 10. Structural model

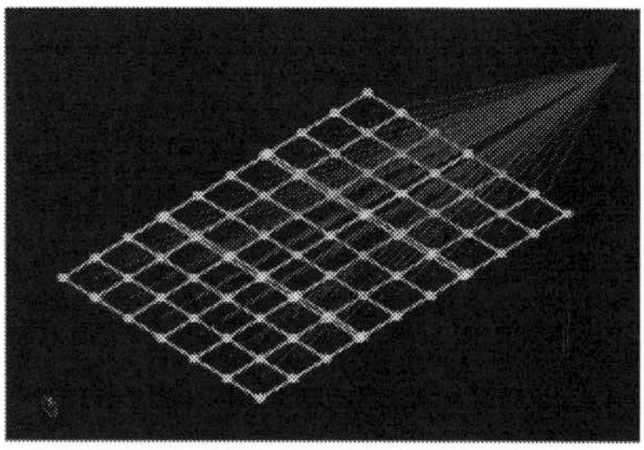

Figure 11. Thermal model

6.2. Improvement of the structural model

In constructing the structural model of GOSAT's two solar cell paddles, we were not allowed to access the detailed satellite design data. Therefore, the structural model was based on partly assumed data. Moreover, it is difficult to estimate the production errors on GOSAT. We thus compared the structural model design data and the observation data. The preliminary observation of solar cell paddle motion was made as GOSAT conducted orbit-raising maneuvers using the 20 Newton Gas Jet thrusters. This maneuvering applied relatively large force to the satellite's main body, thereby causing large bending of the solar cell paddles, followed by the induced vibration thereof. An earlier paper in this series reported the measurement results.

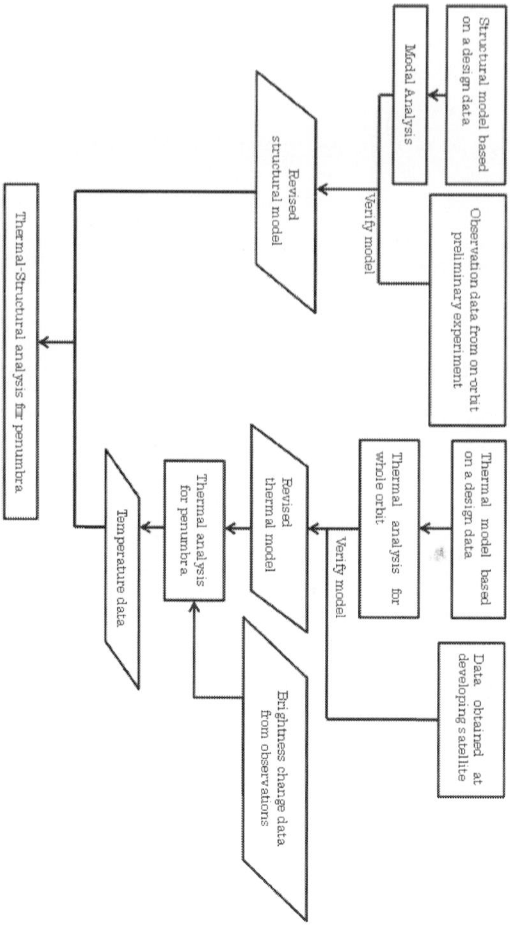

Figure 12. Flowchart of thermal snap analysis using observed data

In this experiment, we severely shook the solar paddle with the 20N thrusters, in order to observe its behavior. The FFT data obtained from this experiment on solar paddle vibration is available. Fig. 13 shows an example of the obtained FFT data. From these data, the natural frequencies of the actual solar array paddle were assumed to be 0.215 [Hz] (out-of-plane) and 0.459 [Hz] (in-plane). These natural frequencies are identical to those obtained by analysis during development.

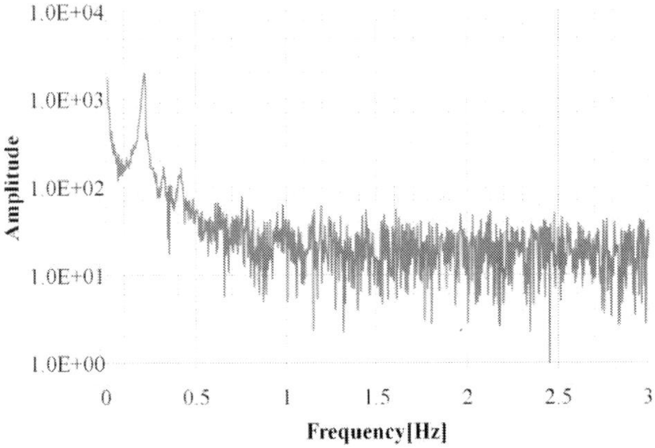

Figure 13. Example of FFT data on out-of-plane solar array paddle vibration as obtained from observation

Next, we conducted modal analysis of the structural model of the solar array paddle to calculate the natural frequency. In case of any differences between the analysis results and observation results, we adjusted the structural data around the root of the solar cell paddles.

By iterating this process, the structural model was revised to generate data similar to that obtained from the solar array paddle in orbit.

Tab. 1 below lists shows the final modal analysis results of natural frequency; Fig. 14 shows the modal shapes of the structural model. From these data, the modal analysis results of the revised structural model match the observed data. Hence, we expect that the revised model could sufficiently simulate the actual solar array paddle.

Mode number	Frequency [Hz]	Modal shapes
1	0.218	First order out-of-plane
2	0.454	First order in-plane
3	1.262	First order twist
...	...	...

Table 1. Modal analysis results

6.3. Detailed thermal input profile from observed data

During the time when passing the boundary between the sunlight area and umbra area, the thermal environment changes gradually. To conduct thermal snap simulation, a detailed thermal input profile is needed, but estimating changes from trajectory information alone is difficult due to atmospheric effects. We thus estimated the value of solar light incident (i.e. dominant factor in thermal input) from the brightness value of images taken in observation. By using gray-scale processed images, we assume that if the average gray scale value of an image is max, then the value of solar light incident will also be max. If it is minimum, then the value will be minimum. Fig. 15 shows the estimated solar light incident and total thermal input profile. From the result, we can see that GOSAT's actual optical environment changes gently three times, as long as the value is forecast based on geometric calculation of the satellite's trajectory.

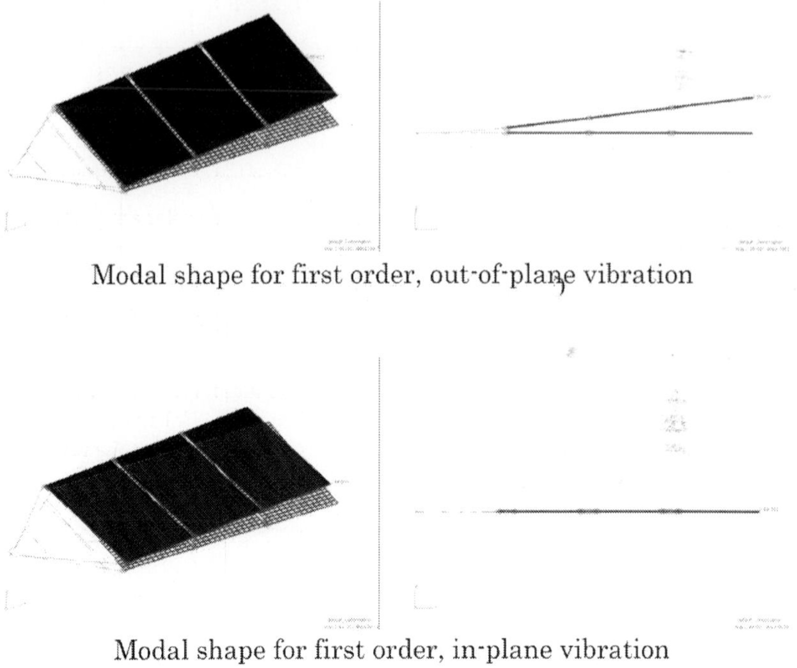

Figure 14. Modal shapes of the structural model (Out of Plane and In-Plane Vibration)

By using the thermal input profile thus revealed, we conducted thermal analysis for the penumbra. Fig. 16 shows the simulation results of solar paddle temperature and the values of thermometers. The thermometers are attached at two points. However, no temperature sensor was attached on the front (solar cell; optical incidence) side of the solar paddle, but one was attached only on a backside plane (radiation plane; indicated as Bottom CFRP in Fig. 16). The tendency of simulated solar paddle temperature is similar, however, to the actual data. We therefore conclude that our thermal analysis can sufficiently simulate actual thermal changes.

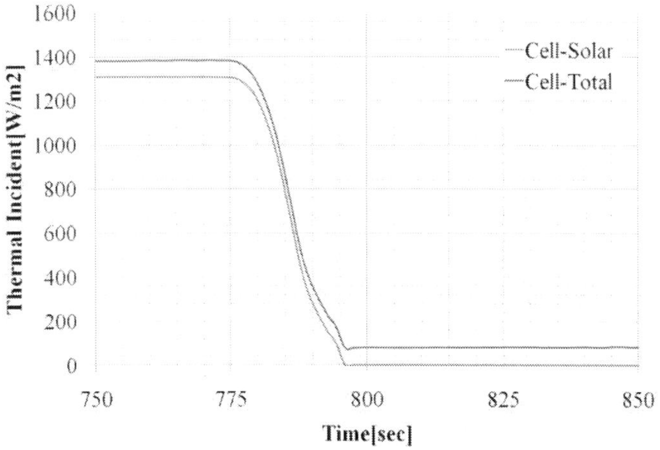

Figure 15. Estimated solar light incident and total thermal input profile from images

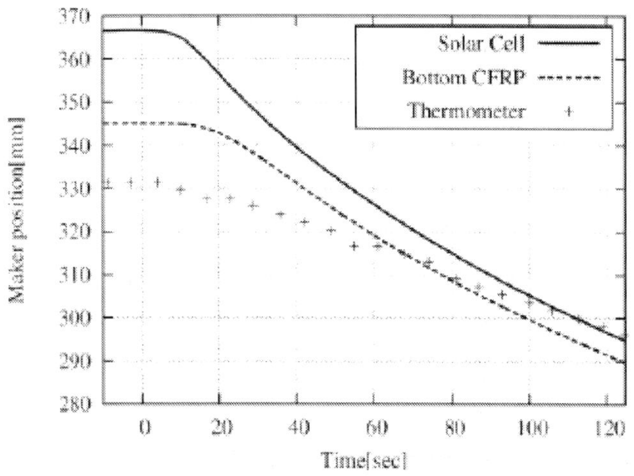

Figure 16. Estimated solar array paddle temperature and values of thermometers attached on a backside plane (Bottom CFRP)

6.4. Results of thermal snap analysis

Thermal snap analysis is conducted using the revised structural model and temperature distribution. Fig. 17 shows the results of thermal snap analysis in the penumbra. The simulation data shows that the solar array paddles will bend in quasi-static while the paddle is illuminated by the Sun. And from the deformation plot, it is assumed that the wires have a large effect on a deformation.

By using the models, we conducted thermal deformation analysis for the whole orbit—a task not possible through camera observation alone. Fig. 18 shows the results. It shows that much larger deformation occurred in the penumbra than at any other time.

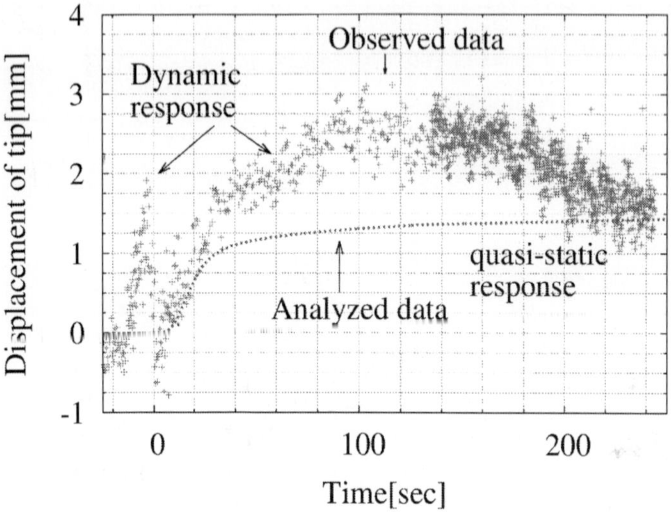

Figure 17. Analysis results of deformation in penumbra

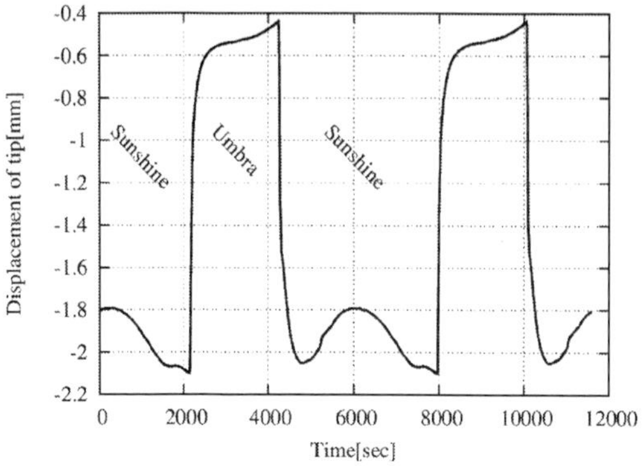

Figure 18. Solar array paddle deformation for two round orbit (The orbital period of GOSAT is 6000 [sec].)

6.5. Detailed modeling for hinges

As dynamic response offers the possibility of adding more awkward disturbances to attitude stability than quasi-static deformation, we decided to check the penumbra data in detail. The observed data obtained in the penumbra indicates that the solar array paddles have a much lower natural frequency. Fig. 19 shows the FFT data on out-of-plane deformation in the penumbra.

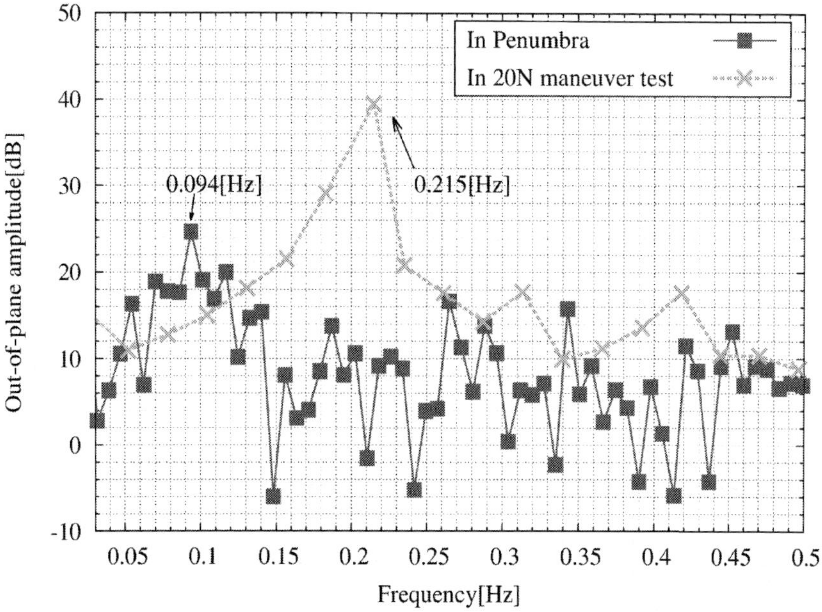

Figure 19. FFT analysis of out-of-plane deformation in the penumbra and same data obtained when maneuvering to raise orbital altitude using powerful 20N thrusters

Fig. 19 shows the data obtained when using the powerful 20N thrusters to raise GOSAT's orbital altitude, with vibration of 0.215 [Hz] being observed. When the satellite goes into eclipse, however, the data obtained did not reveal vibration of 0.215 [Hz], but instead showed a lower frequency of 0.094 [Hz]. Similar data was not reported during the satellite integration and qualification tests.

From these reasons, we assumed that the small load (e.g. solar pressure) acts on the solar panels, and that this load acts as rotation torque at each hinge. As all solar cell panels of GOSAT are linked together to maintain the angle of deployment to the deployed position, the small motion generated at each hinge will also act on all the hinges.

Vibration will occur at each hinge that connects each solar cell panel. Hinges and wires are used to interconnect the solar cell panels, in order to deploy the solar cell panels and maintain the open position of each panel. As each hinge will have some backlash and in order to pull each panel to the deployed position, wires are used to maintain the deployed position after each panel is deployed, and to connect the solar panels, resulting in only a small load under microgravity conditions. Fig. 20 and 21 shows the concept.

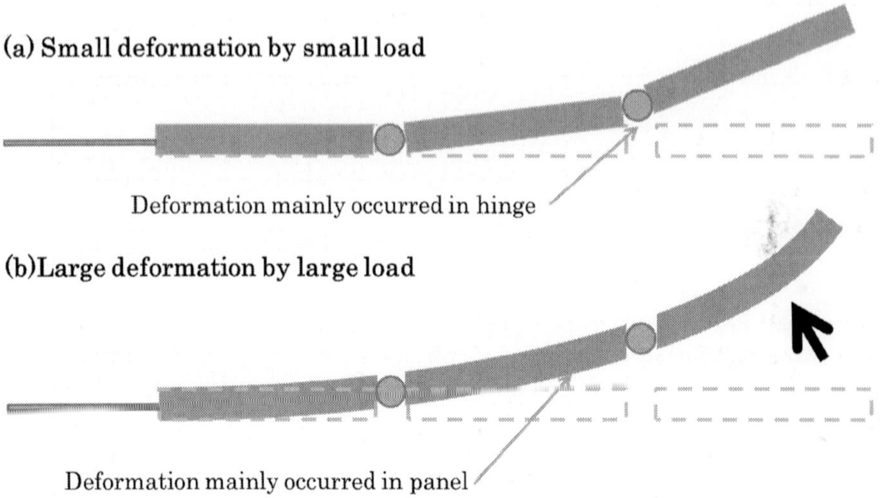

Figure 20. Concept of vibration affected by the hinge gap

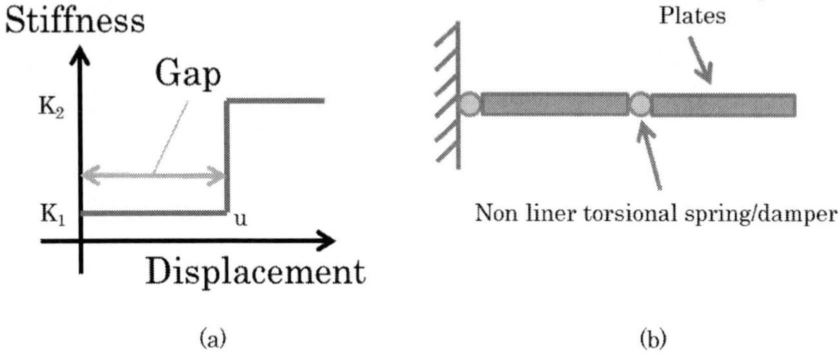

Figure 21. Concept of a hinge modeled using a nonlinear torsional spring. In Fig. 21, (a) shows how to define the stiffness of torsional springs; (b) shows the model for the solar paddle.

Should a small load (e.g. thermal stress) be applied to the solar paddle, deformation will mainly occur in the hinges that have small gaps. Conversely, in case a large load (e.g. inertia force at maneuver) is applied, deformation of the solar paddle is typically larger than the gaps; therefore, such deformation is mainly caused by the elastic deformation of the panels.

To simulate the effect caused by hinges with a gap, we modeled the hinges with nonlinear torsional springs. In Fig. 21, (a) shows how we added nonlinear characteristics to the torsional springs. The length of u is used to define the gap size. And stiffness within the gap is defined by K_1. When the whole dynamic response shown in Fig. 17 is assumed to have occurred within gap area, the gap width is easily defined. Then, the stiffness of the gap is defined as shown below.

To estimate stiffness, the modeling method for the model of spring-connected plates is used (Kojima et al., 2004). The rotational motion equation for the model of spring-connected plates as shown in Fig. 21 (b) is:

$$I\theta + K\theta = 0 \tag{1}$$

where, I denotes the inertia moment of plates and K the stiffness of torsional springs. Here, given harmonic vibration in which frequency is , the stiffness of gap area K is expressed as:

$$K = I\omega^2 \tag{2}$$

1 is 0.094 [Hz]. Therefore, we assumed the stiffness of gap area K as about 1.21 [Nm/rad] at first, and adjust the value until the vibration properties in penumbra coincide with the value of the observation results.

By using the new model with a detailed hinge, we conducted frequency response analysis. The analysis targets were both a small load condition and a large load condition shown as Tab.2. Fig. 22 shows the results. From the results, we can see that the new model could simulate the vibration occurring in the 20N maneuver test and in the penumbra at the same time.

Excitation force	Condition 1	0.4 [m/s²] inertia force
	Condition 2	7.2 [μPa] pressure on top side
Solver		General-purpose non-liner analysis software "MSC. Marc 2011"
Frequency interval		0.1[Hz]
BC		Tip of york : Fixed
Number of nodes		7903
Number of elements		8292

Table 2. Details of frequency response analysis

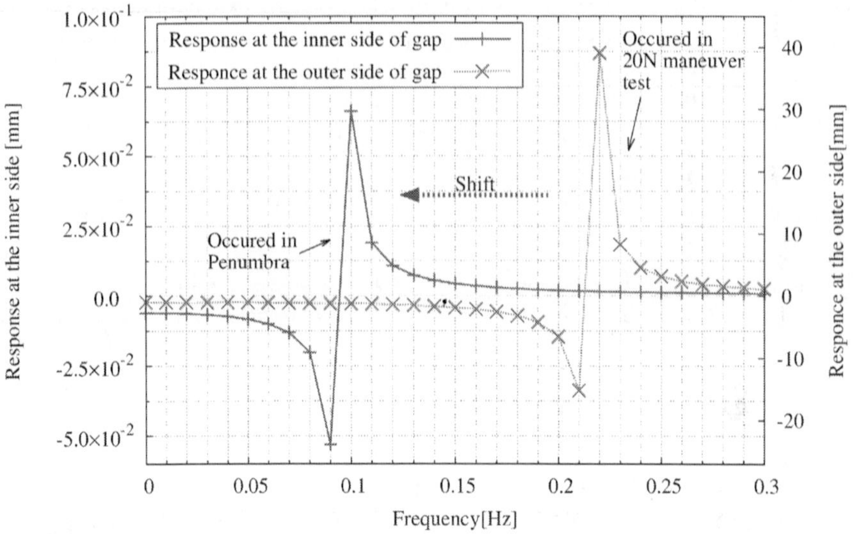

Figure 22. Frequency response analysis results of a structural model with a hinge gap. The analysis of response at the inner side of the gap simulates the dynamic regime in the penumbra. The analysis of response at the outer side of the gap simulates the dynamic regime in the preliminary experiment using the 20N thrusters of GOSAT (see Section 6.2). From the results, the new model with a nonlinear hinge demonstrated that it could simulate the peak shift in response showed in the penumbra (Fig. 19).

7. Modeling of stick-slip motion between wires and pulleys

The conventional simulator also had a problem about amplitude of the dynamic responses shown as Fig.17. The slowness of the temperature change in the paddle might be the reason for that. As shown in Fig.16, the significant decreasing of temperature began 5-10 seconds after going into penumbra. While discussing about the dynamic responses, we take notice of some particular measured data (showed in Fig.23). In the data, the solar array paddle shows the characteristic triangular wave before eclipse and the smaller dynamic responses than ordinary one. These triangular waves are found commonly in cantilever vibration affected "Stick-Slip phenomenon" (Maekawa, et.al., 2008). The Stick-Slip is a phenomenon that it arise the "stick" and "slip" behavior continuously to objects shared sliding surfaces. The phenomenon arise on the ground that the change of friction coefficient. And in slip phase, the stored strain energy is released at once. The Stick-Slip phenomenon occurred at space systems had precedents in Hubble Space Telescope (Thomton, 1993). Then, we assumed that the Stick-Slip phenomenon was occurred at deploy-speed-control wires and pulleys. That is because that the wires have small heat capacity and will be great affected by thermal environment changes. And the wires were empirically-deduced that govern the deformation of solar array paddle. The picture of them is shown in Fig. 24, and the position relation is indicated in Fig. 25. When simulating the thermal snap, sometime these specific dynamic conditions are needed to be considered. For example, when the analysis for Hubble Space Telescope was conducted, the effect from specific cross-section shape of boom was estimated. (Foster, 1995) In case of ADEOS, the tension control mechanism was evaluated in detail. (Taniwaki, 2007)

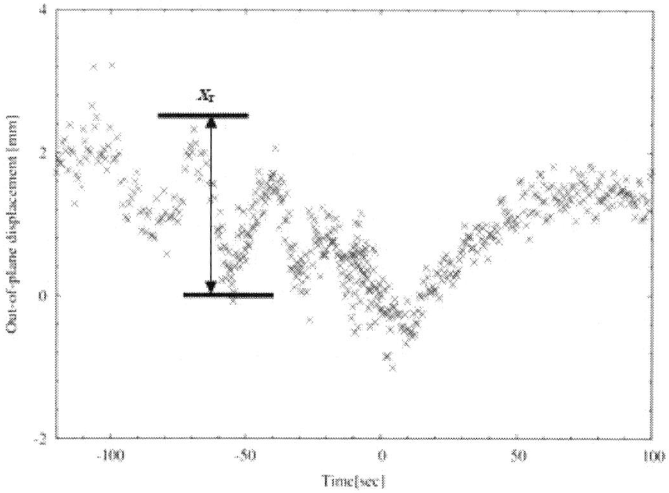

Figure 23. Particular example that solar array paddle shows unexpected behavior.

Figure 24. Similar product of GOSAT's Deploy-speed-control wires and pulleys.

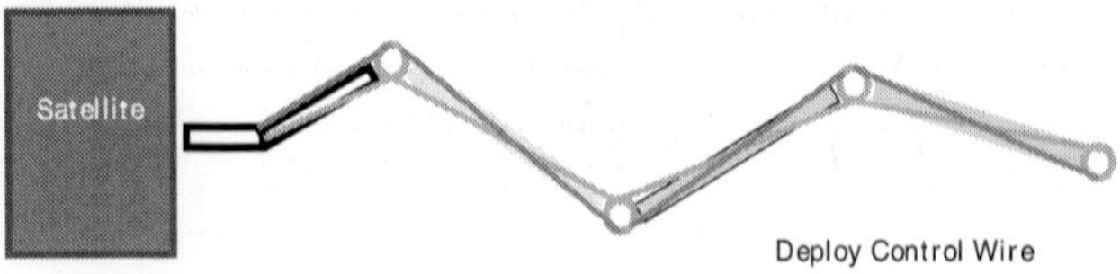

Figure 25. Position relation of GOSAT's Deploy-speed-control wires and pulleys.

To verify the effect of Stick-Slip phenomenon, the phenomenon was introduced to the structural model. In that time, the detailed information of wires and pulleys were not available. Therefore the value from measured data was used to demonstrate the effect of Stick-Slip phenomenon. Using the Bowden and Leben's equation (Bowden & Leben, 1939), the force change while the transition to the slip phase was estimated like below.

$$F_d = x_r k \quad (3)$$

Where the x_r is a displacement defined like showed in Fig.23, and the k is a stiffness of the deformation.

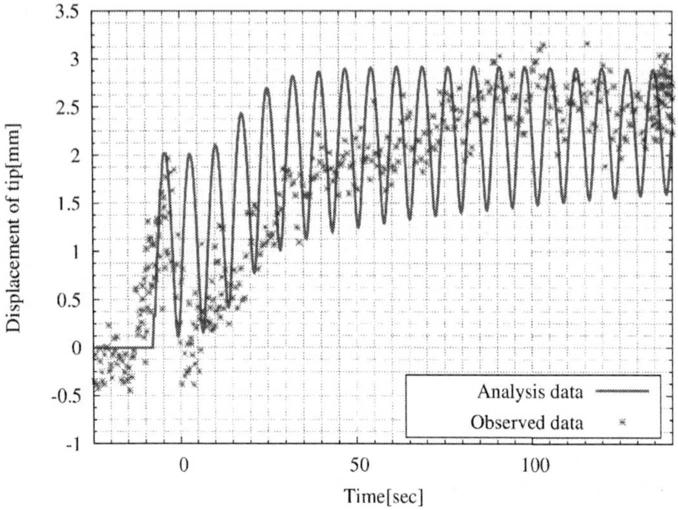

Figure 26. Time transient response simulation with detailed hinge model and stick-slip motion modeling.

From the Fig. 23, the released force was estimated like $F_d \fallingdotseq 0.029$[N]. Using this value, the new simulator was constructed, and verification analysis was conducted. In this analysis, the damping element of paddles was not concerned. The details of analysis are showed in

Tab.3. And the result of analysis is in Fig.26. As shown in Fig.26., the revised simulator could indicate roughly same value of amplitude of dynamic responses with consideration of Stick-Slip phenomenon

Analysis type	Thermal-Structure Time transient
Solver	MSC.Marc 2011
Time step	0.1[sec]
Analysis time	360[sec] (After 770[sec] of static state analysis)
BC	Tip of york : Fixed
	Temperatures of Top panel, Bottom panel, wires. Given by thermal model

Table 3. Details of thermal snap analysis

8. Conclusions

When a satellite in Low Earth Orbit (LEO) goes through an eclipse, sudden changes in the thermal environment will occur. These sudden thermal changes caused by a difference in temperature between the Sun side and the opposite side of the satellite's solar array paddle. This temperature difference causes bending and then vibration of the solar array paddle. However, the bending and vibration amplitude are too small to be observed by conventional satellite attitude sensors. To solve this problem, we used an onboard camera to directly measure solar paddle bending. The previously reported methods of observing the motion of the satellite's solar array paddle were not free from errors in image data processing, especially when the object was weakly illuminated by the Sun.

This chapter paper described the improved image data processing algorism applied to measure bending and deformation of the satellite's solar array paddle that are believed to cause degraded satellite attitude stability, which occurs when the satellite goes into an eclipse. The revised observation data shows both quasi-static deformation and rapid dynamic vibration in the penumbra. We also conducted numerical analysis to verify the observed data and understand the features of thermal snap on the solar array paddle. From this analysis, we found that sudden changes in solar array paddle temperature induce quasi-static deformation, and that the wires controlling solar array paddle deployment have a large influence on solar array paddle vibration. Additionally, we also found that a specific lower frequency vibration appearing in the penumbra can help to explain this vibration mechanism. Developing the detailed hinge model, we succeeded to simulate the vibration property change related to dynamical regime. Then, we focused on the specific results of observation and assumed that Stick-Slip phenomenon have a great influence on behavior of paddles. To simulate the influence, we revised the structural model and conducted the thermal snap analysis. The result indicated the gap of hinges and Stick-Slip phenomenon mainly effect on dynamical response of thermal snap on GOSAT.

As ongoing work, we will model the Stick-Slip phenomenon more in detail and introduce the damping element using observed data.

Author details

Mitsushige Oda[1], Akihiko Honda[2], Satoshi Suzuki[3] and Yusuke Hagiwara[4]

1 JAXA and Tokyo Institute of Technologies, Japan

2 Tokyo Institute of technology, Japan

3 Advanced Engineering Services Co.Lted, Japan

4[4]Mitsubishi Heavy Industries, Ltd, Japan

References

[1] B., A. Boley(1972), Approximate Analyses of Thermal Induced Vibrations of Beam and Plates, *Journal of Applied Mechanics*, vol. 39, no. 1, pp212-216

[2] Chijie Lin, Ramesh B. Malla(2004), Coupled Thermo-Structural Analysis of an Earth Orbiting Flexible Structure, in 45th AIAA/ASME/ASCE/AHS/ASC Structures, Structural Dynamics & Materials Conference, AIAA2004-1793

[3] Earl, A. Thomton, Yool A. Kim(1993): Thermally Induced Bending Vibrations of a Flexible Rolled-Up Solar Array, Journal of Spacecraft and Rockets, Vol.30, No.4, pp438-448

[4] E., A. Thornton (1996), Thermal Structures for Aerospace Application, AIAA, pp. 343-354

[5] F. P. Bowden & L. Leben (1939): The Nature of Sliding and the Analysis of Friction, *Proceedings of Royal Society London*, vol.A169, pp371-391

[6] Foster, C. L., Tinker, G. S., Nurre, W. & Till, W. A. (1995). NASA Technical Paper, The Solar

[7] Array-Induced Disturbance of the Hubble Space Telescope Pointing System, 3556

[8] Japan Society of Mechanical Engineering (February 2007). Mechanical Engineers' Handbook Applications 11: Space Equipment and Systems, Japan Society of Mechanical Engineering, ISBN 978-4-88898-154-5, Japan

[9] John D.Johnston, Earl A. Thornton(1998), Thermally Induced Attitude Dynamics of a Spacecraft with a Flexible Appendage, *Journal of Guidance, Control, and Dynamics*, vol. 21, no. 4, pp. 581-587

[10] Kojima, Y., Taniwaki, S., & Ohkami, Y. (2004): Attitude Vibration Caused by a Stick-Slip Motion for Flexible Solar Array of Advanced Earth Observation Satellite. *Journal of Vibration and Control*, Vol. 10, No. 10, 2004, pp.1459-1472.

[11] Ming-De Xue, Jin Duan, Zhi-Hai Xiang(2007), Thermally-Induced bending-torsion coupling vibration of large space structures, Comput Mech, vol. 40, pp.707-723

[12] Mobara, M. (1994). Introduction to aerospace engineering guidance and control of satellite and rocket, Baifukan, ISBN 978-4-56303-493-1, Japan

[13] Oda, M., Hagiwara, Y., Suzuki, S., Nakamura, T., Inaba, N., Sawada, H., Yoshii, M., & Goto, N. (2011): Measurement of Satellite Solar Array Panel Vibrations Caused by Thermal Snap and Gas Jet Thruster Firing. *Recent Advances in Vibrations Analysis*, In-Tech, ISBN 978-953-307-696-6, Croatia.

[14] S, Maekawa, K, Nakano (2008): Relationship between Schallamach waves in contact surfaces and stick-slip in sliding systems, Manuscript of JAST Tribology Conference 2008, pp81-82

[15] Taniwaki, S., Kojima, Y., Ohkami, Y. (2007), Attitude Stability Analysis for Stick-Slip-Induced Disturbances of Extended Structure with Tension Control Mechanism, Journal of System Design and Dynamics, vol. 1, No. 4, pp714-723

Recent Advances on Force Identification in Structural Dynamics

N. M. M. Maia, Y. E. Lage and M. M. Neves

Additional information is available at the end of the chapter

http://dx.doi.org/10.5772/51650

1. Introduction

This chapter presents recent advances on force identification for structural dynamics that have been developed by the authors using the concept of transmissibility for multiple degree-of-freedom (MDOF) systems.

Being applied for many years only to the single degree-of-freedom (SDOF) system or to MDOF systems in a very limited way, the transmissibility concept has been developed along the last decade or so in a consistent manner, to be applicable in a general and complete way to MDOF systems. Various applications for MDOF systems may now be found, such as evaluation of unmeasured frequency response functions (FRFs), force identification, detection of damage, etc. A review of the multiple applications of the transmissibility concept has been published recently [1].

It is the application of this generalized transmissibility concept to both the direct and inverse force identification that is described along this chapter. The direct problem is understood as the one where one knows the applied forces and wishes to estimate the reactions at the supports; the inverse force identification problem is when one wishes to determine how many forces are applied, where they are applied and which are their magnitudes.

To determine the location and magnitude of the dynamic forces that excite the system is an important issue in structural dynamics [2, 3], especially when operational forces cannot be directly measured, as it happens at inaccessible locations [4, 5]; it is often the case that transducers cannot be introduced in the structure to allow the experimental measurement of the external loads and only a limited number of sensors and positions are available. The identification of forces from vibration measurements at a few accessible locations is a very important problem in various areas, such as vibration control, fatigue life prediction and health monitoring.

Although the force identification problem may be solved from the dynamic responses by simply reversing the direct problem, this is usually ill-posed and sensitive to perturbations in the measured data.

Over the past years, the theory of inverse methods has been actively developed in many research areas presenting in common the effects of matrix ill-conditioning, reflecting the ill-posedness nature of the inverse problem itself. Those problems can often be overcome by methods such as pseudo-inversion for over-determined systems, use of Kalman filters [6, 7], Singular Value Decomposition and Tikhonov regularization [8-10].

Various research works in force identification can be found in the literature, such as those related to the identification of impact forces, implementation of prediction models based on reflected waves or simply from the dynamic responses [11-18], prediction of forces in plates for systems with time dependent properties [11] and identification of harmonic forces [13].

These methods to identify operational loads based on response measurements can be classified into three main categories: deterministic methods, stochastic methods and methods based on artificial intelligence. Two main classes of identification technique are considered in the group of deterministic methods for load identification: frequency-domain methods and time-domain methods. The force identification in time domain has been less studied than its frequency domain equivalent, therefore there are not that many force identification studies in the literature. A review on the state of the art for dynamic load identification may be found in [3, 14].

Although out of the scope of this chapter, some references are here given with respect to recent time-domain force identification developments. One interesting approach based on modal filtering [15] is the Sum of Weighted Accelerations Technique (SWAT), which allows to obtain the time-domain force reconstruction by isolating the rigid body modal accelerations. Another approach for time-domain force reconstruction is the Inverse Structural Filter (ISF) method of Kammer and Steltzner [16] that inverts the discrete-time equations of motion. A variant of this, expected to produce a stable ISF when the standard method fails was recently developed and named as Delayed Multi-step ISF (DMISF). For a more detailed description on these methods (SWAT, ISF and DMISF) see e.g. [17] and for its application to rotordynamics, see [18].

In this chapter, the authors treat the frequency-domain problem from a different perspective, which is based on the MDOF transmissibility concept. As aforementioned, usually the transmissibility of forces is defined in textbooks for SDOF systems, simply as the ratio between the modulus of the transmitted force magnitude to the support and the modulus of the applied force magnitude. For SDOF systems, the expression of either the transmissibility of motion or forces is exactly the same; however, as explained in [1], that is not the case for MDOF systems. On the one hand, the problem of extending the idea of transmissibility of motion to an MDOF system is essentially a problem of how to relate a set of unknown responses to a set of known responses associated to a given set of applied forces; on the other hand, for the transmissibility of forces the question is how to relate a set of reaction forces to a set of applied ones.

Some initial attempts on the generalization of the transmissibility concept are due toVakakis et al. [19-21], Liu et al. [22, 23] and Varoto [24]. Similar efforts can also be found in the indirect measurement of vibration excitation forces [2, 4, 5]. To the best knowledge of the authors, a general answer to the problem is due to Ribeiro [25], and in [26] the experimental evaluation of the transmissibility concept for MDOF systems is presented. The concept of transmissibility of forces for MDOF has been proposed in 2006 [27], where the authors explain the formulation of the transmissibility using both the dynamic stiffness and the receptance matrices.

The use of the transmissibility in conjunction with a two step methodology for force identification is the main novelty of this chapter. For the force identification based on the transmissibility of motion, two steps are taken, (i) firstly the number of forces and their location are obtained, and (ii) secondly the reconstruction of the load vector is performed using some of the responses obtained experimentally together with the updated numerical model. Both have been numerically developed and implemented, as well as experimentally tested in the research group during the last years to access the potential of these new methods.

In section 2, the authors review the generalized transmissibility concepts, both in terms of displacements and forces. They are introduced and deduced from two different perspectives, (i) from the frequency response functions, (ii) from the dynamic stiffness

In section 3, a numerical model and an experimental application are presented to illustrated the transmissibility concept.

In section 4 the methodologies proposed for force identification based on the transmissibility concept are introduced.

Some simulated and experimental results are presented to show how these methodologies are able to help us identifying applied and reaction forces. The authors present a discussion on these proposed methods and on the obtained results.

2. Transmissibility in MDOF systems

The transmissibility concept may be found in any fundamental textbook on mechanical vibrations (e.g. [28]), related to SDOF systems.

The transmissibility of motion is defined as the ratio between the modulus of the response amplitude (output) and the modulus of the imposed base harmonic displacement (input). Depending on the imposed frequency, the result can vary from an amplification to an attenuation in the response amplitude relatively to the input one.

On the other hand, the transmissibility of force is defined as the ratio between the modulus of the transmitted force magnitude to the support and the modulus of the imposed force magnitude.

It happens that for SDOF systems the expression for calculating the transmissibility is the same, either referring to forces or to motion. This is not the case for MDOF systems.

The generalization of these definitions to MDOFs has been developed in the last decade, as mentioned before. In this section a brief review of these generalizations is given, introducing also the concepts and notation used for the force identification problem.

2.1. Transmissibility of motion in MDOF systems

To introduce the problem, the authors follow here as near as possible the notation used in [1]. Let K be the set of n_K co-ordinates where the displacement responses $\boldsymbol{Y}_K$ are known (measured or computed), U the set of n_U co-ordinates where the displacement responses $\boldsymbol{Y}_U$ are unknown, and A the set of co-ordinates where the forces $\boldsymbol{F}_A$ may be applied (Fig. 1).

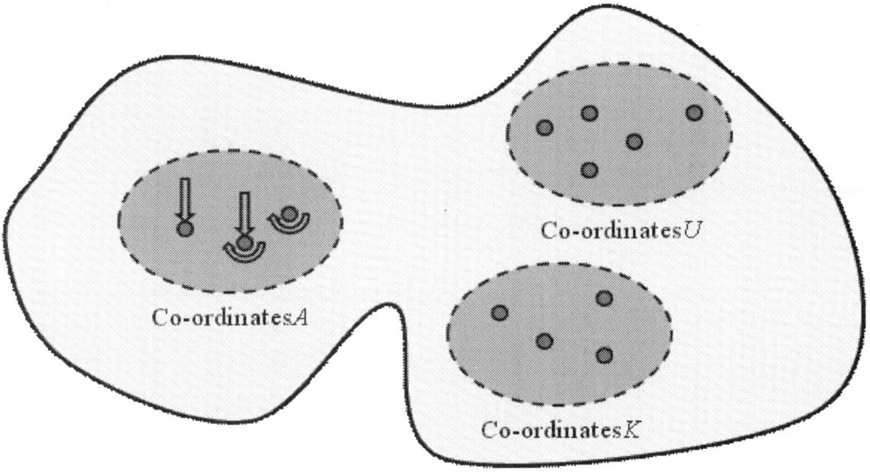

Figure 1. Illustration of an elastic body with the three sets of co-ordinates *K*, *U* and *A*.

To obtain the needed transmissibility of motion one may consider two distinct ways. The first is based on the frequency response function (FRF) matrices $\boldsymbol{H}(\omega)$, known as the fundamental formulation, while the second is based on the dynamic stiffness matrix $\boldsymbol{Z}(\omega)$ and is named alternative formulation.

The receptance frequency response matrix $\boldsymbol{H}(\omega)$ relates the dynamic displacement amplitudes $\boldsymbol{Y}$ with the external force amplitudes $\boldsymbol{F}$ as (using harmonic excitation, in steady-state conditions):

$$\boldsymbol{Y} = \boldsymbol{H}\,\boldsymbol{F} \quad \Leftrightarrow \quad \boldsymbol{Y} = \left(\boldsymbol{K} - \omega^2\,\boldsymbol{M} + i\omega\,\boldsymbol{C}\right)^{-1}\boldsymbol{F} \tag{1}$$

where $\boldsymbol{K}$, $\boldsymbol{M}$ and $\boldsymbol{C}$ are the stiffness, mass and viscous damping matrices, respectively. $\boldsymbol{H}(\omega)$ includes all the degrees of freedom in which the system is discretized and corresponds to the inverse of the dynamic stiffness matrix $\mathbf{Z}(\omega)$. One may underline that the mass-normalized orthogonality properties are observed here:

$$\begin{aligned} \mathbf{\Phi}^T M \mathbf{\Phi} &= \mathbf{I} \\ \mathbf{\Phi}^T K \mathbf{\Phi} &= \mathbf{diag}(\omega_r^2) \end{aligned} \tag{2}$$

Assuming proportional damping, $C=\alpha K+\beta M$ and therefore,

$$Y = H\,F = \Phi\left[diag(\omega_r^2-\omega^2) + i\,\omega\left(\alpha\, diag(\omega_r^2) + \beta\, I\right)\right]^{-1} \Phi^T F \tag{3}$$

where Φ is the mode shape matrix, ω_r is the rth natural frequency and α and β are constants.

From (1) it is easy to understand that if the responses $\boldsymbol{Y}$ at the discretization points are known, then the force reconstruction (in frequency-domain) would be given by:

$$\boldsymbol{F} = \boldsymbol{H}^{-1}\,\boldsymbol{Y} \tag{4}$$

2.1.1. Transmissibility of motion in terms of FRFs

Based on harmonically applied forces at co-ordinates A, one may establish that displacements at co-ordinates U and K are related to the applied forces at co-ordinates A by the following relationships:

$$\boldsymbol{Y}_U = \boldsymbol{H}_{UA}\,\boldsymbol{F}_A \tag{5}$$

$$\boldsymbol{Y}_K = \boldsymbol{H}_{KA}\,\boldsymbol{F}_A \tag{6}$$

Eliminating the external forces $\boldsymbol{F}_A$ between (5) and (6), one obtains

$$\boldsymbol{Y}_U = \boldsymbol{H}_{UA}\left(\boldsymbol{H}_{KA}\right)^+ \boldsymbol{Y}_K = \left(\boldsymbol{T}_d\right)_{UK}^A \boldsymbol{Y}_K \tag{7}$$

where

$$\left(\boldsymbol{T}_d\right)_{UK}^A = \boldsymbol{H}_{UA}\left(\boldsymbol{H}_{KA}\right)^+ \tag{8}$$

is the transmissibility matrix relating both sets of displacements. $(\boldsymbol{H}_{KA})^+$ is the pseudo-inverse of the sub-matrix $\boldsymbol{H}_{KA}$. An important property of the transmissibility matrix to be used here is that it does not depend on the magnitude of the involved forces and only requires the knowledge of a set of co-ordinates that include all the co-ordinates where the forces are applied. Indeed, it is required that n_K be greater or equal to n_A. One important aspect of this definition is that sub-matrices $\boldsymbol{H}_{UA}$ and $\boldsymbol{H}_{KA}$ may be obtained experimentally.

2.1.2. *Transmissibility of motion in terms of dynamic stiffness*

There exists an alternative approach to obtain the transmissibility matrix for the displacements, using the dynamic stiffness matrices introduced in (1). Assuming again harmonic loading and defining two subsets, *A* and *B*, *A* being the set where the dynamic loads may be applied and *B* the set formed by the remaining co-ordinates, where no forces are applied ($\boldsymbol{F}_B = \boldsymbol{0}$), one can obtain (after grouping adequately the degrees of freedom of the problem):

$$\begin{bmatrix} \boldsymbol{Z}_{AK} & \boldsymbol{Z}_{AU} \\ \boldsymbol{Z}_{BK} & \boldsymbol{Z}_{BU} \end{bmatrix} \begin{Bmatrix} \boldsymbol{Y}_K \\ \boldsymbol{Y}_U \end{Bmatrix} = \begin{Bmatrix} \boldsymbol{F}_A \\ \boldsymbol{0} \end{Bmatrix} \tag{9}$$

Developing eq. (9), it follows that

$$\begin{aligned} &\boldsymbol{Z}_{AK}\, \boldsymbol{Y}_K + \boldsymbol{Z}_{AU}\, \boldsymbol{Y}_U = \boldsymbol{F}_A \qquad \text{a} \\ &\boldsymbol{Z}_{BK}\, \boldsymbol{Y}_K + \boldsymbol{Z}_{BU}\, \boldsymbol{Y}_U = \boldsymbol{0} \qquad \text{b} \end{aligned} \tag{10}$$

From (10b) one obtains the transmissibility in terms of the dynamic stiffnesses:

$$\boldsymbol{Y}_U = -\left(\boldsymbol{Z}_{BU}\right)^{+} \boldsymbol{Z}_{BK} \boldsymbol{Y}_K = \left(\boldsymbol{T}_d\right)_{UK}^{A} \boldsymbol{Y}_K \tag{11}$$

where $(\boldsymbol{Z}_{BU})^{+}$ is the pseudo-inverse of $\boldsymbol{Z}_{BU}$.

From (11) it is possible to obtain the response at the unknown co-ordinates, as long as the pseudo-inverse is viable, which requires that n_B is greater or equal to n_U.

Indeed, from all this resulted two conditions:

$$\left(\boldsymbol{T}_d\right)_{UK}^{A} = -\left(\boldsymbol{Z}_{BU}\right)^{+} \boldsymbol{Z}_{BK} = \boldsymbol{H}_{UA}\left(\boldsymbol{H}_{KA}\right)^{+} \qquad n_B \geq n_U \text{ and } n_K \geq n_A \tag{12}$$

2.2. Transmissibility of forces in MDOF systems

To introduce the transmissibility of forces for MDOF systems, the authors follow a similar procedure to the one used in the previous sub-section. The problem consists now of relating the set of known applied forces to a set of unknown reactions (or the other way around), relating the set of known applied forces (set *K*) with a set of unknown reaction forces (set *U*), which are illustrated in Fig.2. At the set *U* it will be assumed that $\boldsymbol{Y}_U = \boldsymbol{0}$. In general, there will be other co-ordinates, where neither there are any applied forces nor there are any reactions, that shall constitute the set *C*.

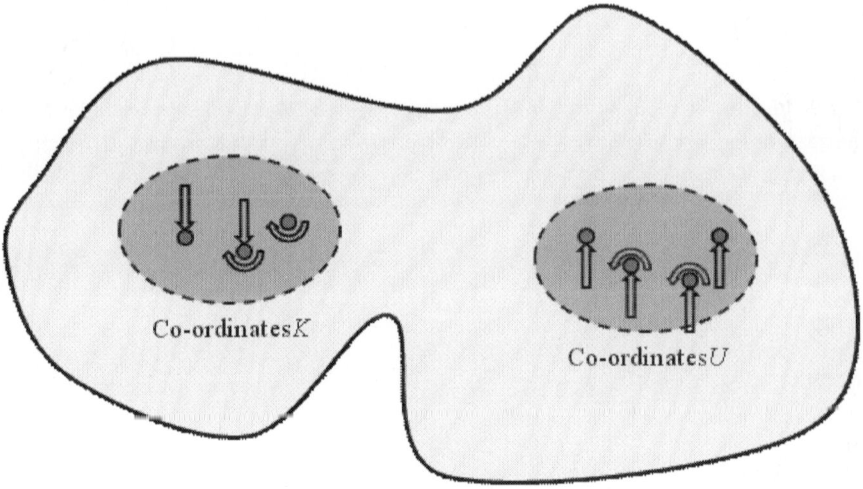

Figure 2. Illustration of both sets of co-ordinates *K* and *U*.

2.2.1. Transmissibility of forces in terms of FRFs

With the definition of the new sets *K*, *U* and *C*, the problem may be defined in the following way:

$$\begin{Bmatrix} \boldsymbol{Y}_K \\ \boldsymbol{Y}_U \\ \boldsymbol{Y}_C \end{Bmatrix} = \begin{bmatrix} \boldsymbol{H}_{KK} & \boldsymbol{H}_{KU} \\ \boldsymbol{H}_{UK} & \boldsymbol{H}_{UU} \\ \boldsymbol{H}_{CK} & \boldsymbol{H}_{CU} \end{bmatrix} \begin{Bmatrix} \boldsymbol{F}_K \\ \boldsymbol{F}_U \end{Bmatrix} \tag{13}$$

Imposing $\boldsymbol{Y}_U = \boldsymbol{0}$, it follows that

$$\boldsymbol{H}_{UK}\ \boldsymbol{F}_K + \boldsymbol{H}_{UU}\ \boldsymbol{F}_U = 0 \tag{14}$$

and so

$$\boldsymbol{F}_U = \left(\boldsymbol{T}_f\right)_{UK}\ \boldsymbol{F}_K \tag{15}$$

where

$$\left(T_f\right)_{UK} = -\left(H_{UU}\right)^{-1} H_{UK} \tag{16}$$

is the force transmissibility matrix.

This is the direct force identification method, i.e., one knows the applied forces and calculate the reactions at the supports, where the displacements are assumed as zero. The inverse problem is also possible, if one is able to measure the reaction forces and if their number is higher than the number of applied forces, in order to calculate the pseudo-inverse of H_{UK}:

$$F_K = \left(\left(T_f\right)_{UK}\right)^{+} F_U \tag{17}$$

where

$$\left(\left(T_f\right)_{UK}\right)^{+} = -\left(H_{UK}\right)^{+} H_{UU} \tag{18}$$

Note that in spite of the fact that here the reaction forces are known, the notation U (that in principle stands for "unknown") is kept.

In the inverse problem, one may not know how many applied force exist and where they are applied. If that is the case, one must follow a different approach, as it will be explained in section 4.1

If the condition $\mathbf{Y}_U = \mathbf{0}$ is relaxed, from eq. (13) it follows that:

$$Y_U = H_{UK} F_K + H_{UU} F_U \tag{19}$$

$$F_U = \left(T_f\right)_{UK} F_K + \left(H_{UU}\right)^{-1} Y_U \qquad \text{a}$$

and

$$F_K = \left(\left(T_f\right)_{UK}\right)^{+} F_U + \left(H_{UK}\right)^{+} Y_U \qquad \text{b} \tag{20}$$

2.2.2. *Transmissibility of forces in terms of dynamic stiffness*

Again, there is an alternative approach to obtain the force transmissibility matrix, using the dynamic stiffness matrices.

Assuming harmonic loading and the mentioned sets *K*, *U* and *C*, one can obtain (after grouping adequately the degrees of freedom of the problem) the following result:

$$\begin{bmatrix} \boldsymbol{Z}_{KK} & \boldsymbol{Z}_{KC} & \boldsymbol{Z}_{KU} \\ \boldsymbol{Z}_{CK} & \boldsymbol{Z}_{CC} & \boldsymbol{Z}_{CU} \\ \boldsymbol{Z}_{UK} & \boldsymbol{Z}_{UC} & \boldsymbol{Z}_{UU} \end{bmatrix} \begin{Bmatrix} \boldsymbol{Y}_K \\ \boldsymbol{Y}_C \\ \boldsymbol{Y}_U \end{Bmatrix} = \begin{Bmatrix} \boldsymbol{F}_K \\ \boldsymbol{F}_C \\ \boldsymbol{F}_U \end{Bmatrix} \tag{21}$$

It is worthwhile noting that joining together the sets *K* and *C* in a new set *E* makes it easier to see that imposing $\boldsymbol{Y}_U = \boldsymbol{0}$ one obtains the following relationships:

$$\begin{bmatrix} \boldsymbol{Z}_{EE} & \boldsymbol{Z}_{EU} \\ \boldsymbol{Z}_{UE} & \boldsymbol{Z}_{UU} \end{bmatrix} \begin{Bmatrix} \boldsymbol{Y}_E \\ \boldsymbol{0} \end{Bmatrix} = \begin{Bmatrix} \boldsymbol{F}_E \\ \boldsymbol{F}_U \end{Bmatrix} \tag{22}$$

from which it is clear that:

$$\begin{aligned} \boldsymbol{Z}_{EE}\,\boldsymbol{Y}_E &= \boldsymbol{F}_E \qquad \text{a} \\ \boldsymbol{Z}_{UE}\,\boldsymbol{Y}_E &= \boldsymbol{F}_U \qquad \text{b} \end{aligned} \tag{23}$$

Eliminating $\boldsymbol{Y}_E$ between (23a) and (23b), it turns out that

$$\boldsymbol{F}_U = \left(\boldsymbol{T}_f\right)_{UE} \boldsymbol{F}_E \tag{24}$$

where

$$\left(\boldsymbol{T}_f\right)_{UE} = \boldsymbol{Z}_{UE}\left(\boldsymbol{Z}_{EE}\right)^{-1} \tag{25}$$

The inverse problem corresponds to

$$\boldsymbol{F}_E = \left(\left(\boldsymbol{T}_f\right)_{UE}\right)^{+} \boldsymbol{F}_U \tag{26}$$

with

$$\left(\left(\boldsymbol{T}_f\right)_{UE}\right)^{+} = \boldsymbol{Z}_{EE}\left(\boldsymbol{Z}_{UE}\right)^{+} \tag{27}$$

It is important to note that only some of the co-ordinates of the set E have applied forces. This means that in (23) some rows of F_E are zero and only the columns (in Z_{EE}) whose co-ordinates have applied forces (set K) are needed for the transmissibility matrix. In other words, from the set E only the co-ordinates corresponding to the K set are used.

2.3. Summary

From sections 2.2.1 and 2.2.2, one can conclude that for the direct problem of transmissibility of forces there is no restrictions in the number of co-ordinates used:

$$\begin{aligned} \left(\boldsymbol{T}_f\right)_{UK} &= -\left(\boldsymbol{H}_{UU}\right)^{-1}\boldsymbol{H}_{UK} \\ \left(\boldsymbol{T}_f\right)_{UE} &= \boldsymbol{Z}_{UE}\left(\boldsymbol{Z}_{EE}\right)^{-1} \end{aligned} \tag{28}$$

whereas in the inverse problem of transmissibility of forces there are some restrictions that can make this option not very useful in practice, especially when using the dynamic stiffnesses, since one needs to calculate the pseudo-inverse matrices:

$$\begin{aligned} \left(\left(\boldsymbol{T}_f\right)_{UK}\right)^{+} &= -\left(\boldsymbol{H}_{UK}\right)^{+}\boldsymbol{H}_{UU} & \quad n_U \geq n_K \\ \left(\left(\boldsymbol{T}_f\right)_{UE}\right)^{+} &= \boldsymbol{Z}_{EE}\left(\boldsymbol{Z}_{UE}\right)^{+} & \quad n_U \geq n_E \end{aligned} \tag{29}$$

3. Numerical and experimental applications

As explained before, the transmissibility matrices may be obtained from a numerical model (which should be updated for the range of frequencies involved) or from results obtained experimentally. In this section, the methodology used in each case is described and illustrated through a comparison example.

3.1. Transmissibility in terms of the numerical model

For the numerical model, one needs the knowledge of the structure within the discretization chosen, to create the receptance matrix $\boldsymbol{H}(\omega)$, which is the inverse of the corresponding dynamic stiffness matrix $\boldsymbol{Z}(\omega)$. Here, the numerical model is created using the Finite Element Method (FEM), although other alternatives may also be used. As seen before, the dynamic stiffness matrix is defined as:

$$\boldsymbol{Z}(\omega) = \boldsymbol{K} - \omega^2\boldsymbol{M} + i\omega\,\boldsymbol{C} \tag{30}$$

where C represents the viscous damping matrix, often of the proportional type, i.e., $C=\alpha K+\beta M$, where α and β are constants to be evaluated experimentally.

To build the dynamic stiffness matrix, a specific structural finite element is chosen according to the approximation considered. For example, in the case of a reasonably long and slender beam one can use the Euler-Bernoulli beam element (instead of a shell or solid structural element). Then, the global matrices are assembled for the chosen discretization of the structure.

In order to improve the accuracy of the numerical model when simulating what is obtained experimentally, concentrated masses are often added at the corresponding nodes to model the effect of the accelerometers used in the testing positions.

Although the receptance matrix $H(\omega)$ is the inverse of the corresponding dynamic stiffness matrix, one should avoid such direct numerical inversion (frequency by frequency). Instead, $H(\omega)$ is calculated from eq. (3), after a modal analysis in free vibration.

Then, using (8) or (16), one can calculate the needed transmissibility matrices $(T_d)_{UK}^{A}$ or $(T_f)_{UK}$. Of course, alternatively one may use the equivalent expressions (11) or (25), respectively.

3.2. Transmissibility in terms of experimental measurements

Depending on the type of transmissibility to obtain, the corresponding experimental setup should be established. Essentially, it is important to observe that for the transmissibility of motion one measures the FRFs relating co-ordinates U and K with co-ordinates A, normally using accelerometers and force transducers. For validation purposes, one may also measure the applied forces. In the examples presented next for the transmissibility of motion, a suspended (free-free) beam is always used.

For the transmissibility of forces, in the direct problem, one measures the applied forces at co-ordinates K (in the inverse problem, one measures the reaction forces at co-ordinates U). For validation purposes, one also measures the ones to be estimated. The test specimen for the transmissibility of forces is always a simply supported beam.

For the experimental setup, the following equipment is used:

- Vibration exciter (Brüel & Kjær Type 4809);
- Power amplifier (Brüel & Kjær Type 2706);
- Force transducers(PCB PIEZOTRONICS Model 208C01);
- Data acquisition equipment (Brüel & Kjær Type 3560-C).

In Fig. 3, a schematic representation of the experimental setup used for the force transmissibility tests is presented, in this case a simply supported beam with a single applied force.

The excitation signal used was a multi-sine transmitted to the exciter, with constant amplitude in the frequency. In reality the signal measured by the force transducer does not exhibits a constant amplitude along the frequency, as it depends on the dynamic response of the structure.

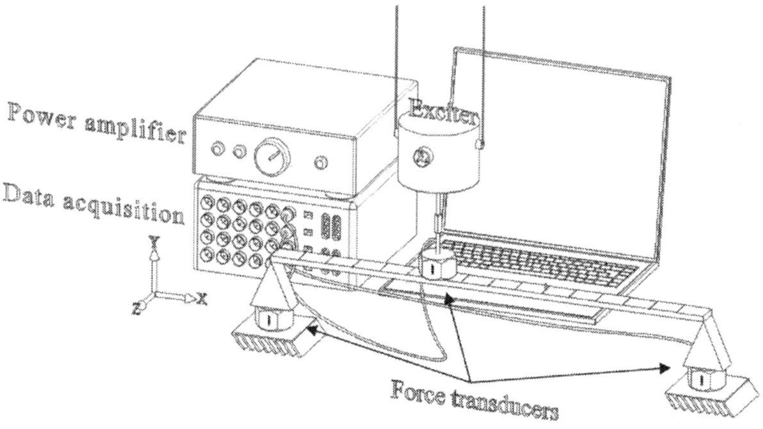

Figure 3. Example of the experimental setup developed for the transmissibility of forces.

For the transmissibility of motion, a beam suspended by nylon strings is used, to simulate free-free conditions. In order to facilitate the interchange of the available accelerometers between the measure positions without affecting the dynamics of the structure, it is important to add equivalent masses (dummies) to model the effect of the sensors.

After obtaining experimentally the needed receptances, by using (8) or (16) one can establish the transmissibility matrices $(T_d)_{UK}^A$ or $(T_f)_{UK}$.

3.3. Examples

The same steel beam was used in all the examples. With the purpose of illustrating the applicability of the presented formulations to obtain the transmissibility plots, the authors used the geometric and material parameters presented in Table 1. Note that these data correspond to the values obtained after updating the FE model.

Young's modulus – E	208 GPa
Density – ρ	7840 kg/m³
Length – L	0.8 m
Section width - b	$5.0\times10^{-3}m$
Section height - h	$20.0\times10^{-3}m$
Section area - A	$1\times10^{-4}m^2$
Second moment of area - I	$2.0883\times10^{-10}m^4$
proportional damping - α	$4\,s$
proportional damping - β	$2.0\times10^{-6}s^{-1}$

Table 1. Beam properties (after updating).

3.3.1. *Numerical model*

The standard two-node Euler-Bernouli bidimensional finite element is used here to build the needed numerical model of the beam.

The beam was discretized into sixteen finite elements, which correspond to $N = 17$ nodes, ordered from 1 up to 17. As the analysis and the model are limited to the plane xOy, each node has three degrees of freedom (which are u_x, u_y, θ). Hence, the matrices of the numerical model have an order of $3xN$ for the free-free beam. In what the measurements are concerned, only the displacements and applied forces along the y direction are used and therefore the numbering of nodes and co-ordinates y coincide.

The model was updated using E, ρ and I as updating parameters and a proportional damping model is included using α and β as updating parameters (Table 1).

3.3.2. *Example 1 — Transmissibility of motion*

In this example, the free-free beam has only one applied force at node 11 along the y direction, denoted as F_{11}, and the measurements are taken at nodes 3, 7, 13 and 17 also along the y direction, denoted as Y_3, Y_7, Y_{13}, and Y_{17} (Fig. 4).

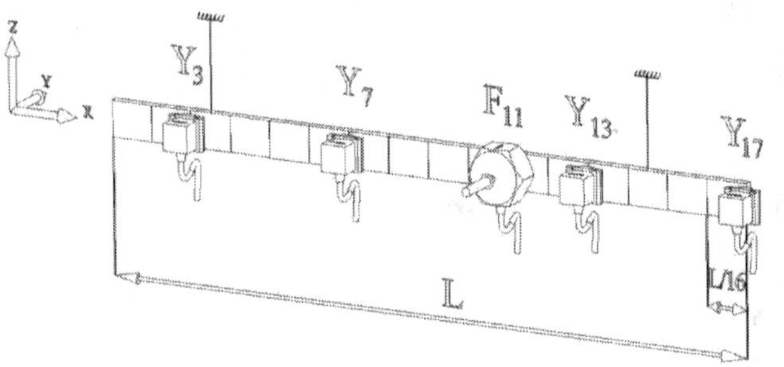

Figure 4. Schematic representation of the accelerometers and force transducer positions.

The transmissibility of motion between the pair of nodes (3,7) and the pair of nodes (13,17) may be expressed as

$$\mathbf{Y}_U = \left(\mathbf{T}_d\right)_{UK}^{A} \mathbf{Y}_K \quad \Leftrightarrow \quad \begin{Bmatrix} Y_{13} \\ Y_{17} \end{Bmatrix} = \begin{bmatrix} T_{13,3} & T_{13,7} \\ T_{17,3} & T_{17,7} \end{bmatrix} \begin{Bmatrix} Y_3 \\ Y_7 \end{Bmatrix} \tag{31}$$

These transmissibilities were obtained from eq. (8), using the FRFs calculated from the numerical model and measured experimentally. Two of them are plotted in Fig. 5, where one can observe that both ways are able to produce the transmissibility response of the structure, as they match reasonably well.

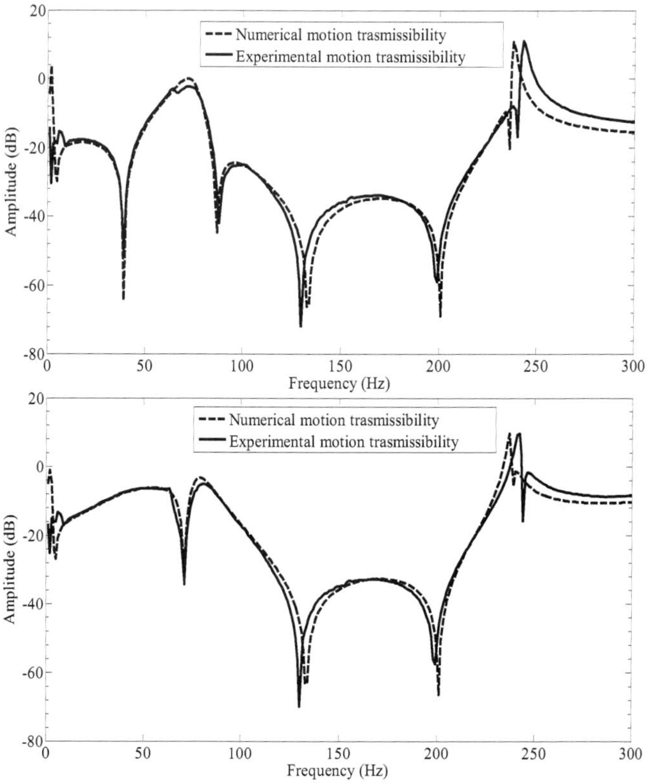

Figure 5. Numerical and experimental transmissibilities (upper plot, $T_{13,3}$; bottom plot, $T_{13,7}$)

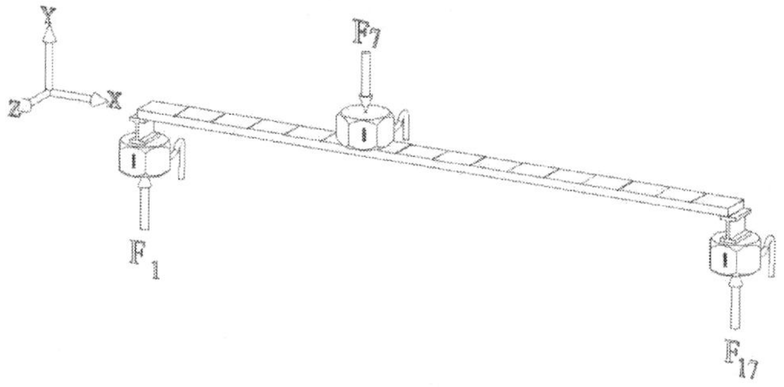

Figure 6. Schematic representation of the positions of the force transducers.

3.3.3. *Example 2 — Transmissibility of forces*

In this case, a simply supported beam is considered with one applied force at node 7 and reactions at nodes 1 and 17. Only the magnitude of the forces is measured, and the transmissibility is obtained directly from the measurements and compared with the numerical results. The experimental setup is illustrated in Fig. 6.

The force transmissibility relation between node 7 and the pair of nodes (1,17) may be expressed as

$$\begin{Bmatrix} F_1 \\ F_{17} \end{Bmatrix} = \begin{bmatrix} T_{1,7} \\ T_{17,7} \end{bmatrix} \{F_7\} \tag{32}$$

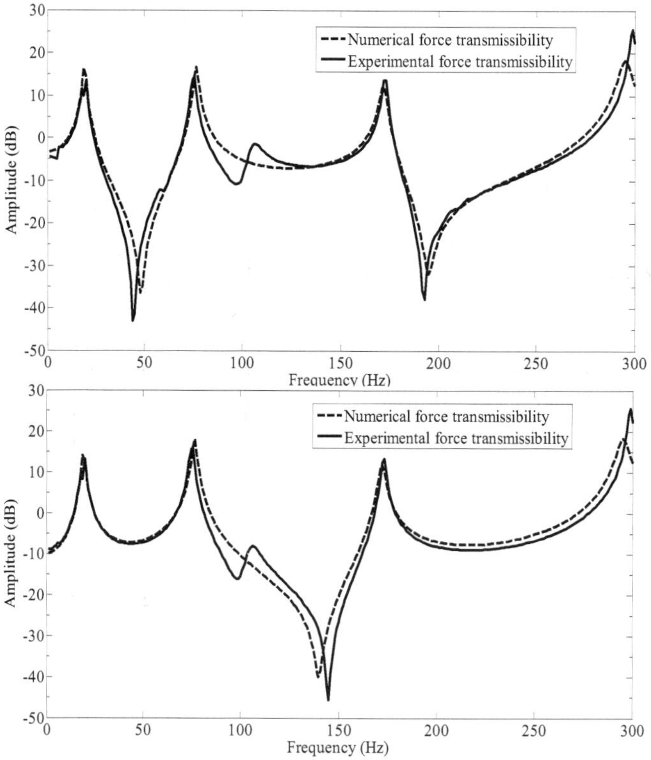

Figure 7. Numerical and experimental transmissibilities (upper plot, $T_{1,7}$; bottom plot, $T_{17,7}$)

The force transmissibilities were obtained using eq. (16) and are plotted in Fig. 7, where it is clear that both numerical and experimental FRFs are able to produce the transmissibility response of the structure. Note that around 100 Hz there is a "bump" in the experimental curve, due to the effect of the supports of the beam themselves; this effect has not been included in the numerical model because it was not important, as these results are only of an illustrative type.

4. Force identification

This section shall be divided into (i) part one for the force localization algorithm based on the transmissibility of motion and reconstruction using the measured responses and the updated numerical model, and (ii) part two for the force reconstruction using the transmissibility of forces.

4.1. Force localization based on the transmissibility of motion and force reconstruction

The force identification problem is a difficult matter, as one has a limited knowledge of the measured responses, due to the complexity of the structure, lack of access to some locations, etc. In other words, there are difficulties due to the incompleteness of the model.

Due to this difficulty in calculating the load vector directly, the authors propose to divide the process into two distinct steps:

1. the localization of the forces, i.e. the identification of the number and position of the applied forces using the concept of transmissibility of motion;
2. the load vector reconstruction.

For the first step, a search for the number and position of forces using the transmissibility of motion is performed. Essentially, this step consists of searching for the transmissibility matrix correspondent to the dynamics of the system and using the available measured data and the numerical model involved.

Once the corresponding transmissibility matrix is found, one has a solution for the number and position of the forces applied to the structure.

The second step consists of reconstructing the load vector with the results obtained in the first step. A more detailed description about this methodology is given in the following sections.

4.1.1. Force localization

In a first stage, to apply the method proposed in the previous section, one finds the transmissibility matrix that converts the dynamic responses $\boldsymbol{Y}_K$ into $\boldsymbol{Y}_U$. As one does not know the position of the applied forces, it was decided to cover all the possibilities until the calculated responses ($\boldsymbol{Y}_U$) match the measured ones $\tilde{\boldsymbol{Y}}$, over a range of frequencies. To calculate the vector $\boldsymbol{Y}_U$ one may use either eq. (7) or (11).

The maximum number of forces must be less or equal to the dimension of the known dynamic response vector $\boldsymbol{Y}$.

The successive combinations of the tested nodes are obtained according to the following scheme:

$$\text{For one force:} \quad \{(1), \ldots \ (\text{N});$$

$$\text{For two forces:} \quad \begin{cases} (1,2), \ldots \ (1,\text{N}); \\ (2,3), \ldots \ (2,\text{N}); \\ (3,4), \ldots \ (3,\text{N}); \\ \quad \vdots \\ \quad (N-1,N); \end{cases}$$

$$\text{For three forces:} \quad \begin{cases} (1,2,3), \ldots \ (1,2,N) \\ \quad \vdots \end{cases}$$

$$\vdots$$

The error in each combination is kept in a vector to identify the combination with the least associated error (in absolute value). Firstly, the algorithm scrolls through the possible combinations of position and number of forces. For each combination, the associated error between the calculated vector $\boldsymbol{Y}_U$ and the measured response vector $\tilde{\boldsymbol{Y}}$ is calculated; this is carried out over a frequency range defined by the user. The error between the predicted and the measured dynamic response at each co-ordinate i can be defined as:

$$\text{error}_i = \sum_{\omega} \left(\log\left(\text{abs}\left(\tilde{Y}_{U_i}(\omega) \right) \right) - \log\left(\text{abs}\left(Y_{U_i}(\omega) \right) \right) \right)^2 \tag{33}$$

For each combination, the calculated error is kept in an entry of the error vector and analyzed later on:

$$\boldsymbol{\varepsilon} = \{\text{error}_i\} \tag{34}$$

The accumulated error for a given combination of co-ordinates where $\boldsymbol{F}$ can be located is the norm of $\boldsymbol{\varepsilon}$. The calculations are repeated for sucessive combinations of number and position of forces. The combination of the force locations that gives the lowest error leads to the number and position of the forces applied to the structure. As already mentioned, the maximum number of forces that can be found is equal to the dimension of the known dynamic response vector.

As one does not know *a priori* how many forces exist, one has to follow a trial and error procedure that consists basically in assuming an increasing number of forces and the corresponding number of measurements; if the right number of forces is N_f, one has a minimum

error ε for a certain set of co-ordinates. When one proceeds and assumes N_f +1 forces and measurements, the error will be higher then ε, telling us that the right answer was effectively N_f at a certain set of co-ordinates.

It is clear that all the combinations of the N_f +1 forces that contain the right combination of the N_f forces should exhibit a local minimum, though not the absolute one.

The method was implemented computationally (in MatLab®).

4.1.2. Force reconstruction

In a second step, the reconstruction of the force amplitudes consists of solving an inverse problem using the measured dynamic responses $\mathbf{Y}_K$:

$$\boldsymbol{F}_A = \left(\boldsymbol{H}_{KA}\right)^+ \boldsymbol{Y}_K \tag{35}$$

Note that for the given system to be invertible, the number of dynamic responses to be used (set K) must be higher or equal than the number of applied forces (set A). However, this is always verified, as in the first step one has already imposed it.

4.1.3. Example 3 — Localization of the applied forces

This is a numerical example, illustrated in Fig.8, where a set of uncorrelated forces is applied at co-ordinates 1 and 5 (set *A*), and one uses the three known responses (set *K*) to identify the number and location of forces.

A set of simulated results (to mimic the experimental measurements) are obtained at nodes 1, 3, 5, 11 and 17 (see Fig. 8); they define the following sets:

$$\mathbf{Y}_K = \{Y_3 \quad Y_5 \quad Y_{17}\}^T \quad \text{and} \quad \mathbf{Y}_U = \{Y_1 \quad Y_{11}\}^T \tag{36}$$

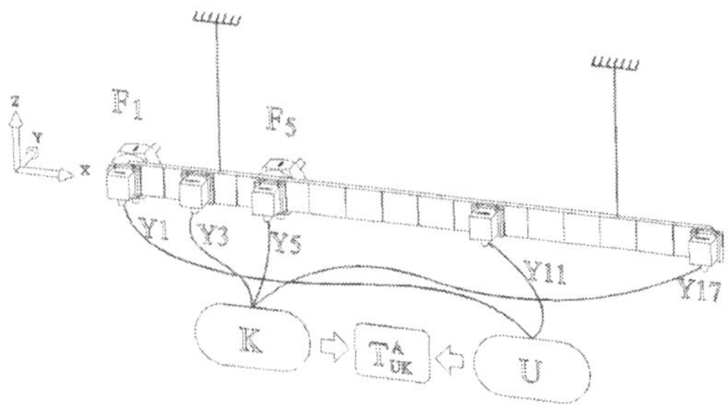

Figure 8. Illustration with the responses and applied force locations for example 3.

Considering these responses, the maximum number of identifiable applied forces is three, as explained before. The forces are uncorrelated and applied to the structure at co-ordinates 1 and 5. A series of force combinations have to be systematically generated as follows. In this case, all combinations up to three forces have been considered:

$$\text{one force}: \quad \{ \ (1),\ (2),\ (3),\dots\ (17); \qquad \#\text{combination number } [1 \text{ to } 17]$$

$$\text{two forces}: \quad \begin{cases} (1,2),\ (1,3),\ (1,4),\ (1,5)\ \dots(1,17); \\ \vdots \\ (16,17); \end{cases} \qquad \#\text{combination number } [18 \text{ to } 153]$$

$$\text{three forces}: \quad \begin{cases} (1,2,3)\dots \\ \vdots \end{cases} \qquad \#\text{combination number } [154 \text{ to } 833]$$

Applying the localization method described in the previous subsection 4.1.1, one obtains the plot of the error defined in eq. (33), as shown in Fig.9.

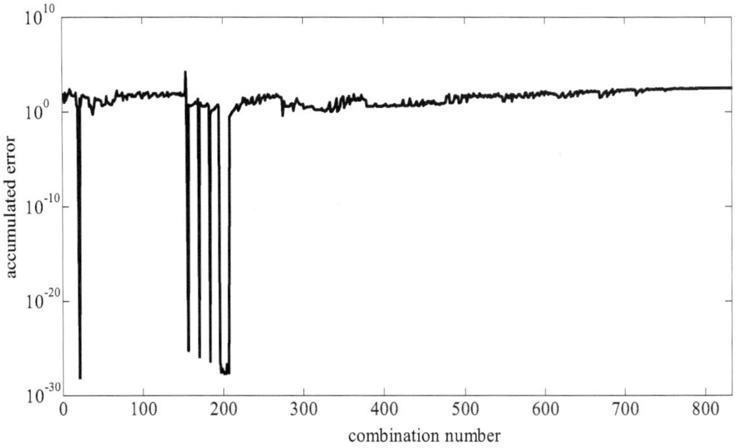

Figure 9. Accumulated error in frequency for each combination of forces.

It is clear that there are several situations (combinations) where the error is close to zero and other where is not.

The minimum error happens with the combination number 21, corresponding to two forces applied at co-ordinates 1 and 5, thus identifying the correct positions and number of forces (Table 2).

Combination	Number of forces	Position of the forces	Number of identified forces	Identified positions	Absolute error
21	2	1, 5	2	1, 5	2,69e-29

Table 2. Data of the combination with minimum error.

To better understand why there exist more combinations with small errors, Table 3 shows these combinations with its corresponding error value. All of them have a common group of co-ordinates, corresponding to the correct combination of number and positions of the forces. In this case the correct positions are obtained with success through the minimum error.

Combination	Real position of the forces	Absolute error
21	1,5	2,69e-29
156	1,2,5	7,99e-26
170	1,3,5	2,85e-27
183	1,4,5	6,99e-28
197	1,5,7	0,70e-29
198	1,5,8	1,86e-28
199	1,5,9	1,82e-28
200	1,5,10	5,54e-28

Table 3. Some combinations and their respective error.

This illustrates the localization step performed with two forces, whose number and location were not known at the beginning. From these results, it can be stated that the transmissibility of motion can be considered as adequate to perform this task. Note that, in spite of the high numbers of combinations that exist, the computations are relatively quick, as they involve only sub-matrices, and for the first permutations they are of a small order.

4.1.4. Example 4 — Localization and reconstruction 1

This is an experimental example, where a multisine signal is fed into the shaker, attached to the beam at co-ordinate 13. Later on, the applied force is compared with the reconstructed one.

The experimental measurements are obtained at nodes 5, 7, 11 and 15. The measured vectors are as follows:

$$\mathbf{Y}_K = \{Y_7 \quad Y_{15}\}^T \quad \text{and} \quad \mathbf{Y}_U = \{Y_5 \quad Y_{11}\}^T \tag{37}$$

Considering these responses, the maximum number of identifiable applied forces is two, as explained before.

A series of force combinations was systematically generated as described before. Applying the localization method, one obtains the graph of Fig. 10.

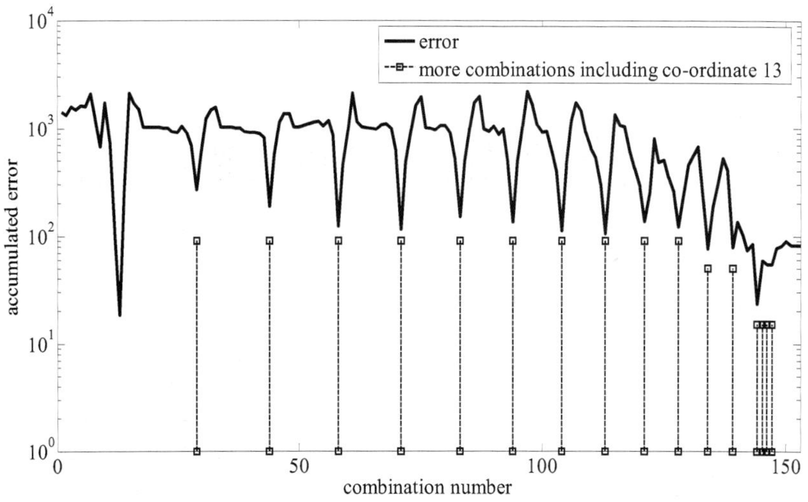

Figure 10. Accumulated error in frequency for each force combination.

As one can see, the absolute minimum corresponds to combination no. 13, which is right because this combination represents the force applied at co-ordinate 13. So, the method could localize correctly the position of the force at co-ordinate 13.

Once the localization of the force is accomplished, its reconstruction is a simple calculation, using the measured displacements relating those co-ordinates to the force location. As the force is located at node 13, taking the measurement at co-ordinates 5, 7, 11 and 15, for instance, it follows that:

$$\begin{Bmatrix} \tilde{Y}_5 \\ \tilde{Y}_7 \\ \tilde{Y}_{11} \\ \tilde{Y}_{15} \end{Bmatrix} = \begin{Bmatrix} H_{5,13} \\ H_{7,13} \\ H_{11,13} \\ H_{15,13} \end{Bmatrix} F_{13} \Leftrightarrow F_{13} = \begin{Bmatrix} H_{5,13} \\ H_{7,13} \\ H_{11,13} \\ H_{15,13} \end{Bmatrix}^{+} \begin{Bmatrix} \tilde{Y}_5 \\ \tilde{Y}_7 \\ \tilde{Y}_{11} \\ \tilde{Y}_{15} \end{Bmatrix} \quad (38)$$

Using the information from the measured responses, the reconstruction is now immediate. To validate this methodology, the result was ploted against its experimentally measured curve, as in Fig. 11. One may affirm that the method is able to predict the applied force. It is possible that a better matching of the curves may be obtained with a finer updated FE model.

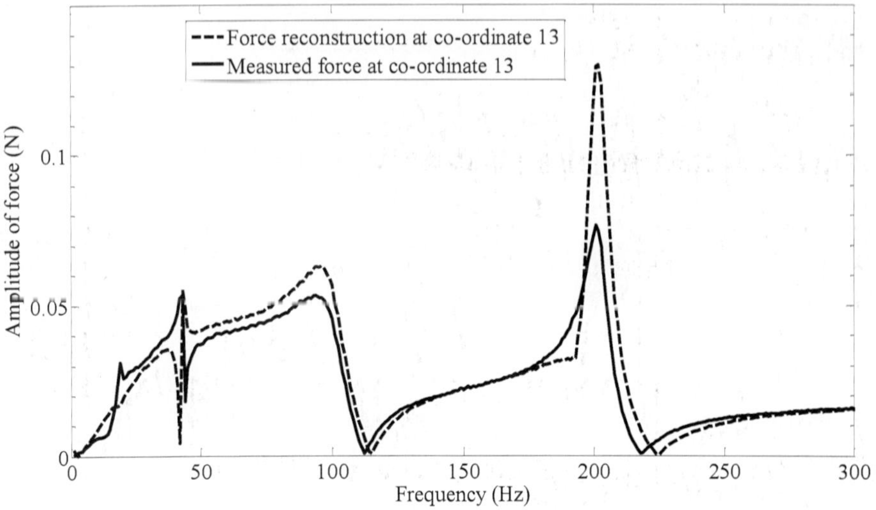

Figure 11. Comparison between the experimental and the reconstructed forces at co-ordinate 13.

4.1.5. Example 5 — Localization and reconstruction 2

Here, the same multisine signal is fed into two shakers, attached to the free-free beam at the co-ordinates 1 and 11. Later on, the applied forces are compared with the reconstructed ones.

The experimental measures are obtained at nodes 3, 7, 17 and 13. The measured vectors are as follows.

$$\mathbf{Y}_K = \{Y_3, Y_7, Y_{17}\}^T \quad \text{and} \quad \mathbf{Y}_U = \{Y_{13}\} \tag{39}$$

Considering these responses, the maximum number of identifiable applied forces is three. Applying the localization method, one obtains the plot shown in Fig.12.

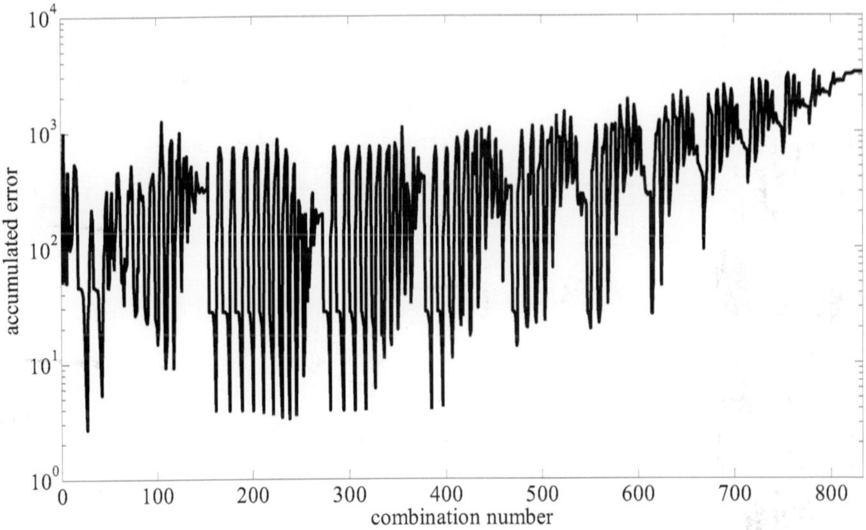

Figure 12. Accumulated error in frequency for each force combination.

As one can see, the absolute minimum corresponds to the right combination (no. 27), representing the forces applied at co-ordinates 1 and 11. So, the method located correctly the position of the forces. Note that, as there are two forces, a high number of combinations with small errors appear; observing the co-ordinates of those combinations, the best of them have in common the correct co-ordinates where the forces are applied and the others include co-ordinates physically close to them. Again, the force reconstruction is obtained using the measured displacements:

$$\begin{Bmatrix} \tilde{Y}_3 \\ \tilde{Y}_7 \\ \tilde{Y}_{13} \\ \tilde{Y}_{17} \end{Bmatrix} = \begin{bmatrix} H_{3,1} & H_{3,11} \\ H_{7,1} & H_{7,11} \\ H_{13,1} & H_{13,11} \\ H_{17,1} & H_{17,11} \end{bmatrix} \begin{Bmatrix} F_1 \\ F_{11} \end{Bmatrix} \Leftrightarrow \begin{Bmatrix} F_1 \\ F_{11} \end{Bmatrix} = \begin{bmatrix} H_{3,1} & H_{3,11} \\ H_{7,1} & H_{7,11} \\ H_{13,1} & H_{13,11} \\ H_{17,1} & H_{17,11} \end{bmatrix}^{+} \begin{Bmatrix} \tilde{Y}_3 \\ \tilde{Y}_7 \\ \tilde{Y}_{13} \\ \tilde{Y}_{17} \end{Bmatrix} \tag{40}$$

Figs. 13 and 14 present the reconstructed forces versus the measured ones. Again, one can state that the method is able to predict the applied force.

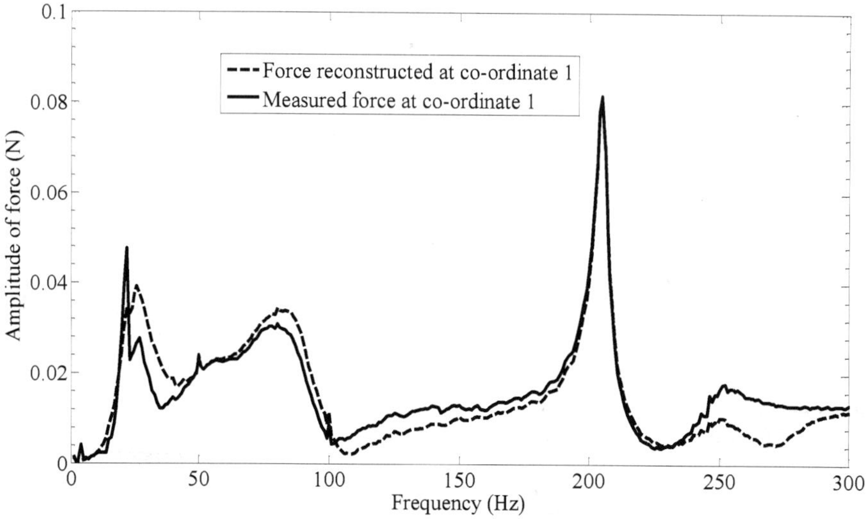

Figure 13. Comparison between the experimental and the reconstructed forces at co-ordinate1

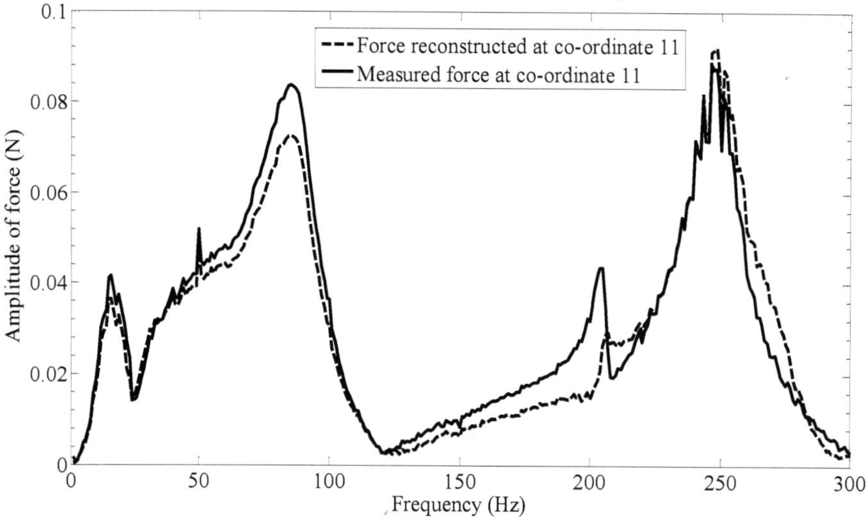

Figure 14. Comparison between the experimental and the reconstructed forces at co-ordinate11

4.2. Force reconstruction based on the transmissibility of forces

The main objective of this section is the estimation of the existing forces (reactions or applied forces) in the structure using the MDOF concept of transmissibility of forces. Two types of problems involving estimation of forces are here considered:

1. Reaction forces estimation, with the objective of calculating a set of unknown reactions from a set of known applied loads, as expressed by equation (15);

2. Applied forces estimation, with the objective of calculating a set of applied forces from a set of known reactions, as expressed by equation (17).

The method to estimate the applied forces is limited by the number of reactions, as it is not possible to perform the needed pseudo-inverse if the number of applied forces is greater than the number of reactions. So, it is a required condition that $n_K \le n_U$.

4.2.1. Example 6 — Reaction forces estimation knowing the applied ones

The first experimental reconstruction case was carried out with the configuration presented in Fig. 6 (simple supported beam). One has a single applied force at node 7 (set *K*) and two reactions at nodes 1 and 17 (set *U*).

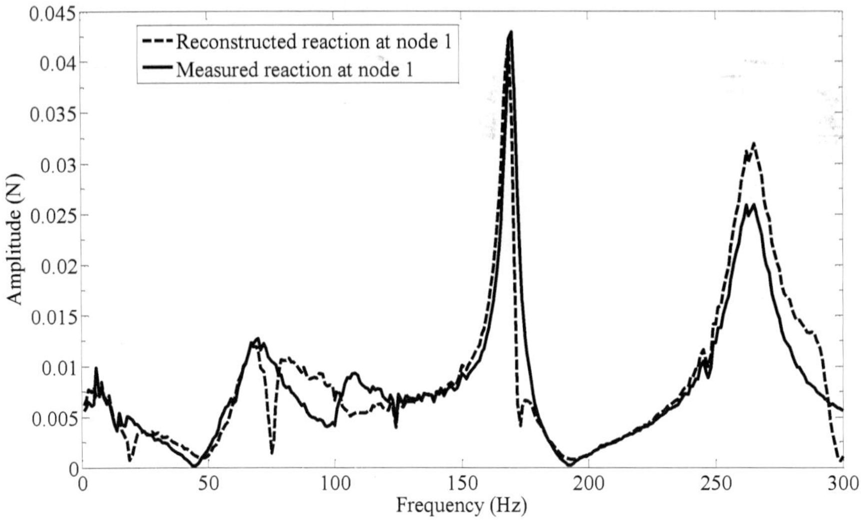

Figure 15. Comparison between the experimental and estimated force reaction F_1

In this case, with one applied force and two reactions, the transmissibility has a dimension of 2x1 and can be obtained either from the receptance matrix or from the dynamic stiffness matrix, as proposed in this work. The two different formulations are equivalent and are very close to the experimental results.

As the objective is to estimate the reaction forces, one needs the numerical model for the transmissibility matrix and to know the experimental vector of applied forces, which in this case has only one component. The calculation of the reactions is then reduced to the following form:

$$\begin{Bmatrix} F_1 \\ F_{17} \end{Bmatrix} = \begin{bmatrix} T_{1,7} \\ T_{17,7} \end{bmatrix} \{F_7\} \tag{41}$$

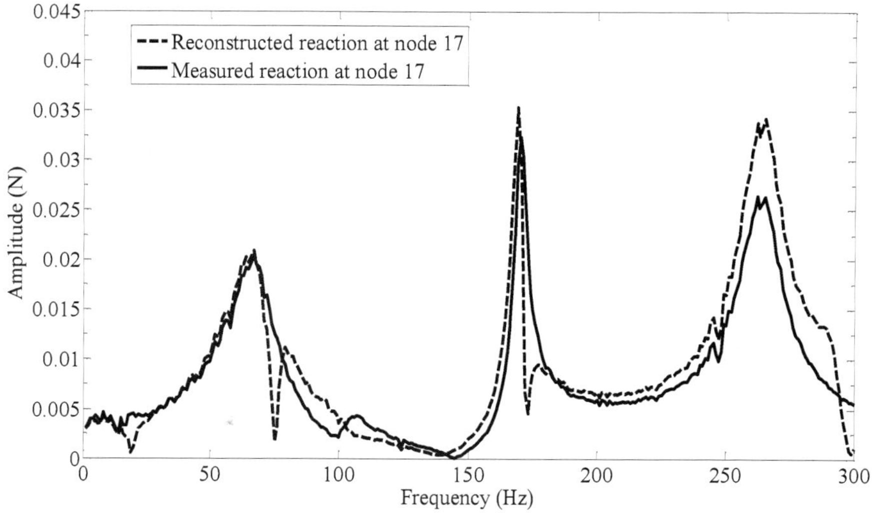

Figure 16. Comparison between the experimental and estimated force reaction F_{17}

From Figs. 15 and 16, it is clear that the reconstructed reactions match reasonably well the experimentally measured ones. Better results may even be possible if a finer updating procedure on the FE model is achieved.

4.2.2. Example 7 — Applied force reconstruction knowing the reaction forces

For the reconstruction of the applied forces (the inverse problem), one needs to known the vector of the reaction forces $\{F_U\}$ and the inverse transmissibility matrix that can be obtained from the numerical model.

In this case the same configuration presented in Fig. 6 was used (simple supported beam), with one applied force at node 7 (set K) and two reaction forces at nodes 1 and 17 (set U).

Knowing the reaction forces, the reconstruction of the applied force follows, in this case, the following expression:

$$\{F_7\} = \begin{bmatrix} T_{1,7} \\ T_{17,7} \end{bmatrix}^+ \begin{Bmatrix} F_1 \\ F_{17} \end{Bmatrix} \tag{42}$$

The reconstructed values are compared with the experimentally measured ones, in Fig. 17.

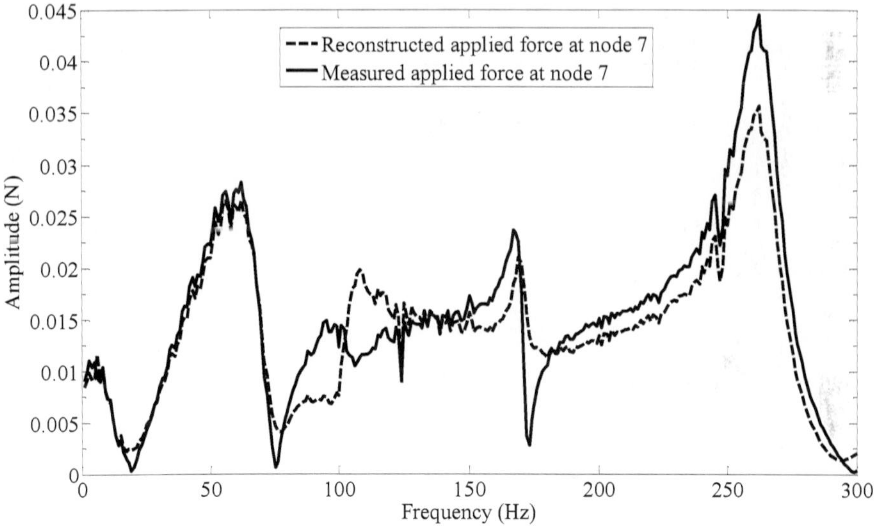

Figure 17. Comparison between the experimental and estimated applied load F7

In all the tested cases a good approach of the reconstructed forces was verified, as the values obtained by the direct and inverse problems are close enough to the experimentally measured ones.

5. Conclusions

In this work, the authors reviewed recent advances in the application of MDOF transmissibility-based methods for the identification of forces.

From these developments, one can draw the following main conclusions:

i. it is possible to localize forces acting on a structure, based on the motion transmissibility matrix, comparing the expected responses with the ones measured along the structure;

ii. finding where the forces are applied corresponds to finding the transmissibility matrix related to the smallest error between the expected responses and measured ones along the frequency range;

iii. in all the examples the identification of the number of forces and their localization have been accomplished;

iv. the magnitude of the estimated forces exhibits a very good correlation with the measured ones for most of the frequency range.

Acknowledgements

The current investigation had the support of IDMEC/IST and FCT, under the project PTDC/EME-PME/71488/2006.

Author details

N. M. M. Maia, Y. E. Lage and M. M. Neves

*Address all correspondence to: nmaia@dem.ist.utl.pt

IDMEC-IST, Technical University of Lisbon, Department Mechanical Engineering, Lisboa, Portugal

References

[1] Maia, N. M. M, Urgueira, A. P. V, & Almeida, R. A. B. Whys and Wherefores of Transmissibility. In: Dr. Francisco Beltran-Carbajal (ed.) Vibration Analysis and Control- New Trends and Developments, InTech; (2011). http://www.intechopen.com/books/vibration-analysis-and-control-new-trends-and-developments/whys-and-wherefores-of-transmissibility., 187-216.

[2] Hillary, B. Indirect Measurement of Vibration Excitation Forces. PhD Thesis. Imperial College of Science, Technology and Medicine, Dynamics Section, London, UK; (1983).

[3] Stevens, K. K. Force Identification Problems- an overview. Proceedings of the (1987). SEM Spring Conference on Experimental Mechanics. Houston, TX, USA; 1987.

[4] Mas, P, Sas, P, & Wyckaert, K. Indirect force identification based on impedance matrix inversion: a study on statistical and deterministic accuracy. Proceedings of 19th International Seminar on Modal Analysis, Leuven. Belgium, (1994). , 1049-1065.

[5] Dobson, B. J, & Rider, E. A review of the indirect calculation of excitation forces from measured structural response data. Proceedings of the Institution Mechanical Engineers, Part C, Journal of Mechanical Engineering Science,(1990). , 69-75.

[6] Ma, C. K, Chang, J. M, & Lin, D. C. Input Forces Estimation of Beam Structures by an Inverse Method. Journal of Sound and Vibration, (2003). , 387-407.

[7] Ma, C. K, & Ho, C. C. An Inverse Method for the Estimation of Input Forces Acting On Non-Linear Structural Systems. Journal of Sound and Vibration, (2004). , 953-971.

[8] Thite, A. N, & Thompson, D. J. The quantification of structure-borne transmission paths by inverse methods. Part 1: improved singular value rejection methods. Journal of Sound and Vibration, (2003). , 411-431.

[9] Thite, A. N, & Thompson, D. J. The quantification of structure-borne transmission paths by inverse methods. Part 2: use of regularization methods, Journal of Sound and Vibration, (2003). , 433-451.

[10] Choi, H, Thite, G, & Thompson, A. N. D. J. A threshold for the use of Tikhonov regularization in inverse force determination. Applied Acoustics, (2006). , 700-719.

[11] Michaels, J. E, & Pao, Y. H. The inverse source problem for an oblique force on an elastic plate. Journal of the Acoustic. Society of America, June (1985). , 2005-2011.

[12] Martin, M. T, & Doyle, J. F. Impact force identification from wave propagation responses. International Journal of Impact Engineering, January (1996). , 65-77.

[13] Huang, C. H. An inverse non-linear force vibration problem of estimating the external forces in a damped system with time-dependent system parameters. Journal of Sound and Vibration, (2001). , 749-765.

[14] Tadeusz UhlThe inverse identification problem and its technical application, Arch Appl Mech, (2007). , 325-337.

[15] Zhang, Q, Allemang, R. J, & Brown, D. L. Modal Filter: Concept and Applications, 8th International Modal Analysis Conference (IMAC VIII), Kissimmee, Florida, (1990). , 487-496.

[16] Kammer, D. C, & Steltzner, A. D. Structural identification of Mir using inverse system dynamics and Mir/shuttle docking data, Journal of Vibration and Acoustics, (2001). , 230-237.

[17] Allen, M. S, & Carne, T. G. Comparison of Inverse Structural Filter (ISF) and Sum of Weighted Accelerations (SWAT) Time Domain Force Identification Methods, Mechanical Systems and Signal Processing,(2008). , 1036-1054.

[18] Paulo, P. A Time-Domain Methodology For Rotor Dynamics: Analysis and Force Identification. MSc thesis. Instituto Superior Técnico Lisbon; (2011).

[19] Vakakis, A. F. Dynamic Analysis of a Unidirectional Periodic Isolator, Consisting of Identical Masses and Intermediate Distributed Resilient Blocks. Journal of Sound andVibration, (1985). , 25-33.

[20] Vakakis, A. F, & Paipetis, S. A. Transient Response of Unidirectional Vibration Isola-torswith Many Degrees of Freedom, Journal of Sound and Vibration, (1985). 0002-2460X., 99(4), 557-562.

[21] Vakakis, A. F, & Paipetis, S. A. The Effect of a Viscously Damped Dynamic Absorber ona Linear Multi-Degree-of-Freedom System, Journal of Sound and Vibration, (1986). (1), 49-60.

[22] Liu, W. Structural Dynamic Analysis and Testing of Coupled Structures. PhD thesis. Imperial College London; (2000).

[23] Liu, W, & Ewins, D. J. Transmissibility Properties of MDOF Systems, Proceedings of the16th International Operational Modal Analysis Conference (IMAC XVI), Santa Barbara, California, (1998). , 847-854.

[24] Varoto, P. S, & Mcconnell, K. G. Single Point vs Multi Point AccelerationTransmissi-bility Concepts in Vibration Testing, Proceedings of the 12th International Modal Analysis Conference (IMAC XVI), Santa Barbara, California, USA,(1998). , 83-90.

[25] Ribeiro, A. M. R. On the Generalization of the Transmissibility Concept, Proceedings of the NATO/ASI Conference on Modal Analysis and Testing, Sesimbra, Portugal, (1998). , 757-764.

[26] Fontul, M, Ribeiro, A. M. R, Silva, J. M. M, & Maia, N. M. M. Transmissibility Matrix in Harmonic and Random Processes, Shock and Vibration, (2004). , 563-571.

[27] Maia, N. M. M, Fontul, M, & Ribeiro, A. M. R. Transmissibility of Forces in Multiple-Degree-of-Freedom Systems, Proceedings of ISMA (2006). Noise and Vibration Engi-neering, Leuven, Belgium 2006.

[28] Rao, S. S. Mechanical Vibrations. Fourth International Edition. Prentice-Hall; (2004).

RECENT ADVANCES IN VIBRATIONS ANALYSIS

Edited by **Natalie Baddour**

Exact Transfer Function Analysis of Distributed Parameter Systems by Wave Propagation Techniques

Bongsu Kang
Indiana University – Purdue University Fort Wayne
USA

1. Introduction

The vibrations of elastic structures such as strings, beams, and plates can be described in terms of waves traveling in waveguides (Cremer et al., 1973; Graff, 1975; Fahy, 1987). While the subject of wave propagation has been extensively studied in the fields of acoustics in fluids and solids rather than vibrations of elastic structures, wave analysis techniques have been employed to reveal physical characteristics associated with structural vibrations of elastic media (Argento & Scott, 1995; Kang & Tan, 1998). One of the advantages of the wave analysis technique, when applied to the structural vibration analysis, is its compact and systematic approach to analyze complex structures with discontinuities (Mace, 1984; Yong & Lin, 1989; Kang et al., 2003; Mei & Mace, 2005). Applying the concept of wave reflection and transmission, Mace (1984) obtained the frequency equations of Euler-Bernoulli beams including waves of both propagating and near-field types. By the phase-closure principle, also referred to as the wave-train closure principle (Cremer et al., 1973), Mead (1994) determined natural frequencies of Euler-Bernoulli beams. This principle states that if the phase difference between incident and reflected waves is an integer multiple of 2π, then the waves propagate at a natural frequency and their motions constitute a vibration mode. Based on the same principle, Kang (2007) presented a systematic approach to the free and forced vibration analysis of multi-span beams.

The classical method, known as the normal mode or eigenfunction expansion, of solving the forced vibration problem of a distributed parameter system involves expansion of the forcing function into the eigenfunctions of the associated free vibration problem. While this method is theoretically sound and powerful, the method is difficult to implement when the problem to be solved is a non-self-adjoint system typically due to complicating effects such as damping, discontinuities, or non-classical boundary conditions, for which case obtaining the exact eigensolutions is not often feasible. Although approximate eigensolutions may be used instead of exact ones, the problem still persists in the form of poorly convergent solution and/or significant error in the solution. As an alternative approach to solve forced vibration problems, Yang and Tan (1992) presented a method for evaluating exact closed-form transfer functions for a class of one-dimensional distributed parameter systems. Applying the energy functionals of constrained and combined damped systems, Yang (1996a, 1996b) presented a method to obtain a closed-form transient response solution in eigenfunction series for a distributed damped system.

The dynamic displacement of any point in an elastic waveguide can be determined by superimposing the amplitudes of the constituent waves traveling along the waveguide, which is a basis of wave propagation. Based on this simple fact, an alternative technique for the free and forced vibration analysis of one-dimensional distributed parameter systems is presented. The method of normal mode expansion is often difficult to implement for nonself-adjoint systems with complicating effects such as mode couplings, non-proportional damping, discontinuities, or arbitrary boundary conditions, since the method requires eigensolutions or assumed normal modes as a priori. However, this alternative analysis technique based on the elastic wave propagation does not pose such a requirement and leads to the exact, closed-form, distributed transfer function of a distributed parameter system. The general wave solution of the equation of motion governing the dynamics of a waveguide is cast into a matrix form in terms of the constituent waves defined in the Laplace domain. The spatial amplitude variation of the traveling wave is represented by the field transfer matrix and the amplitude distortion of the traveling wave incident upon a discontinuity due to geometric or kinetic constraints is described by the local wave reflection and transmission matrices. Combining these matrices in a progressive manner along the waveguide by applying the concepts of global wave reflection and transmission matrices leads to the exact characteristic equation and corresponding mode shapes for the free response analysis and the transfer function of the system for the forced response analysis. The transient response solution for a complex system can be obtained through the Laplace inversion of the transfer function using numerical inversion algorithms. The exact frequency response solution, which includes infinite normal modes of the system, can be obtained in terms of the complex frequency response function from the transfer function. One of the main advantages of this analysis technique is its systematic formulation resulting in a recursive computational algorithm which can be implemented into highly efficient computer codes. This systematic approach also allows modular formulation which can be readily expandable to include additional discontinuities with little alteration to the existing formulation. In addition, it is also computationally advantageous that the technique always results in operating matrices of a fixed size regardless of the number of discontinuities in a waveguide. This analysis technique is applicable to any one-dimensional waveguides (strings, axial rods, torsional bars, beams, and frame structures), in particular systems with multiple point discontinuities such as viscoelastic supports, attached inertias, and geometric/material property changes. The analysis technique is demonstrated using the second order wave equation, fourth order beam equation, and sixth order curved beam equation.

2. Second order systems

The free transverse vibration of a taut string, longitudinal vibration of a thin bar, and the torsional vibration of a shaft are governed by the equation of motion in the same form, the wave equation, and thus they are mathematically analogous. Therefore, with no loss of generality, the transverse vibration of the string is taken as a representative problem for the description of the present analysis technique based on wave propagation. The equation governing the transverse motion of a uniformly damped string of span length L is

$$T\frac{\partial^2 W}{\partial X^2}+C_e\frac{\partial W}{\partial t}=m\frac{\partial^2 W}{\partial t^2} \tag{2.1}$$

where W is the transverse displacement, X the spatial variable, t the temporal variable; and T denotes the constant tension, C_e the damping coefficient, and m the mass per unit length of the string. With introduction of the following non-dimensional variables and parameters

$$w=W/L \quad x=X/L \quad \tau=t/t_0 \quad c_e=C_eL/t_0 \quad t_0=L/c_0 \quad c_0=\sqrt{T/m} \tag{2.2}$$

the equation of motion takes the non-dimensional form of

$$w''+c_e\dot{w}=\ddot{w} \quad (0\le x\le 1) \tag{2.3}$$

where the prime (′) and dot (·) denote the differentiation with respect to x and τ, respectively. Applying the Laplace transform to Eq. (2.3) yields

$$\overline{w}''(x;s)+c_es\overline{w}(x;s)=s^2\overline{w}(x;s) \tag{2.4}$$

where s denotes the Laplace variable and zero initial conditions are assumed.

2.1 Wave solution

Denoting C as the amplitude of the wave traveling along the string, the solution of Eq. (2.4) can be assumed in wave form

$$\overline{w}(x;s)=Ce^{i\gamma x} \tag{2.5}$$

where γ is the non-dimensional wavenumber normalized against span length L. Applying the above wave solution to Eq. (2.4) gives the frequency equation of the problem

$$\gamma^2+s^2-c_es=0 \text{ or } \gamma^2=c_es-s^2 \tag{2.6}$$

from which the general wave solution can be found as the sum of two constituent waves

$$\overline{w}(x;s)=C^+e^{-i\gamma x}+C^-e^{i\gamma x} \tag{2.7}$$

where the coefficient C represents the amplitude of each wave component with its traveling direction indicated by the plus (+) or (−) sign. Note that γ is complex valued for nonzero c_e, hence the classification of the wave into propagating and attenuating waves does not apply to this case. Defining $f(x;s)=e^{-i\gamma x}$ as the *field transfer function* which relates the wave amplitudes by

$$C^+(x+x_0)=fC^+(x_0) \text{ or } C^-(x+x_0)=f^{-1}C^-(x_0) \tag{2.8}$$

where $f^{-1}(x;s)=e^{i\gamma x}$, the wave solution in Eq. (2.7) can be re-written as

$$\overline{w}(x;s)=fC^++f^{-1}C^- \tag{2.9}$$

2.2 Wave reflection and transmission

When a wave traveling along a string is incident upon a discontinuity such as an elastic support, geometric/material property change, or boundary, it is reflected and transmitted at different rates depending on the properties of the discontinuity. The rates of wave reflection and transmission can be determined in terms of *wave reflection* and *transmission coefficients*. For example, consider an infinitely long string constrained at a local coordinate $\xi = 0$ as shown in Fig. 1, where the constraint is a point support consisting of an attached mass (M_c), a transverse spring (K_c), and a viscous damper (C_c).

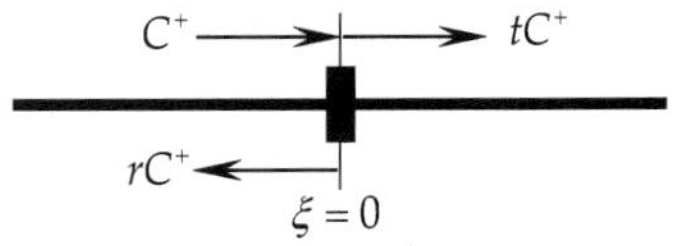

Fig. 1. Wave reflection and transmission at a discontinuity.

When a positive-traveling wave C^+ is incident upon the support, it gives rise to a reflected wave rC^+ and a transmitted wave tC^+, where r and t represent the wave reflection and transmission coefficients, respectively. The resulting transverse displacements at the left and right of the constraint are

$$\bar{w}_l = C^+ + rC^+ \quad \bar{w}_r = tC^+ \tag{2.10}$$

Since $\bar{w}_l = \bar{w}_r$ at ξ=0, one can find

$$1 + r = t \tag{2.11}$$

In addition, the kinetic equilibrium condition at ξ=0 states that

$$\bar{w}_r' - \bar{w}_l' = \varphi \bar{w}_r \tag{2.12}$$

where φ represents the kinetic properties of the constraint as

$$\varphi(s) = k_c + c_c s + m_c s^2 \tag{2.13}$$

Note that k_c, c_c, and m_c in Eq. (2.13) are the non-dimensional spring constant, damping coefficient, and the attached mass, respectively, defined by

$$k_c = K_c L/T \quad c_c = C_c c_0/T \quad m_c = M_c c_0^2/TL \tag{2.14}$$

Equation (2.12) leads to

$$\gamma(r + t - 1) = i\varphi t \tag{2.15}$$

Combining Eqs. (2.11) and (2.15), the wave reflection and transmission coefficients can be found as

$$r = \frac{i\varphi}{2\gamma - i\varphi} \text{ and } t = \frac{2\gamma}{2\gamma - i\varphi} \tag{2.16}$$

When a wave is incident upon a boundary at ξ=0, it is only reflected, therefore

$$\overline{w}_l' = -\varphi \overline{w}_l \tag{2.17}$$

which leads to

$$r = \frac{\gamma + i\varphi}{\gamma - i\varphi} \tag{2.18}$$

In the limiting case where $\varphi=\infty$, it can be seen that $r=-1$ which is the wave reflection coefficient for the classical fixed boundary. When the wave is incident upon a series of discontinuities along its traveling path, it is more computationally efficient to employ the concepts of *global wave reflection and transmission coefficients,* in particular when the free or forced vibration analysis of a multi-span string is sought. These coefficients relate the amplitudes of incoming and outgoing waves at a discontinuity. Consider wave motion in a multi-span string as illustrated in Fig. 2. Define R_{ir} as the global wave reflection coefficient which relates the amplitudes of negative- and positive-traveling waves on the right side of discontinuity i such that

$$C_{ir}^- = R_{ir} C_{ir}^+ \tag{2.19}$$

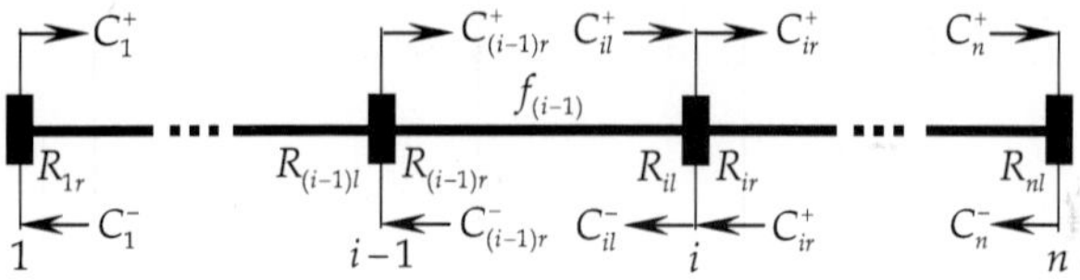

Fig. 2. Waves traveling along a multi-span string.

Since $C_{ir}^- = f_i R_{(i+1)l} C_{ir}^+$, one can find R_{ir} in terms of the global wave reflection coefficient on the left side of discontinuity i+1; i.e.,

$$R_{ir} = f_i^2 R_{(i+1)l} \tag{2.20}$$

In addition, by combining the following wave equations at discontinuity i

$$C_{ir}^+ = t_i C_{il}^+ + r_i C_{ir}^- \text{ and } C_{il}^- = t_i C_{ir}^- + r_i C_{il}^+ \tag{2.21-22}$$

the relationship between the global wave reflection coefficients on the left and right sides of discontinuity i can be found as

$$R_{il} = r_i + t_i^2 (R_{ir}^{-1} - r_i)^{-1} \tag{2.23}$$

R_{ir} and R_{il} progressively expand to include all the global wave reflection coefficients of discontinuities along the string before terminating its expansion at the boundaries where

$$C_1^+ = r_1 C_1^- \quad C_n^- = r_n C_n^+ \tag{2.24}$$

While the global wave reflection coefficient relates the amplitudes of waves traveling in the opposite direction of each other within a subspan *i*, there is a need for another coefficient which relates the amplitudes of waves traveling in the same direction in two adjacent subspans. This inter-span coefficient is particularly useful when the mode shape or forced response of a string with several or more subspans needs to be determined since it allows an intuitive and systematic formulation of the system's transfer function. Denoting this inter-span wave transfer coefficient as the *global wave transmission coefficient* T_i, define

$$C_{ir}^{+} = T_i C_{(i-1)r}^{+} \tag{2.25}$$

Rewriting Eq. (2.21) by applying $C_{il}^{+} = f_{(i-1)} C_{(i-1)r}^{+}$ and $C_{ir}^{-} = R_{ir} C_{ir}^{+}$, and then comparing it with Eq. (2.25), the global wave transmission coefficient at discontinuity *i* can be found as

$$T_i = (1 - r_i R_{ir})^{-1} t_i f_{(i-1)} \tag{2.26}$$

The global wave reflection and transmission coefficients are the key elements in determining the exact transfer function of a multi-span string as discussed in the following sections.

2.3 Free response analysis

The global reflection and transmission coefficients of waves traveling along a multi-span sting are now combined with the field transfer function to analyze the free response of a multi-span sting. With reference to Fig. 2, consider wave motion in the first span. Based on the definition of the global wave reflection coefficient, at the boundary

$$C_1^{-} = R_{1r} C_1^{+} \tag{2.27}$$

However, recalling $C_1^{+} = r_1 C_1^{-}$ from Eq. (2.24), it can be found that

$$(r_1 R_{1r} - 1) C_1^{+} = 0 \tag{2.28}$$

For nontrivial solutions,

$$F(s) = r_1 R_{1r} - 1 = 0 \tag{2.29}$$

which is the characteristic equation in terms of the Laplace variable *s* for the multi-span string with arbitrary discontinuities and boundaries. This remarkably simple expression for the characteristic equation is due to the fact that R_{1r} recursively expands to include all the effects of constraints in the remaining side of the string until its expansion terminates at the rightmost boundary which yields $R_{nl}=r_n$. This equation simply states that when the string system vibrates at one of its natural frequencies, $r_1R_{1r}=1$. As a simple example, for a single span uniformly damped string fixed at both ends, $r_1=r_2=-1$ from Eq. (2.18) and $R_{1r}=e^{-2i\gamma}$ from Eq. (2.20). Therefore, $F(s)=e^{-2i\gamma}-1=0$ gives the natural wavenumbers $\gamma_n=n\pi$ (n=1,2,3,...), or in terms of the non-dimensional frequency $\omega_n^2 + ic_e\omega_n = (n\pi)^2$ by Eq. (2.6) with *s* replaced by $i\omega$.

The mode shapes of the multi-span string system can be found in a systematic manner by relating wave amplitudes between two adjacent subspans. Let ξ_i denote the local coordinate within span *i*. The transverse displacement at any point in span *i* can be expressed as

$$\bar{w}_i(\xi_i) = f_i(\xi_i)C_i^+ + f_i^{-1}(\xi_i)C_i^- \tag{2.30}$$

However, due to Eq. (2.19)

$$\bar{w}_i(\xi_i) = [f_i(\xi_i) + f_i^{-1}(\xi_i)R_{ir}]C_i^+ \tag{2.31}$$

Now by applying the global wave transmission coefficient defined in Eq. (2.25) to C_i^+ in the above equation, the displacement at any point in span i of the string can be expressed in terms of the wave amplitude in span i–1 as

$$\bar{w}_i(\xi_i) = [f_i(\xi_i) + f_i^{-1}(\xi_i)R_{ir}]T_iC_{i-1}^+ \tag{2.32}$$

Assume a disturbance arise in span 1; i.e., a wave originates and starts traveling from the leftmost boundary of the string. Then, by successively applying the global transmission coefficient of each discontinuity on the way up to the first span, the mode shape of span i can be found in terms of wave amplitude C_1^+; i.e.,

$$\bar{w}_i(\xi_i) = [f_i(\xi_i) + f_i^{-1}(\xi_i)R_{ir}]\prod_{j=i}^{1} T_jC_1^+ \quad C_1^+ \quad 0 \le \xi_i \le l_i \tag{2.33}$$

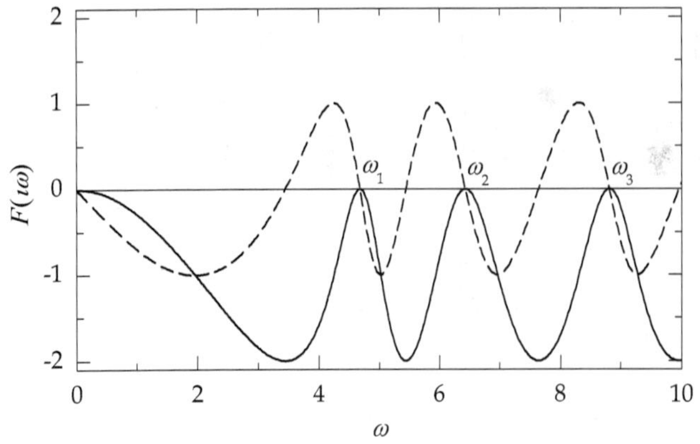

Fig. 3. Plot of the characteristic equation. The solid and dashed curves represent the real and imaginary parts, respectively.

For example, consider a fixed-fixed undamped string with three supports specified by φ_2=5+0.1s^2, φ_3=7+0.1s^2, and φ_4=4+0.1s^2 according to Eq. (2.13). l_1=0.25, l_2=0.3, l_3=0.25, and l_4=0.2 are assumed. Once the global wave reflection coefficient at each discontinuity has been determined, one can apply Eq. (2.29) to find the natural frequencies. Shown in Fig. 3 is the plot of the characteristic equation, where the first three natural frequencies are indicated. The mode shapes can be found from Eq. (2.33) in a systematic way once the global wave transmission coefficient at each discontinuity has been determined. Figure 4 shows the mode shapes for the first three modes.

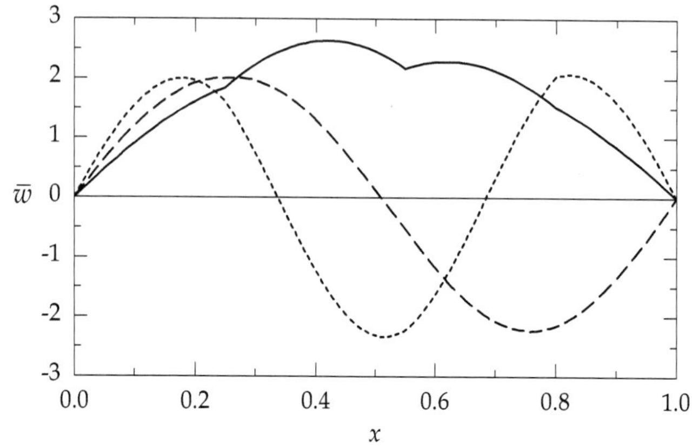

Fig. 4. First three mode shapes obtained from Eq. (2.33). The solid, dashed, and dotted curves represent the 1st, 2nd, and 3rd modes, respectively.

2.4 Transfer function analysis

Consider a multi-span string subjected to an external point load $\bar{p}(s)$, normalized against tension T, applied at $x=x_0$ as shown in Fig. 5. Since the waves injected by the load travel in both direction from the point of loading, a set of local coordinates $\{\xi, \xi^*\}$ is introduced such that the wave traveling toward each boundary of the string is considered positive as indicated in Fig. 5. Let $C_1^{\pm}$ and $D_1^{\pm}$ denote the injected waves that travel in the region $x<x_0$ and $x\geq x_0$ within span 1, respectively. The transverse displacement of span 1 of the string can be expressed as

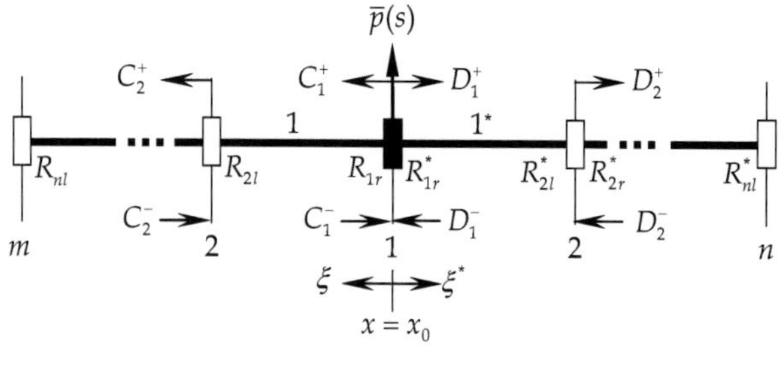

Fig. 5. Wave motion in a multi-span string due to a point load.

$$\bar{w}_1(\xi_1, x_0; s) = f_1(\xi_1; s)C_1^+ + f_1^{-1}(\xi_1; s)C_1^- \quad 0 \leq \xi_1 \leq l_1 \tag{2.34}$$

Defining v as the wave injected by the applied load, the difference in amplitudes between the incoming and outgoing waves at the loading point is

$$C_1^+ - D_1^- = v \quad D_1^+ - C_1^- = v \tag{2.35}$$

where v can be determined from the geometric and kinetic continuity conditions at ξ=0 as

$$v = \frac{\sqrt{-1}}{2\gamma}\overline{p}(s) \tag{2.36}$$

Applying the global wave reflection coefficient on each side of the loading point gives

$$C_1^- = R_{1r}C_1^+ \quad D_1^- = R_{1r}^*D_1^+ \tag{2.37}$$

where the asterisk (*) signifies that the global wave reflection coefficient is defined in the region $x \geq x_0$ to distinguish it from the one defined in the region $x<x_0$. Combining Eqs. (2.35) and (37), one can determine the ampltitude of the wave that rises at the loading point in each direction

$$C_1^+ = T_1 v \quad T_1 = (1 - R_{1r}^* R_{1r})(1 + R_{1r}^*) \tag{2.38.1}$$

$$D_1^+ = T_1^* v \quad T_1^* = (1 - R_{1r} R_{1r}^*)(1 + R_{1r}) \tag{2.38.2}$$

Note that T_1 and T_1^* and can be considered as the global wave transmission coefficients that characterize the transmissibility of the wave injected by the external force in the region $x<x_0$ and $x \geq x_0$, respectively. It is evident that and T_1 and T_1^* are different unless the string system is symmetric about the loading point. Applying the results in Eqs. (2.37) and (2.38) to Eq. (2.34), the wave motion in either side of $x=x_0$ can be found; i.e., in the region $x<x_0$

$$\overline{w}_1(\xi_1, x_0; s) = [f_1(\xi_1;s) + f_1^{-1}(\xi_1;s)R_{1r}]T_1 v \quad 0 \leq \xi_1 \leq l_1 \tag{2.39}$$

Now, in the same manner as for the mode shape analysis, since $C_{i+1}^+ = T_i C_i^+$ and $C_1^+ = T_1 v$, the wave motion in span i on either side of the loading point can be found. For the region $x<x_0$

$$\overline{w}_i(\xi_i, x_0; s) = [f_i(\xi_i;s) + f_i^{-1}(\xi_i;s)R_{ir}]\prod_{k=i}^{1} T_k v \quad 0 \leq \xi_i \leq l_i \tag{2.40}$$

Note that the ratio

$$G_i(\xi_i, x_0; s) = \overline{w}_i(\xi_i, x_0; s)/\overline{p}(s) \tag{2.41}$$

is the transfer function governing the forced response of any point in span i due to the point loading at $x=x_0$. The Laplace inversion of $G_i(\xi_i, x_0; s)$ is the Green's function of the problem. Thefore, denoting L^{-1} as the inverse Laplace transform operator, the forced response at any point within any subspan of the multi-span string can be determined from the following convolution integral; e.g., for span i in the region $x<x_0$

$$w_i(\xi_i, x_0; \tau) = \int_0^\tau G_i(\xi_i, x_0; \tau - \overline{\tau})p(\tau)d\overline{\tau} \quad (i = 1, 2, \cdots, m) \tag{2.42.1}$$

$$G_i(\xi_i, x_0; \tau) = L^{-1}[\frac{\sqrt{-1}}{2\gamma}[f_i(\xi_i;s) + f_i^{-1}(\xi_i;s)R_{ir}]\prod_{k=i}^{1} T_1] \quad 0 \leq \xi_i \leq l_i \tag{2.42.2}$$

The exact Laplace inversion of $G_i(\xi_i, x_0; s)$ in close form is not feasible in general, in particular for multi-span string systems. One may have to resort to the numerical inversion of Laplace transforms. It is found that the algorithm known as the fixed Talbot method (Abate & Valko, 2004), which is based on the contour of the Bromwich inversion integral, appears to perform the numerical Laplace inversion in Eq. (2.42) with a satisfactory accuracy and reasonable computation time. In this method, the accuracy of the results depends only on the number of precision decimal digits (denoted with M in the algorithm) carried out during the inversion. It is found that for well-damped second- and higher order distributed parameter systems, M=32 and for lightly damped or undamped systems, M=64 gives acceptable results. To demonstrate the effectiveness of the present analysis approach, consider the unit impulse response of a single span undamped string, in particular the response near the point of loading immediately after the loading, which is well known for its deficiency in numerical convergence. The response solution by the method of normal mode expansion is

$$w(x,x_0;\tau)=\sum_{n=1}^{N}\frac{2\sin n\pi x_0}{n\pi}\sin n\pi x\sin n\pi\tau \tag{2.43}$$

The corresponding transfer function from Eq. (2.40) is

$$\bar{w}(x,x_0;s)=\frac{1}{2s(e^{2s}-1)}\begin{cases} e^{-sx}(e^{2s(1-x_0)}-1)(e^{2sx}-e^{2sx_0}) & \text{for } 0\le x\le x_0 \\ e^{-sx}(e^{2sx_0}-1)(e^{2sx}+e^{2s(1-x_0)}) & \text{for } x_0<x\le 1 \end{cases} \tag{2.44}$$

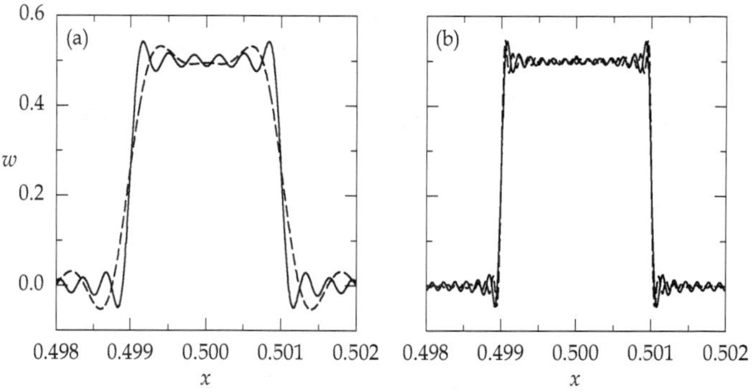

Fig. 6. Unit impuse response of a single span string near the point of loading at τ=0.001: (a) M=32 and N=2,500; (b) M=64 and N=20,000. The solid and dashed curves represent the solutions from Eqs. (2.43) and (2.44), respectively.

Shown in Fig. 6 is the comparison of the response solutions given in Eqs. (2.43) and (2.44) when τ=0.001 near $x=x_0$=0.5. N=2,500 and N=20,000 are used for the evaluation of the series solution, while M=32 and M=64 are used for the numerical Laplace inversion of the transfer function in Eq. (2.44). It can be seen that the series solution in Eq. (2.43) fails to accurately represent the actual impulse response behavior with N=2,500. This is expected for the series solution since it would take a large number of harmonic terms ($N\geq$10,000 for this example) to represent such a sharp spike due to the impulse. It can be seen that the result with M=32 reasonably represents the actual behavior, and the result with M=64 is almost comparable to the series solution with N=20,000. However, if one tries to obtain the response at a time very close to the moment of impact, the numerical Laplace inversion becomes extremely strenuous or beyond the machine precision of the computing machine. This is because the expected response would consist of waves that have unrealistically short wavelengths. This is not a unique problem for the present wave approach since the same problem would manifest itself in the series solution given in Eq. (2.43), requiring an impractically large number of harmonics terms for a convergent solution.

If $p(\tau) = p_0 e^{i\omega\tau}$; i.e., a harmonic forcing function, the steady-state response of the problem can be readily found in terms of the complex frequency function defined as

$$H(\xi_i, x_0; \omega) = G(\xi_i, x_0; s)|_{s=i\omega} \tag{2.45}$$

Therefore the frequency response at any point within any subspan of the string can be obtained by; e.g., for span i in the region $x<x_0$

$$w_i(\xi_i, x_0; \tau) = |H_i(\xi_i, x_0; \omega)| p_0 e^{i(\omega\tau + \phi_i)} \quad 0 \le \xi_i \le l_i \tag{2.46.1}$$

$$H_i(\xi_i, x_0, \omega) = \frac{\sqrt{-1}}{2\gamma}[f_i(\xi_i, \omega) + f_i^{-1}(\xi_i, \omega) R_{ir}] \prod_{k=i}^{1} T_k \quad \phi_i = \angle H_i \tag{2.46.2}$$

One of the main advantages of this approach is its systematic formulation resulting in a recursive computational algorithm which can be implemented into highly efficient computer codes consuming less computer resources. This systematic approach also allows modular formulation which can be easily expandable to include additional subspans with very minor alteration to the existing formulation. Another significant advantage of the present wave-based approach to the forced response analysis of a multi-span string is that the eigensolutions of the system is not required as a priori as in the method of normal mode expansion which assumes the forced response solution in terms of an infinite series of the system eigenfunctions - truncated later for numerical computations. However, exact eigensolutions are often difficult to obtain particularly for non-self-adjoint systems, and also approximated eigensolutions can result in large error. In contrast, the current analysis technique renders closed-form transfer functions from which exact frequency response solutions can be obtained.

3. Fourth order systems

The analysis techniques described by using the vibration of a string can be applied to the transverse vibration of a beam of which equation of motion is typically a fourth order partial differential equation. Denoting X and t as the spatial and temporal variables, respectively, the equation of motion governing the transverse displacement $W(X,t)$ of a damped uniform Euler-Bernoulli beam of length L subjected to an external load $P(X,t)$ is

$$m\frac{\partial^2 W}{\partial t^2} + C_e\frac{\partial W}{\partial t} + EI\frac{\partial^4 W}{\partial X^4} = P(X,t) \tag{3.1}$$

where m denotes the mass per unit length, EI the flexural rigidity, C_e the external damping coefficient of the beam. With introduction of the following non-dimensional variables and parameters

$$w = W/L\,,\ x = X/L\,,\ \tau = t/t_0\,,\ t_0 = mL^4/EI\ \ c_e = C_e t_0/m\,,\ p(x,t) = P(X,t)L^3/EI \tag{3.2}$$

the equation of motion takes the non-dimensional form of

$$\ddot{w} + c_e\dot{w} + w^{(4)} = p(x,t) \quad (0 \le x \le 1) \tag{3.3}$$

Applying the Laplace transform to Eq. (3.3) yields

$$s^2\bar{w}(x;s) + c_e s\bar{w}(x;s) + \bar{w}^{(4)}(s;x) = \bar{p}(x;s) \tag{3.4}$$

Letting $\bar{p}(x;s) = 0$, the homogeneous wave solution of Eq. (3.4) can be assumed as:

$$\bar{w}(x;s) = Ce^{i\gamma x} \tag{3.5}$$

where γ is the non-dimensional wavenumber normalized against span length L. Applying the above wave solution to Eq. (3.4) gives the frequency equation of the problem

$$\gamma^4 = -(s^2 + c_e s) \tag{3.6}$$

from which the general wave solution can be found as the sum of four constituent waves

$$\bar{w}(x;s) = C_a^+ e^{-i\gamma x} + C_a^- e^{i\gamma x} + C_b^+ e^{-\gamma x} + C_b^- e^{\gamma x} \tag{3.7}$$

where the coefficient C represents the amplitude of each wave with its traveling direction indicated by the superscript; plus (+) and minus (-) signs denote the wave's traveling directions with respect to the x-coordinate. The subscripts a and b signify the spatial wave motion of the same type traveling in the opposite direction. Note that γ is complex valued in general. The general wave solution in Eq. (3.7) may be re-expressed in vector form by grouping the wave components (wave packet) traveling in the same direction; i.e.,

$$\mathbf{C}^+ = \begin{Bmatrix} C_a^+ \\ C_b^+ \end{Bmatrix} \quad \mathbf{C}^- = \begin{Bmatrix} C_a^- \\ C_b^- \end{Bmatrix} \tag{3.8}$$

and then

$$\bar{w}(x;s) = [1 \quad 1][\mathbf{f}(x;s)\mathbf{C}^+ + \mathbf{f}^{-1}(x;s)\mathbf{C}^-] \tag{3.9}$$

where $\mathbf{f}(x;s)$ is the diagonal *field transfer matrix* (Mace, 1984) given by

$$\mathbf{f}(x;s) = \begin{bmatrix} e^{-i\gamma x} & 0 \\ 0 & e^{-\gamma x} \end{bmatrix} \tag{3.10}$$

which relates the wave amplitudes by

$$\mathbf{C}^+(x + x_0) = \mathbf{f}\mathbf{C}^+(x_0) \quad \text{or} \quad \mathbf{C}^-(x + x_0) = \mathbf{f}^{-1}\mathbf{C}^-(x_0) \tag{3.11}$$

3.1 Wave reflection and transmission matrices

For a wave packet with multiple wave components, the rates of wave reflection and transmission at a point discontinuity can be found in terms of the *wave reflection matrix* **r** and *wave transmission matrix* **t**, in the same manner described in Section 2.2. When the flexural wave packet in Eq. (3.8) travels along a beam and is incident upon a kinetic constraint (ξ=0) which consists of, for example, transverse (K_t) and rotational (K_r) springs and transverse damper (C_t), **r** and **t** at the discontinuity can be found by applying the geometric continuity kinetic equilibrium conditions at ξ=0; i.e., with reference to Fig. 7, one can establish the following matrix equations

$$\begin{bmatrix} 1 & 1 \\ -i & -1 \end{bmatrix}\mathbf{C}^+ + \begin{bmatrix} 1 & 1 \\ i & 1 \end{bmatrix}\mathbf{rC}^+ = \begin{bmatrix} 1 & 1 \\ -i & -1 \end{bmatrix}\mathbf{tC}^+ \tag{3.12.1}$$

$$\begin{bmatrix} 1 & -1 \\ i & -1 \end{bmatrix}\mathbf{C}^+ + \begin{bmatrix} 1 & -1 \\ -i & 1 \end{bmatrix}\mathbf{rC}^+ = \begin{bmatrix} 1 - ik_r/\gamma & -1 - k_r/\gamma \\ i + (k_t + c_t s)/\gamma^3 & -1 + (k_t + c_t s)/\gamma^3 \end{bmatrix}\mathbf{tC}^+ \tag{3.12.2}$$

where $k_t = K_t L^3/EI$, $k_r = K_r L/EI$, and $c_t = C_t t_0/m$. Solving the above equations gives the local wave reflection and transmission matrices as:

$$\mathbf{r} = \frac{1}{2}\begin{bmatrix} (1-i)(\alpha - \beta i) & -(1+i)(\alpha + \beta) \\ -(1-i)(\alpha + \beta) & (1+i)(\alpha + \beta i) \end{bmatrix} \tag{3.13.1}$$

$$\mathbf{t} = \frac{1}{2}\begin{bmatrix} 2 - (1-i)\alpha - (1+i)\beta & (1+i)(\alpha - \beta) \\ (1-i)(\alpha - \beta) & 2 - (1+i)\alpha - (1-i)\beta \end{bmatrix} \tag{3.13.2}$$

$$\alpha = \frac{k_r + c_r s}{k_r + c_r s + (2+2i)\gamma} \qquad \beta = \frac{k_t + c_t s}{k_t + c_t s - (2-2i)\gamma^3}$$

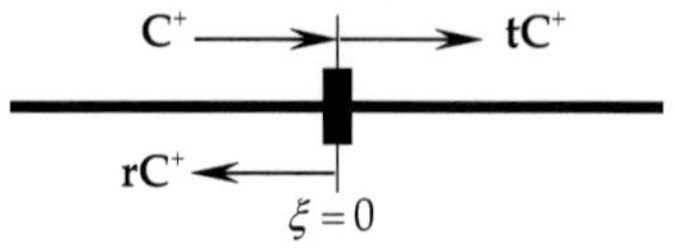

Fig. 7. Wave reflection and transmission at a discontinuity.

However, as previously discussed in Section 2.2, when the wave packet is incident upon a series of discontinuities along its traveling path, it is more computationally efficient to employ the concepts of *global wave reflection* and *transmission matrices.* These matrices relate the amplitudes of incoming and outgoing waves at a point discontinuity. When compared to the string problem, the only difference in formulating these matrices for the beam problem is to use vectors and matrices instead of single coefficients. Therefore, with reference to Fig. 2, let $\mathbf{R}_{ir}$ as the global wave reflection matrix which relates the amplitudes of negative- and positive-traveling waves on the right side of discontinuity i such that

$$\mathbf{C}_{ir}^- = \mathbf{R}_{ir}\mathbf{C}_{ir}^+ \tag{3.14}$$

Since $\mathbf{C}_{ir}^{-} = \mathbf{f}_i \mathbf{R}_{(i+1)l} \mathbf{C}_{ir}^{+}$, one can find $\mathbf{R}_{ir}$ in terms of the global wave reflection matrix on the left side of discontinuity i+1; i.e.,

$$\mathbf{R}_{ir} = \mathbf{f}_i \mathbf{R}_{(i+1)l} \mathbf{f}_i \tag{3.15}$$

In addition, by combining the following wave equations at discontinuity i

$$\mathbf{C}_{ir}^{+} = \mathbf{t}_i \mathbf{C}_{il}^{+} + \mathbf{r}_i \mathbf{C}_{ir}^{-} \quad \mathbf{C}_{il}^{-} = \mathbf{t}_i \mathbf{C}_{ir}^{-} + \mathbf{r}_i \mathbf{C}_{il}^{+} \tag{3.16-17}$$

the relationship between the global wave reflection matrices on the left and right sides of discontinuity i can be found as

$$\mathbf{R}_{il} = \mathbf{r}_i + \mathbf{t}_i (\mathbf{R}_{ir}^{-1} - \mathbf{r}_i)^{-1} \mathbf{t}_i \tag{3.18}$$

$\mathbf{R}_{ir}$ and $\mathbf{R}_{il}$ progressively expand to include all the global wave reflection matrices of intermediate discontinuities along the beam before terminating its expansion at the boundaries. Since incident waves are only reflected at the boundaries, one can find the following wave equations

$$\mathbf{C}_{1}^{+} = \mathbf{r}_1 \mathbf{C}_{1}^{-} \quad \mathbf{C}_{n}^{-} = \mathbf{r}_n \mathbf{C}_{n}^{+} \tag{3.19.1-2}$$

where $\mathbf{r}$ can be found by imposing the force and moment equilibrium conditions at the boundary; e.g., for the same kinetic constraint previously considered

$$\mathbf{r} = \begin{bmatrix} \gamma - ik_r & -\gamma - k_r \\ -i\gamma^3 - (k_t + c_t s) & \gamma^3 - (k_t + c_t s) \end{bmatrix}^{-1} \begin{bmatrix} \gamma + ik_r & -\gamma + k_r \\ i\gamma^3 - (k_t + c_t s) & -\gamma^3 - (k_t + c_t s) \end{bmatrix} \tag{3.20}$$

Now, to determine the global wave transmission matrix $\mathbf{T}_i$, define

$$\mathbf{C}_{ir}^{+} = \mathbf{T}_i \mathbf{C}_{(i-1)r}^{+} \tag{3.21}$$

Rewriting Eq. (3.16) by applying $\mathbf{C}_{il}^{+} = \mathbf{f}_{(i-1)} \mathbf{C}_{(i-1)r}^{+}$ and $\mathbf{C}_{ir}^{-} = \mathbf{R}_{ir} \mathbf{C}_{ir}^{+}$, and then comparing it with Eq. (3.21), one can find that

$$\mathbf{T}_i = (\mathbf{I}_{2\times2} - \mathbf{r}_i \mathbf{R}_{ir})^{-1} \mathbf{t}_i \mathbf{f}_{(i-1)} \tag{3.22}$$

where $\mathbf{I}_{2\times2}$ is the 2×2 identity matrix.

3.2 Free response analysis

The global reflection and transmission matrices of waves traveling along a multi-span beam are now combined with the field transfer matrix to analyze the free vibration of the beam. With reference to Fig. 2, where $C_i^{\pm}$, R_i, and f_i are now replaced by $\mathbf{C}_i^{\pm}$, $\mathbf{R}_i$, and $\mathbf{f}_i$, respectively, at the left boundary

$$\mathbf{C}_{1}^{-} = \mathbf{R}_{1r} \mathbf{C}_{1}^{+} \tag{3.23}$$

However, due to Eq. (3.19), it can be found that

$$(\mathbf{r}_1\mathbf{R}_{1r} - \mathbf{I}_{2\times2})\mathbf{C}_1^+ = \mathbf{0} \tag{3.24}$$

Applying the condition for non-trivial solutions to the above matrix equation, one can obtain the following characteristic equation in terms of the Laplace variable s

$$F(s) = \text{Det}[\mathbf{r}_1\mathbf{R}_{1r} - \mathbf{I}_{2\times2}] = 0 \tag{3.25}$$

By applying a standard root search technique (e.g., Newton-Raphson method or secant method) to Eq. (3.25), one can obtain the natural frequencies of the multi-span beam.
The mode shapes of the multi-span beam can be systematically found by relating wave amplitudes between two adjacent subspans, in the same way described in Section 2.3. Defining ξ_i as the local coordinate in span i, the transverse displacement of any point in span i can be found as

$$\overline{w}_i(\xi_i) = [1 \quad 1][\mathbf{f}_i(\xi_i)\mathbf{C}_i^+ + \mathbf{f}_i^{-1}(\xi_i)\mathbf{C}_i^-] \tag{3.26}$$

However, due to Eq. (3.14)

$$\overline{w}_i(\xi_i) = [1 \quad 1][\mathbf{f}_i(\xi_i) + \mathbf{f}_i^{-1}(\xi_i)\mathbf{R}_{ir}]\mathbf{C}_i^+ \tag{3.27}$$

Since $\mathbf{C}_i^+ = \mathbf{T}_i\mathbf{C}_{i-1}^+$ from Eq. (3.21),

$$\overline{w}_i(\xi_i) = [1 \quad 1][\mathbf{f}_i(\xi_i) + \mathbf{f}_i^{-1}(\xi_i)\mathbf{R}_{ir}]\mathbf{T}_i\mathbf{C}_{i-1}^+ \tag{3.28}$$

Assume a wave packet originates and starts traveling from the leftmost boundary of the beam. By successively applying the global transmission matrix of each discontinuity on the way up to the first span, the mode shape of span i can be found in terms of wave amplitude $\mathbf{C}_1^+$; i.e.,

$$\overline{w}_i(\xi_i) = [1 \quad 1][\mathbf{f}_i(\xi_i) + \mathbf{f}_i^{-1}(\xi_i)\mathbf{R}_{ir}]\prod_{j=i}^{1}\mathbf{T}_j\mathbf{C}_1^+ \quad \mathbf{T}_1 = \mathbf{I}_{2\times2} \quad 0 \le \xi_i \le l_i \tag{3.29}$$

	Discontinuity					
Constraint	1	2	3	4	5	6
Location	0	0.15	0.35	0.6	0.82	1
k_t	∞	3,000	2,500	1,500	3,500	∞
k_r	0	250	150	100	300	∞
c_t	0	0	0	0	0	0

Table 1. Nondimensional system parameter used for Fig. 8.

where note that the amplitude ratio between the two wave components C_a^+ and C_b^+ can be found from Eq. (3.24). For example, shown in Fig. 8 are the first three mode shapes of a uniformly damped five-span beam with system parameters specified in Table 1. Once the wave reflection and transmission matrices at each discontinuity and the boundary are determined, one can apply Eq. (3.25) to find the first three natural wavenumbers γ_1=10.294, γ_2=12.038, and γ_3=14.148, from which the damped natural frequencies of the beam can be determined by using Eq. (3.6). It should be noted from the computational point of view that the present wave approach always results in operationg matrices of a fixed size regardless of the number of subspans. However if the classical method of separation of variables is applied to solve a multi-span beam problem, the size of matrix that determines the eigensolutions of the problem increases as the number of subspans increases, which may cause strenuous computations associated with large-order matrices.

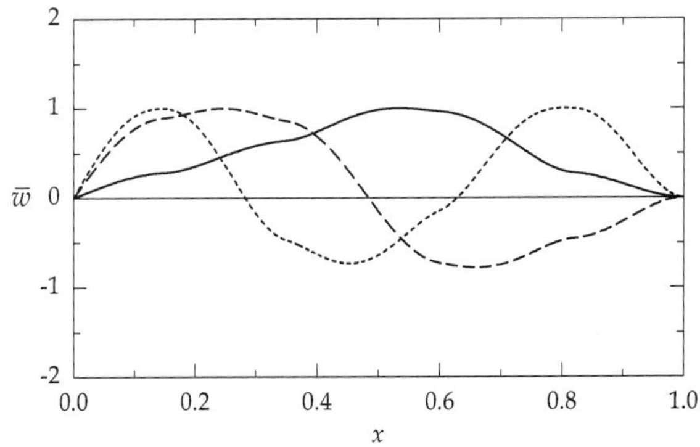

Fig. 8. First three mode shapes obtained from Eq. (3.25). The solid, dashed, and dotted curves represent the 1st , 2nd, and 3rd modes, respectively.

3.3 Transfer function analysis

Consider a multi-span beam subjected to an external force applied at $x=x_0$ where x_0 is located in subspan m. Let $\mathbf{C}^{\pm}$ be the amplitudes of the waves rise in sub-span n as a result of injected waves due to the applied force, and also assume that $\mathbf{C}^{\pm}$ satisfy all the continuity conditions at intermediate discontinuities and boundary conditions of the multi-span beam system. The transverse displacement $\bar{w}_n(x;s)$ of span n can be expressed in wave form

$$\bar{w}_n(x;s) = [1 \quad 1][\mathbf{f}(x;s)\mathbf{C}^{+} + \mathbf{f}^{-1}(x;s)\mathbf{C}^{-}] \tag{3.30}$$

Now, in order to determine the actual wave amplitudes $\mathbf{C}^{\pm}$, consider the multi-span beam with arbitrary supports and boundary conditions under a concentrated applied force of $\bar{p}(s)\delta(x-x_0)$, where $\bar{p}(s)$ is the Laplace transform of $p(\tau)$, as schematically depicted in Fig. 5 with $C_i^{\pm}$, $D_i^{\pm}$, and R_i replaced by $\mathbf{C}_i^{\pm}$, $\mathbf{D}_i^{\pm}$, and $\mathbf{R}_i$, respectively. Since the waves injected at $x=x_0$ travel in both directions, a new set of local coordinates $\{\xi, \xi^{*}\}$ is defined such that the waves traveling towards each end of the beam are measured positive as indicated in the Fig. 5. Let $\mathbf{C}_i^{\pm}$ and $\mathbf{D}_i^{\pm}$ be the amplitudes of the waves traveling within subspan i in the region $x<x_0$ and $x\geq x_0$, respectively. The discontinuity in the shear force at $x=x_0$ can be expressed in term of the difference in amplitudes between the incoming and outgoing waves at the discontinuity such that

$$\mathbf{D}_1^{+} - \mathbf{C}_1^{-} = \mathbf{v} \quad \mathbf{C}_1^{+} - \mathbf{D}_1^{-} = \mathbf{v} \tag{3.31}$$

where $\mathbf{v}$ is the wave vector injected by the applied force which can be determined by the geometric and kinetic continuity conditions at $\xi=0$ as

$$\mathbf{v} = \frac{\overline{p}(s)}{4\gamma^3}\begin{Bmatrix} \sqrt{-1} \\ 1 \end{Bmatrix} \tag{3.32}$$

Applying the global wave reflection matrices on each side of discontinuity 1 gives

$$\mathbf{C}_1^- = \mathbf{R}_{1r}\mathbf{C}_1^+ \quad \mathbf{D}_1^- = \mathbf{R}_{1r}^*\mathbf{D}_1^+ \tag{3.33.1-2}$$

Combining Eqs. (3.31) and (3.33), one can find the wave amplitudes that give rise at the point of loading

$$\mathbf{C}_1^+ = \mathbf{T}_1\mathbf{v} \quad \mathbf{T}_1 = (\mathbf{I} - \mathbf{R}_{1r}^*\mathbf{R}_{1r})^{-1}(\mathbf{I} + \mathbf{R}_{1r}^*) \tag{3.34}$$

$$\mathbf{D}_1^+ = \mathbf{T}_1^*\mathbf{v} \quad \mathbf{T}_1^* = (\mathbf{I} - \mathbf{R}_{1r}\mathbf{R}_{1r}^*)^{-1}(\mathbf{I} + \mathbf{R}_{1r}) \tag{3.35}$$

$\mathbf{T}_1$ and $\mathbf{T}_1^*$ in the above equations can be considered as the generalized wave transmission matrices that characterize the transmissibility of the waves injected by the applied force in the region $x<x_0$ and $x\geq x_0$, respectively. It is evident that $\mathbf{T}_1$ and $\mathbf{T}_1^*$ are different unless the system is symmetric about $x=x_0$. Applying the results in Eqs. (3.34) and (3.35) to Eq. (3.30), the transverse displacement of any point in span 1 on either side of the point of loading can be found. For example, for span 1 in the region $x<x_0$

$$\overline{w}_1(\xi_1, x_0; s) = [1 \quad 1][\mathbf{f}_1(\xi_1; s) + \mathbf{f}_1^{-1}(\xi_1; s)\mathbf{R}_{1r}]\mathbf{T}_1\mathbf{v} \quad 0 \leq \xi_1 \leq l_1 \tag{3.36}$$

In the same manner as for free response analysis, the forced response solution in Eq. (3.36) can be generalized for a multi-span beam by applying the global wave transmission matrix. For $x<x_0$, $\mathbf{C}_1^+ = \mathbf{T}_1\mathbf{v}$ and $\mathbf{C}_{i+1}^+ = \mathbf{T}_i\mathbf{C}_i^+$. Therefore one can progressively expand the solution in Eq. (3.36) until the expansion terminates at the leftmost boundary. Thus, the transverse displacement of any point within span i in the region $x<x_0$ due to external loading at $x=x_0$ can be determined from

$$\overline{w}_i(\xi_i, x_0; s) = \frac{\overline{p}(s)}{4\gamma^3}[1 \quad 1][\mathbf{f}_i(\xi_i; s) + \mathbf{f}_i^{-1}(\xi_i; s)\mathbf{R}_{ir}]\prod_{k=i}^{1}\mathbf{T}_k\begin{Bmatrix} \sqrt{-1} \\ 1 \end{Bmatrix} \quad 0 \leq \xi_i \leq l_i \tag{3.37}$$

One can find the same expression in the region $x\geq x_0$, but in terms of $\mathbf{R}_{ir}^*$ and $\mathbf{T}_k^*$ instead. Note that the ratio $G_i(\xi, x_0; s) = \overline{w}(\xi_i, x_0; s)/\overline{p}(s)$ in Eq. (3.37) is the transfer function governing the forced response of anypoint in span i due to the point loading at $x=x_0$. The Laplace inversion of $G_i(\xi, x_0; s)$ is the Green's function of the problem. Therefore, denoting L^{-1} as the inverse Laplace transform operator, the forced response at any point within any span of the multi-span beam can be found from; e.g., for span i in the region $x<x_0$,

$$w_i(\xi_i, x_0; \tau) = \int_0^t G_i(\xi_i, x_0; \tau - \overline{\tau})p(\tau)d\overline{\tau} \quad (i = 1, 2, \cdots, m) \tag{3.38}$$

$$G_i(\xi_i, x_0; \tau) = L^{-1}\left[\frac{1}{4\gamma^3}[1 \quad 1][\mathbf{f}_i(\xi_i; s) + \mathbf{f}_i^{-1}(\xi_i; s)\mathbf{R}_{ir}]\prod_{k=i}^{1}\mathbf{T}_k\begin{Bmatrix} \sqrt{-1} \\ 1 \end{Bmatrix}\right] \quad 0 \leq \xi_i \leq l_i \tag{3.39}$$

The exact Laplace inversion of $G_i(\xi, x_0; s)$ in close form is prohibitively difficult for the beam problems considered in this study, in particular multi-span beams with non-classical intermediate supports and boundary conditions. One may have to resort to the numerical inversion of Laplace transforms. Shown in Fig. 9 is the unit impuse response curve of a non-uniformly damped three-span beam, whose system parameters are listed in Table 2, sampled at x=0.15 and x=0.825 for x_0= 0.4. The unit impulse response curve obtained from the method of normal mode expansion is also shown for comparison. For the results obtained from the method of normal mode expansion, the mode shape function of $\sin n\pi x$ (n=1, ..., 10) are used for the expansion and the constraint forces at the intermediate supports are represented by using the Dirac-Delta function. The resulting set of modal equations are numerically integrated using a fixed-step Runge-Kutta algorithm since the beam is nonproportionally damped hence the modal equations cannot be decoupled. It can be seen that both results show an excellent agreement.

	Discontinuity				
Constraint	3	2	1=1*	2	3
Location	0	0.15	0.4	0.65	1
k_t	∞	2,000	0	3,000	∞
k_r	0	0	0	0	0
c_t	0	15	0	10	0

Table 2. Nondimensional system parameter used for Fig. 9, where c_e=0.

Regarding the numerical Laplace inversion to obtain the transient response of a Euler-Bernoulli beam model due to an impulse, it should be noted that there exists a singular behavior of the response for small values of time; i.e., the initial response. This is not due to the algorithm of the numerical Laplace inversion, but due to the inherent deficiency of the classical Euler-Bernoulli beam theory that neglects the effects of rotary inertia and shear

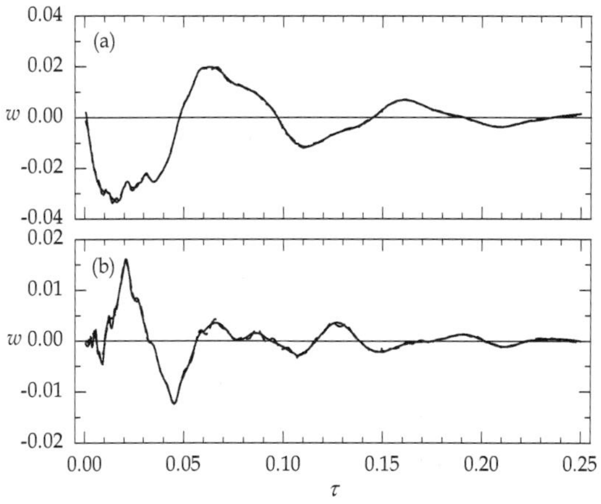

Fig. 9. Unit impulse response of the three-span beam sampled at (a) x=0.15 and (b) x=0.825. The solid and dashed curves represent the results from the present analysis and the method of normal mode expansion, respectively.

deformations. According to this simple beam theory, a wave of very high frequency (or very short wavelength) would be predicted instantaneously at remote locations, which is physically unacceptable since this can be interpreted as that such a wave travels at an infinity velocity. This abnormal behavior of the Euler-Bernoulli beam can be readily verified by observing the phase velocity c of the dispersion equation (Eq. (3.6) with $s=i\omega$ and $\omega=c\gamma$) continues to increase with increasing γ. Therefore, for a meaningful initial transient response to an impulsive load, one must employ Rayleigh or Timoshenko beam models. Studies on the wave reflection and transmission behavior at various point discontinuities and boundaries using a Timoshenko beam model can be found in Refs. (Tan & Kang, 1999; Mei and Mace, 2005).

If $p(\tau)=p_0e^{i\omega\tau}$; i.e., a harmonic forcing function, the steady-state response of the problem can be readily found in terms of the complex frequency function defined as

$$H(\xi_i,x_0;\omega)=G(\xi_i,x_0;s)\,|_{s=i\omega} \tag{3.40}$$

	Discontinuity								
Constraint	4	3	2	1=1*	2*	3*	4*	5*
Location	0	0.12	0.26	0.39	0.56	0.69	0.84	1
k_t	5,000	4,000	3,500	0	3,000	5,000	∞	0
k_r	100	0	100	0	100	50	0	0
c_t	0.5	0.1	0	0	0	0.1	0.1	0.2

Table 3. Nondimensional system parameter used for Fig. 8, where c_e=0.

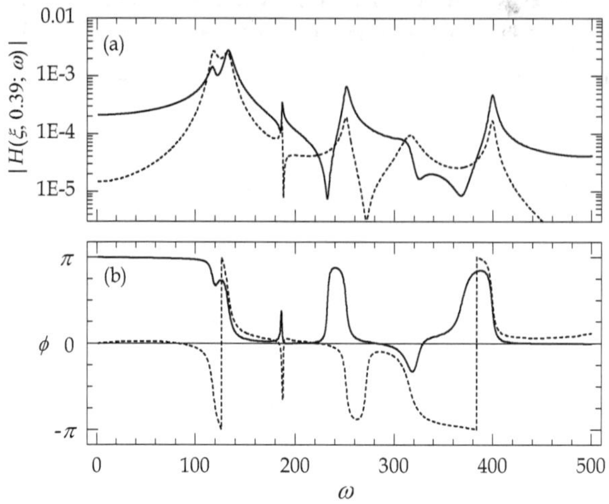

Fig. 10. Frequency response of the six-span beam due to harmonic excitation applied at x=0.39: (a) amplitudes sampled at x=0.325 (solided curve) and x=0.92 (dashed curve); and (b) corresponding phase angle.

Therefore the frequency response at any point within any subspan of the beam can be obtained by; e.g., for span i in the region $x<x_0$

$$w_i(\xi_i, x_0; \tau) = |H_i(\xi_i, x_0; \omega)| p_0 e^{i(\omega\tau + \phi_i)} \quad 0 \le \xi_i \le l_i \tag{3.41}$$

$$H_i(\xi_i, x_0; \omega) = \frac{1}{4\gamma^3}[1 \quad 1][\mathbf{f}_i(\xi_i;\omega) + \mathbf{f}_i^{-1}(\xi_i;\omega)\mathbf{R}_{ir}]\prod_{k=i}^{1}\mathbf{T}_k \begin{Bmatrix} \sqrt{-1} \\ 1 \end{Bmatrix} \quad \phi_i = \angle H_i \tag{3.42}$$

Shown in Fig.10 is the frequency response sampled at x=0.325 and x=0.92 for the non-uniformly damped six-span beam, whose system parameters are listed in Table 3, under harmonic excitation at x=0.39.

4. Sixth order systems

The analysis techniques described in the previous sections can be applied to the vibrations of higher order, one-dimensional, distributed parameter systems. For example, the in-plane vibration of a planar curved beam is governed by a sixth order partial differential equation. Consider a thin planar curved bam. Neglecting the effects of rotary inertia and shear deformations, the coupled equations of motion governing the flexural displacement W and the tangential displacement U of the centroidal axis are

$$\frac{EI}{R^3}\frac{\partial^3}{\partial\theta^3}\left(U - \frac{\partial W}{\partial\theta}\right) - \frac{EA}{R}\left(W + \frac{\partial U}{\partial\theta}\right) = \rho AR\frac{\partial^2 W}{\partial t^2} \tag{4.1.1}$$

$$\frac{EI}{R^3}\frac{\partial^2}{\partial\theta^2}\left(U - \frac{\partial W}{\partial\theta}\right) + \frac{EA}{R}\frac{\partial}{\partial\theta}\left(W + \frac{\partial U}{\partial\theta}\right) = \rho AR\frac{\partial^2 U}{\partial t^2} \tag{4.1.2}$$

where E denotes the Young's modulus, I the second moment of inertia of the cross-section about the centroid, θ the angular coordinate, R the constant radius of curvature for the given range of angle θ, A the cross-sectional area, ρ the mass density, and t the time variable. Details of deriving these equations of motion can be found in Ref. (Graff, 1975). With the following non-dimensional variables and parameters:

$$u = \frac{U}{R} \quad w = \frac{W}{R} \quad \tau = \frac{t}{t_0} \quad t_0 = R^2\sqrt{\frac{\rho A}{EI}} \quad k^2 = \frac{I}{AR^2} \tag{4.2}$$

where t_0 is a characteristic time constant and k is the curvature parameter, Eq. (4.1) takes the following non-dimensional form in the Laplace domain

$$k^2\frac{\partial^3}{\partial\theta^3}\left(\bar{u}(\theta;s) - \frac{\partial\bar{w}(\theta;s)}{\partial\theta}\right) - \left(\bar{w}(\theta;s) + \frac{\partial u(\theta;s)}{\partial\theta}\right) = k^2 s^2 \bar{w}(\theta;s) \tag{4.3.1}$$

$$k^2\frac{\partial^2}{\partial\theta^2}\left(\bar{u}(\theta;s) - \frac{\partial\bar{w}(\theta;s)}{\partial\theta}\right) + \frac{\partial}{\partial\theta}\left(\bar{w}(\theta;s) + \frac{\partial\bar{u}(\theta;s)}{\partial\theta}\right) = k^2 s^2 \bar{u}(\theta;s) \tag{4.3.2}$$

where s represents the Laplace variable. In order to determine the general wave solutions for the equations of motion in Eq. (4.3), the following form of harmonic wave solutions are first assumed

$$\begin{Bmatrix} \bar{w}(\theta;s) \\ \bar{u}(\theta;s) \end{Bmatrix} = \begin{Bmatrix} C_w \\ C_u \end{Bmatrix} e^{i\gamma\theta} \tag{4.4}$$

where γ denotes the wavenumber, normalized against R, of the wave traveling along the centroidal axis. Substituting the above wave solutions into Eq. (4.3) leads to

$$\begin{bmatrix} k^2(\gamma^4+s^2)+1 & i\gamma(1+\gamma^2k^2) \\ i\gamma(1+\gamma^2k^2) & -k^2(\gamma^2+s^2)-\gamma^2 \end{bmatrix} \begin{Bmatrix} C_w \\ C_u \end{Bmatrix} = 0 \tag{4.5}$$

Since the determinant of the matrix in Eq. (4.5) must vanish for nontrivial solutions, one can obtain following frequency equation

$$\gamma^6 + (k^2s^2-2)\gamma^4 + [1+(1+k^2)s^2]\gamma^2 + (k^2s^2+1)s^2 = 0 \tag{4.6}$$

The six roots (essentially three complex conjugates, namely $\pm\gamma_n(s)$, n=1 ,2, 3) of Eq. (4.6) can be obtained in a closed form by transforming Eq. (4.6) into a cubic equation. Each conjugate pair is the wavenumber of the constituent wave components corresponding to each of the two wave modes of the curved beam. Therefore, it can be found that

$$\bar{w}(\theta;s) = \sum_{n=1}^{3}(C_n^+e^{-i\gamma_n\theta} + C_n^-e^{i\gamma_n\theta}) \tag{4.7.1}$$

$$\bar{u}(\theta;s) = \sum_{n=1}^{3}\alpha_n(C_n^+e^{-i\gamma_n\theta} - C_n^-e^{i\gamma_n\theta}) \tag{4.7.1}$$

where the coefficient C represents the amplitude of each flexural wave component with its traveling direction indicated by the +/− signs. The subscript n signifies the spatial wave motion of the same type traveling in the opposite direction. α_n is the amplitude ratio between the flexural and tangential waves which can be found from Eq. (4.5) that

$$\alpha_n = \frac{i\gamma_n(k^2\gamma_n^2+1)}{(1+k^2)\gamma_n^2+k^2s^2} \tag{4.8}$$

The above wave solutions may be recast in vector form by grouping the waves traveling in the same direction in a wave packet; i.e., $\mathbf{C}^+ = \{C_1^+ \quad C_2^+ \quad C_3^+\}^T$ and $\mathbf{C}^- = \{C_1^- \quad C_2^- \quad C_3^-\}^T$, then

$$\bar{w}(\theta;s) = \mathbf{u}[\mathbf{f}\mathbf{C}^+ + \mathbf{f}^{-1}\mathbf{C}^-] \tag{4.9.1}$$

$$\bar{u}(\theta;s) = \boldsymbol{\alpha}[\mathbf{f}\mathbf{C}^+ - \mathbf{f}^{-1}\mathbf{C}^-] \tag{4.9.2}$$

where $u = [1 \quad 1 \quad 1]$, $\boldsymbol{\alpha} = [\alpha_1 \quad \alpha_2 \quad \alpha_3]$, and $\mathbf{f}$ is the field transfer matrix defined by

$$\mathbf{f}(\theta;s) = \begin{bmatrix} e^{-i\gamma_1\theta} & 0 & 0 \\ 0 & e^{-i\gamma_2\theta} & 0 \\ 0 & 0 & e^{-i\gamma_3\theta} \end{bmatrix} \tag{4.10}$$

One now can see that the analysis techniques described in the previous sections for the multi-span beam can be applied to a multi-span curved beam in the same manner to determine the local wave reflection and transmission matrices, global wave reflection and transmission matrices, and the transfer function (Kang et al., 2003). The only difference is the size of matrix, three by three for the curved beam. It should be also noted that the present analysis techniques are applicable to one-dimensional distributed parameter systems involving other types of discontinuities such as geometric/material changes and cracks, as long as the properties of those discontinuities can be characterized by wave reflection and transmission coefficients or matrices. In what follows, an example of free vibration analysis of a three-span curved beam with curvature changes is presented.

4.1 Wave reflection and transmission at a curvature change

Consider a curved beam with two subspans of different curvatures joined at ξ=0. Assuming that the wavenumbers of the waves traveling in each subspan are γ_n and γ'_n (n=1, 2, 3), the geometric continuity conditions at ξ=0 give

$$\begin{bmatrix} 1 & 1 & 1 \\ \alpha_1 & \alpha_2 & \alpha_3 \\ \alpha_1 - i\gamma_1 & \alpha_2 - i\gamma_2 & \alpha_3 - i\gamma_3 \end{bmatrix} \mathbf{C}^+ + \begin{bmatrix} 1 & 1 & 1 \\ -\alpha_1 & -\alpha_2 & -\alpha_3 \\ -\alpha_1 + i\gamma_1 & -\alpha_2 + i\gamma_2 & -\alpha_3 + i\gamma_3 \end{bmatrix} \mathbf{rC}^+ = \begin{bmatrix} 1 & 1 & 1 \\ \alpha'_1 & \alpha'_2 & \alpha'_3 \\ \alpha'_1 - i\gamma'_1 & \alpha_2 - i\gamma'_2 & \alpha'_3 - i\gamma'_3 \end{bmatrix} \mathbf{tC}^+ \tag{4.11.1}$$

and the kinetic continuity conditions give

$$\begin{bmatrix} 1 - i\gamma_1\alpha_1 & 1 - i\gamma_2\alpha_2 & 1 - i\gamma_3\alpha_3 \\ -\gamma_1^2(i\gamma_1 + \alpha_1) & -\gamma_2^2(i\gamma_2 + \alpha_2) & -\gamma_3^2(i\gamma_3 + \alpha_3) \\ \gamma_1(\gamma_1 - i\alpha_1) & \gamma_2(\gamma_2 - i\alpha_2) & \gamma_3(\gamma_3 - i\alpha_3) \end{bmatrix} \mathbf{C}^+ + \begin{bmatrix} 1 - i\gamma_1\alpha_1 & 1 - i\gamma_2\alpha_2 & 1 - i\gamma_3\alpha_3 \\ \gamma_1^2(i\gamma_1 + \alpha_1) & \gamma_2^2(i\gamma_2 + \alpha_2) & \gamma_3^2(i\gamma_3 + \alpha_3) \\ \gamma_1(\gamma_1 - i\alpha_1) & \gamma_2(\gamma_2 - i\alpha_2) & \gamma_3(\gamma_3 - i\alpha_3) \end{bmatrix} \mathbf{rC}^+ = \begin{bmatrix} 1 - i\gamma'_1\alpha'_1 & 1 - i\gamma'_2\alpha'_2 & 1 - i\gamma'_3\alpha'_3 \\ -\gamma'^2_1(i\gamma'_1 + \alpha'_1) & -\gamma'^2_2(i\gamma'_2 + \alpha'_2) & -\gamma'^2_3(i\gamma'_3 + \alpha'_3) \\ \gamma'_1(\gamma'_1 - i\alpha'_1) & \gamma'_2(\gamma'_2 - i\alpha'_2) & \gamma'_3(\gamma'_3 - i\alpha'_3) \end{bmatrix} \mathbf{tC}^+ \tag{4.11.2}$$

$$\alpha'_n = \frac{i\gamma'_n(k'^2\gamma'^2_n + 1)}{(1 + k'^2)\gamma'^2_n + k'^2 s^2} \tag{4.11.3}$$

The local wave reflection and transmission matrices can be found by solving the above equations. The curved beam system shown in Fig. 11 has three subspans of equal span angle of 60° but different radius of curvature. Both ends are clamped. The natural frequencies of the system can be obtained from

$$F(s) = \mathrm{Det}[\mathbf{r}_1 \mathbf{R}_{1r} - \mathbf{I}_{3\times 3}] = 0 \tag{4.12}$$

where $\mathbf{r}_1$ is the wave reflection matrix at the clamed boundary, which is

$$\mathbf{r}_1 = \begin{bmatrix} -1 & -1 & -1 \\ -\alpha_1 & -\alpha_2 & -\alpha_3 \\ \alpha_1 - i\gamma_1 & \alpha_2 - i\gamma_2 & \alpha_3 - i\gamma_3 \end{bmatrix}^{-1} \begin{bmatrix} -1 & -1 & -1 \\ -\alpha_1 & -\alpha_2 & -\alpha_3 \\ \alpha_1 - i\gamma_1 & \alpha_2 - i\gamma_2 & \alpha_3 - i\gamma_3 \end{bmatrix} \tag{4.13}$$

and $\mathbf{R}_{1r}$ is the global wave reflection matrix that includes the effects of the curvature changes and the other clamped boundary. The natural frequencies are compared in Table 4 with the results obtained by the finite element method (FEM) (Maurizi et al., 1993; Krishnan and Suresh, 1998) and the cell discretization method (CDM) (Maurizi et al., 1993). Note that the natural frequencies obtained from the present wave analysis are exact since both propagating and attenuating wave components are considered in the formulation.

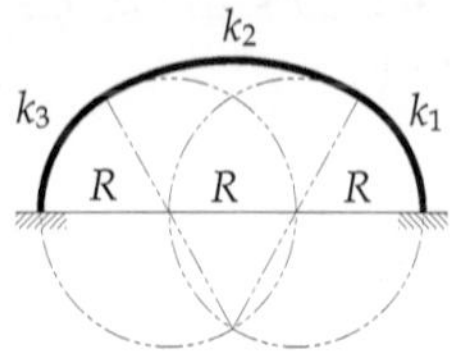

Fig. 11. Curved beam with three subspan of equal span angle of 60°, where k_1=k_3=0.01 and k_2=0.02.

Mode	Present	FEM 12 curved elements Ref. 1	FEM 12 curved elements Ref. 2	FEM 24 straight elements Ref. 1	CDM 100 Degrees of freedom Ref. 1
1	2.683	2.680	2.701	2.685	2.671
2	4.834	4.824	4.828	4.828	4.780
3	9.565	9.536	9.543	9.543	9.413
4	14.585	14.527	14.535	14.535	14.309
5	21.865	21.749	21.751	21.751	21.265

Table 4. Non-dimensional natural frequencies of the curved beam in Fig. 11.

5. Summary

An alternative approach to the dynamic response analysis of one-dimensional distributed parameter systems by applying the concepts of wave motions in elastic waveguides is presented. The analysis techniques are demonstrated using the vibration of multi-span strings, straight beams, and curved beams with general support and boundary conditions, however other one-dimensional systems such as rods and higher order beam models can be treated in the same manner. The present approach allows a systematic formulation that leads to exact, closed-form, distributed transfer functions from which the transient response and frequency response solutions can be obtained. In addition, the present analysis approach results in recursive computational algorithms that always involve computations of fixed-size matrices regardless of the number of subspans, which can be implemented into highly efficient computer codes. Since neither exact nor approximate eigensolutions are required as a priori, the present wave-based approach is suitable for the forced response analysis of non-self-adjoint systems. There are two limiting cases that may affact the analysis accuracy and numerical efficiency of the present wave approach: (1) when the waveguide contains a *very* small amount of inertia or flexibilty (such as massless elements or rigid bodies), which results in making the wavelenght of the constituent waves larger than the span length of the waveguide; and (2) when the waveguide is extremely flexible or its response contains a very sharp impulsive spike (such as the impulse response example shown in Section 2.4), which result in making the wavelength of the constituent waves unrealistically small. In paractice, however, most engineering structures have resonable amounts of inertia, flxibility, and also damping such that these two limiting cases will be rarely encountered.

6. References

Abate, J. & Valkó, P. P. (2004). Multi-Precision Laplace Transform Inversion, *International Journal of Numerical Methods in Engineering*, Vol.60, pp.979-993

Argento, A. & Scott, R. A. (1995). Elastic Wave Propagation in a Timoshenko Beam Spinning about Its Logitudinal Axis, *Wave Motion*, Vol.21, pp. 67-74

Cremer, L.; Heckl, M. & Ungar, E. E. (1973). *Structure-Borne Sound*, Springer, Berlin, Germany

Fahy, F. (1987). *Sound and Structural Vibration*, Academic Press, Orlando, USA

Graff, K. F. (1975). *Wave Motion in Elastic Solids*, Ohio State University Press, Columbus, Ohio, USA

Kang, B. & Tan, C. A. (1998). Elastic Wave Motions in an Axially Strained, Infinitely Long Rotating Timoshenko Shaft, *Journal of Sound and Vibration*, Vol.213, No.3, pp.467-482

Kang, B. (2007). Transfer Functions of One-Dimensional Distributed Parameter Systems by Wave Approach, *ASME Journal of Vibration and Acoustics*, Vol. 129, pp.193-201

Kang, B.; Riedel, R. H. & Tan, C. A. (2003). Free Vibration Analysis of Planar Curved Beams by Wave Propagation, *Journal of Sound and Vibration*, Vol.260, pp.19-44

Krishnan, A. & Suresh, Y. J. (1998) A Simple Cubic Linear Elements for Static and Free Vibration Analyses of Curved Beams, *Computer and Structures*, Vol.68, pp.473-489

Mace, B. R. (1984). Wave Reflection and Transmission in Beams, *Journal of Sound and Vibration*, Vol.97, No.2, pp.237-246

Maurizi, M. J.; Belles, R. E. & Rossi, M. A. (1993). Free Vibration of a Three-Centered Arc Clamped at the Ends, *Journal of Sound and Vibration*, Vol.213, pp.483-510

Mead, D. J. (1994). Wave and Modes in Finite Beams: Application of the Phase-Closure Principle, *Journal of Sound and Vibration*, Vol.171, pp.695-702

Mei, C. & Mace, B. R. (2005). Wave Reflection and Transmission in Timoshenko Beams and Wave Analysis of Timoshenko Beam Structures, *ASME Journal of Vibration and Acoustics*, Vol.127, pp.382-394

Tan, C. A. & Kang, B. (1999). Wave Reflection and Transmission in an Axially Strained, Rotating Timoshenko Shaft, *Journal of Sound and Vibration*, Vol.213, pp.483-510

Yang, B. & Tan, C. A. (1992). Transfer Functions of One-Dimensional Distributed Parameter Systems, *ASME Journal of Applied Mechanics*, Vol. 58, pp.1009-1014

Yang, B. (1996). Closed-Form Transient Response of Distributed Damped Systems, Part I: Modal Analysis and Green's Function Formula, *ASME Journal of Applied Mechanics*, Vol. 63, pp.997-1003

Yang, B. (1996). Closed-Form Transient Response of Distributed Damped Systems, Part II: Energy Formulation fo Constrained and Combined Systems, *ASME Journal of Applied Mechanics*, Vol. 63, pp.1004-1010

Yong, Y. & Lin, Y. K. (1989). Propagation of Decaying Waves in Periodic and Pieccewise Periodic Structures of Finite Length, *Journal of Sound and Vibration*, Vol.129, pp.99-118

Maintenance of Reducers with an Unbalanced Load Through Vibration and Oil Analysis Predictive Techniques

Aparecido Carlos Gonçalves, Daniel Fabiano Lago and
Maria da Consolação Fonseca de Albuquerque
*University of State of Sao Paulo, College of Engineering of Ilha Solteira,
Department of Mechanical Engineering
Brazil*

1. Introduction

Among the techniques for Predicitive Maintenance of Reducers, the most used are the oil analysis and vibration analysis, and the simultaneous use of both of these provides accurate results about the condition of a system under test.
The principle of the vibration analysis is based on the idea that the structures of the machines excited by the dynamic efforts (power action) give vibrational signs, whose frequency is equal to the frequency of the exciting agents. An imbalance in a machine component will cause increased vibration, once it causes an imbalance in the system and consequent increase in the power. Thus, observing the progression of the level of vibration, it is possible to obtain information on the state of the machine (Gonçalves and Campos-Silva, 2011).
The oil analysis enables identifying the first signs of wear of a component. The identification begins with studying the amount of particles, size, shape and composition, which provide accurate information on the conditions of the moving surfaces without having to disassemble the entire set that these parties belong to. These solid particles are generated by the dynamic friction between the parts in contact. According to the study of these particles the wear situations of the set can be related and attributed to physical and chemical conditions, (Barraclough et al, 1999), (Anderson et al, 1999). The oil analysis is achieved through laboratory techniques involving, reagents, instruments and equipment..

2. Analysis of the lubricants

The lubrication is introduced between two sliding solids by the addition of a lubricant at the interface, in order to reduce friction and wear, remove heat and particles generated by the contact.
Normally, the lubricants are liquids, but can also be solid, and pasty and gaseous, according to its physical state. Liquid lubricants are characterized by their liquid viscosity, but other properties are also important. Lubricating oils have names designating these properties. This type of lubricant can be subdivided into: mineral oils, fatty oils, composite oil, aditivated oils and synthetic oils..

2.1 Viscosity

Viscosity is the most important property of lubricating oils and essentially it can be defined as the flow resistance that fluids present. It is defined as the shear stress in a plane of the fluid per unit of normal velocity gradient to the plane. Viscosity can be expressed in terms of kinematical viscosity (mm^2/s or cSt) or absolute viscosity (dynamics) whose unit is the Pa.s. Hutchings (1992) defines the Newtonian viscosity of fluids, in terms of shear deformation by:

$$\tau = \eta \cdot \frac{\partial \gamma}{\partial t} \tag{1}$$

τ = shear stress [Pa],
η= dynamic viscosity [Pa.s],
$\frac{\partial \gamma}{\partial t}$ = shear deformation rate
The kinematical viscosity Z is defined according to Equation 2.

$$Z = \frac{\eta}{\rho} \tag{2}$$

Where:
Z = viscosity in cSt or mm^2/s;
ρ = specific mass.

2.2 Viscosity monitoring

Monitoring viscosity is an important component of many programs for the analysis of oil. Even small changes in viscosity can cause major damage to the lubrication. The Limits of typical industrial oils are set at 5% for precaution, and 10% for critical situations, although applications at high loads and extremely critical systems should also have an alarm system (POA, 2002).

Significant reduction of viscosity	Significant increasing of viscosity
• Loss of the oil film causing excessive wear; • Increase of mechanical friction causing excessive consumption of energy; • Generation of heat due to mechanical friction; • Internal or external leaks; • Increased sensitivity for contamination of particle due to the reduction of the oil film; • Failure of oil film for high temperatures, high loads or during starting and stopping.	• Excessive generation of heat resulting in the oxidation of oil, sludge and development of varnish; • Gaseous cavitation due to inadequate flow of oil to pumps and bearings; • Lack of lubrication due to inadequate oil flow; • Oil whipping in the radial bearing; • Excessive consumption of energy to overcome the friction of fluid; • Poor demulsibility; • Weak pumping during cold starting.

Table 1. Depicts the effects of using a lubricant with inappropriate viscosity.

When a significant change in viscosity is observed, the root cause of the problem should always be investigated and corrected. Changes in the viscosity can result from a change in the basic chemistry of the oil (a change in the molecular structure of the oil), or due to ingression of contaminants. Change in viscosity requires additional testing, such as: number of acids (AN) and infrared spectroscopy with the Fourier transform (FTIR) to confirm the incipient oxidation; contaminant tests to identify the presence of water or soot, or another less commonly used assay, which is the gas chromatography test (GC) to identify any changes in the basic chemistry of the oil.

2.3 Wear in lubricated system

Wear can be defined as a progressive loss of material, as a result of mechanical interaction between two surfaces in contact, lubricated or not. In general these areas are in relative movement (sliding or slipping) and with applied loads.
Various authors characterize wear mechanisms differently but, according to the Modern Tribology Handbook (2001) there are 4 main forms of wear: adhesive, abrasive, corrosive and by fatigue, as well as some sideline cases often classified as wear forms. Oxidation, erosion, erosion by cavitation and impact, are sometimes classified as types of wear, although Rabinowicz (1995) finds that in fact none of them are forms of wear.
These four types of damage are shown in Figure 1.

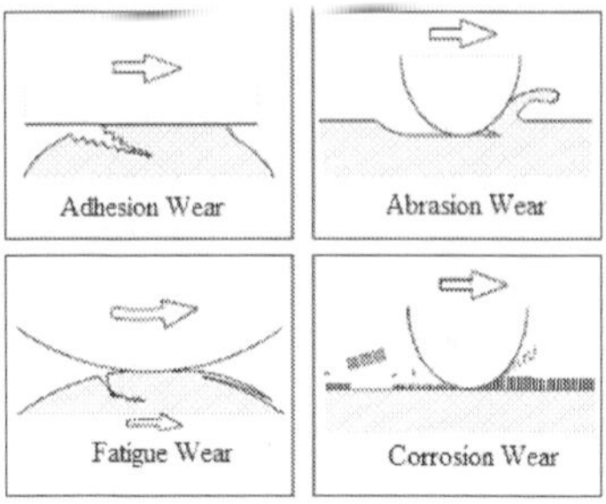

Fig. 1. Schematic Figures of four types of wear (Kato and Adashi, 2001).

These types of wear will provide precise information during the lubricant's analysis, as the particles generated mix with this analyzed lubricant.

2.4 Physical tests

The most common physical tests used with spectrographic programs and analysis programs of wear metals are: viscosity, total acid number (TAN) and determination of the water rate.
The method ASTM D445 is used for identification of viscosity, the method ASTM D 974 or D 664 is to determine the total number of acids and ASTM D 1744 to determine the water concentration by titration(Gonçalves, Padovese, 2010).
In cases where the water is at levels above 0.05 vol.%, infrared spectrography can be used. Although the limits of control for each of these parameters need to be adjusted depending on the type of lubricant and equipment, variations in viscosity of ± 10%, TAN greater than 3 mg/g, and water exceeding 100 to 500 ppm are usually sufficient for an intervention or at least for further investigation.

2.5 Spectrographic analysis of metals

The spectrographic methods include atomic absorption (AA), atomic emission spectrography (AES), inductive coupled plasma emission spectroscopy (ICPE), and X ray fluorescence (XRF). Of these methods, AES and ICPE, which are based on the detection of light emitted by the elements, are the most popular because of cost, speed, and other factors (Kimura and Gonçalves, 2009).

The spectrographic analysis of metals determines the concentration of metals and particles of up to 10 microns in size, such as moderate wear (benign sliding) and the advanced stages of fatigue, since in these wear modalities the predominant distribution of particles is within the detectable scale (<10 microns).

2.6 Particle count

The particle count is the monitoring of the number of particles of a given size by fluid volume, it is used as a preliminary monitoring tool combined with other analytical methods. The counting of particles and direct reading ferrography of direct-reading detect the onset of severe wear with a rapid increase in the quantity and size of the particles. The counting of particles detects all the particles, given that the direct reading ferrography indicates only particles of ferrous wear.

Many sensitive optical instruments are used in the counting of the number of particles in different size ranges. This counting informs the number of particles larger than a given size found in a specified volume of fluid.

For the counting of particles the standard SAE AS 4059, NAS 1638 (National Aerospace Agency), or ISO can be used (Gonçalves and Padovese, 2010).

The ISO 4406 norm (International Standard Organization) is the most widely used. Most of the versions commonly used of this technical standard refer to the number of particles larger than 4, 6, and 14 micrometers in 1 ml of fluid. The number of particles 4+ and 6+ are used as a particles point of reference. The 14+ size indicates the amount of large particles present, which contribute greatly to the possible catastrophic failure of the machine.

Figure 2 and Table 2 below represent a measurement example, where the result is obtained by an ISO code. Table 2.6 shows the various numbers of ISO 4406 code.

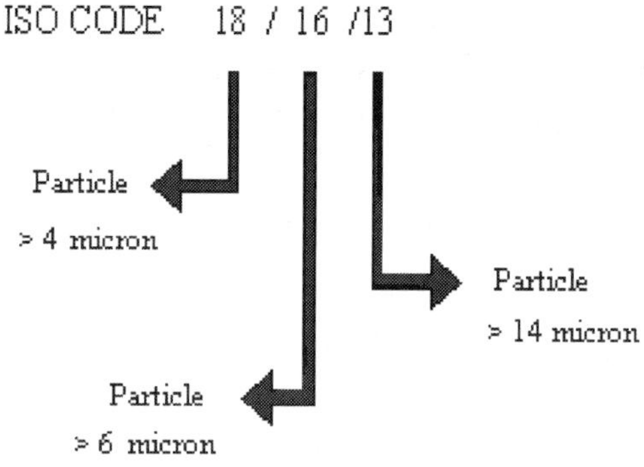

Fig. 2. Example of an obtained ISO Code, (Parker Catalog, 2000).

Range Number	Micron	Actual particle Count range (por ml)
18	4+	1,300 - 2,500
16	6+	320 - 640
13	14+	40 - 80

Table 2. Classification and definition by ISO (Parker Catalog,2000).

2.7 Ferrography

The ferrographic techniques are divided into two levels of analysis. A quantitative one, which is an evaluation technique of the wear conditions of the components of a machine, by quantifying the particles in suspension in the lubricant, and an analytical one, which uses the observation of particles in suspension in the lubricant.

The analytical ferrography is also known as direct reading ferrography. This one measures the concentration of ferrous particles in a fluid sample. With this technique one can get information on the degree of severity of wear in the machine under analysis.

To establish exact guidelines for the oil condition, samples are regularly taken from carefully selected positions of the machine system, preferably during normal operation (Lockwood and Dalley, 1992). In the ferrographic examination of direct reading the optical density is used to quantitatively measure the concentration of the wear particles in lubricating oil or in a hydraulic fluid. The particles are classified according to their sizes in DL (large particles, larger than 5 mm) (large particles), and DS (small particles, smaller than 5mm) (small particles). The wear particle concentration (WPC) and the percentage of large particles (PLP) are derived like this.

The following are equations to calculate the WPC and PBP:

$$WPC = DL+DS \tag{3}$$

$$PLP = [(DL - DS) / (DL + DS)] * 100 \tag{4}$$

The RPD (rotary particle depositor) is a type of device for direct ferrography where an index that represents the magnetic density in a given volume of lubricant is measured.

When the quantitative ferrography indicates an abnormal wear tendency, the analytical ferrographic techniques can be used to specifically identify the nature of the machine's potential problems. This enables a thorough study of particles whose size is between 1 and 250 μm (Arato, 2004). A ferrogram is built and then it can be analyzed aided by optical and electron microscopy, identifying the morphology of the particles, hence identifying any anomaly.

The ferrograms consist of transparent slides where the magnetic particles are deposited, they are separated by sizes by means of a magnetic field, other particles are randomly placed in the "barriers" formed by magnetic particles.

A first type of ferrogram is obtained by passing a flow of diluted lubricant on a plaque by gravity (the plaque is inclined). The plaque is placed on a magnet that attracts the ferrous particles and allows its adherence to the plaque, as illustrated in Figure 3.

Due to the magnetic field the particles are lined up in horizontal chains along the plaque, the larger particles are deposited first and there is a size decrease along the plaque. The non-ferrous particles are randomly placed throughout the plaque. The absence of ferrous particles actually reduces the efficiency of the analysis of non-ferrous particles.

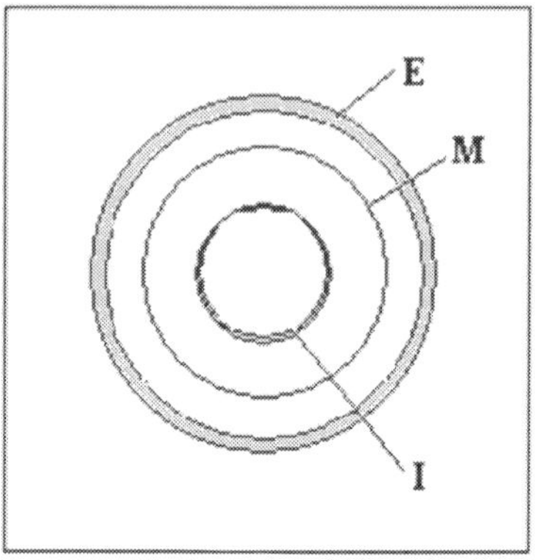

Fig. 3. Obtaining the ferrogram.

3. Analysis of vibrations

Vibration analysis is based on the idea that machine structures, excited by dynamics efforts, give vibratory signs whose frequency is identical to those of the efforts that provoked them; and the global measure in some point is the sum of vibratory answers of the structure to the different excitator efforts (Wang and Willing, 1995).

It is possible, with captors placed in private points, to register the transmitted vibration to the machine by the structures, and with its analysis, to identify the origin of the efforts which it is submitted. Vibration monitoring provides information about the macroscopic behavior inside the machine (Diana and Chelif, 2005).

In that way, as soon as it is obtained, the "vibration signature" of the machine when it is new or reputed as in good state, it will be possible for comparison , to appreciate the evolution of its state and to identify the appearance of new dynamic efforts, serial to a degradation in development process (Azoutsev, 1998).

The measure of a vibration transmitted by a structure under the effect of dynamic efforts will be function of multiple parameters:

a. Mass, rigidity and damper coefficient of the structure that transmits the vibrations;
b. Characteristics of machine fixation on the ground that opposes reactions to vibrations and modifies the intensity;
c. Position of measure point;
d. Position and fixation of the sensor (captor) on the machine;
e. Characteristic of sensor;
f. Pre amplification and transmission of the sign;
g. Performance of analyzed apparels;
h. Rotation and absorbed potency;
i. State of the connection of the cinematic chain (alignment, balancement, engagements, rolling, etc);

In a spectrum all the components of a vibratory level are represented under the form of "picks" and it can follow individually a variation of amplitude, without the masking effect that occurs in a global measure. Figure 4 shows an example of a spectrum.

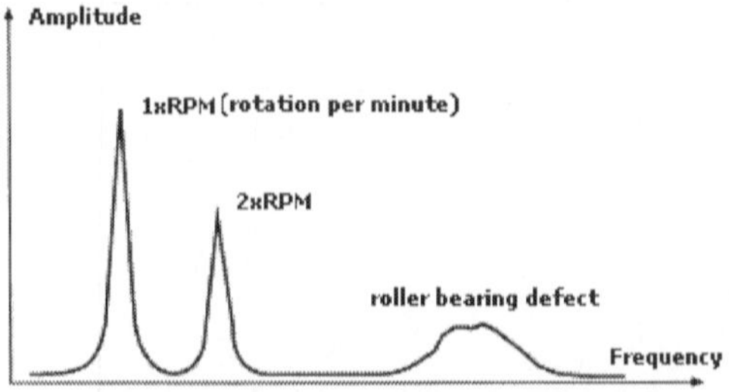

Fig. 4. Typical example of a roller bearing with problem (Mirshawaka, 1991).

The measure of the global value is an approximate analysis method of the sign that makes abstractions of the frequential parameters, expressing the evaluated amplitude in different ways. That takes into account:

a. The pick to pick value. It is measured the maximum amplitude of the fundamental wave that is useful, for example, when the vibratory displacement of a machine is critical in relation to restrictions of maximum load or clearance;
b. Crest value (or pick value). It is an important measure to indicate, for example, the level of a shock of short duration; and
c. Effective value. Measurement that takes into account the evaluation of the value of harmonic components directly related to the vibration energy content.

From the analysis of vibration signals it is possible to make decisions whether or not to intervene in the operation of the machine, so that it can be "available" as much as possible, reducing maintenance costs, time-stop of the machine, reducing the stock and improving safety, among others.

Thus, it is possible from the vibration signs at certain points of the equipment, to identify the emergence of new dynamic effects or the abrupt increase of the magnitude of the response, which are indicators of the appearance of defects or deterioration of the operation.

Vibration data can be measured with accelerometers and processed by a signal analyzer. The signal can be analyzed in the time domain or in the frequency.

3.1 Description of time domain

A vibration signal can be presented by constructing a graph for magnitude values of the signal as a function of time, from a given moment regarded as zero time. The vibration magnitude can be represented by acceleration, velocity or movement.

The wave forms are analyzed by comparison with the wave forms previously collected, and observed by repetitive impulses that can report the frequencies of the bearings, the gears or other components. In Figure 5, the vibration signal caused by imbalance is the dominant signal. It has high magnitude when compared with the defects of bearing or gears. For this reason, the lower magnitude of the wave form is superimposed on the waves caused by imbalance (Green, 2003).

The vibration elements can be divided depending on the period of repetition, finite or infinitely long, in: periodic vibrations, random vibrations and transient vibrations.

- Periodic Vibrations - Vibrations that are repeated according to a given period of time.
- Random Vibrations - Vibrations that are unpredictable as to its instant value, for any future moment.

- Transitional Vibrations - Vibrations that exist only in a limited space in time, and null at any other time.

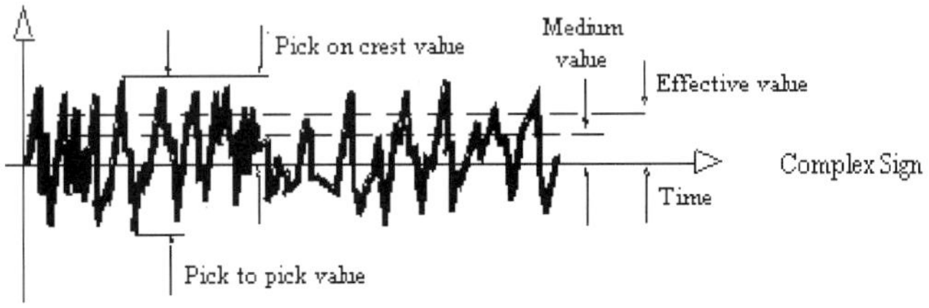

Fig. 5. Vibration Signal in time domain (Arato,2004).

3.2 Description of the frequency domain

The fast Fourier transform (FFT) can derive a wave form in time and present it in the frequency domain as shown in Figure 6. This process is the breaking of all vibrational signals into individual components of the vibration signal and plotting it in a frequency scale. This signal in the frequency domain is called frequency spectrum and provides valuable information about the condition of a machine.

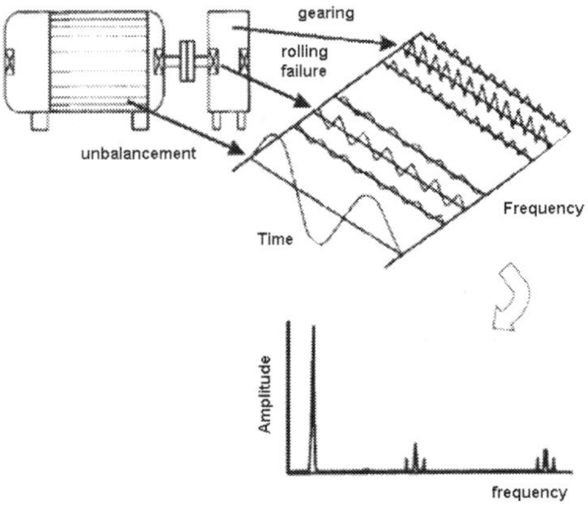

Fig. 6. Vibratory Signal in the frequency domain(Green, 2003)

The frequency spectra are used to obtain information that help determine the location of the problem, the cause of the problem and the time for the problem to become critical. This depends on the type of machine and is always relative to the level of vibration of the machine's proper operation. The frequency at which the vibration occurs indicates the type of failure and it provides an indication of what is causing the failure.

In FFT transformation, a small section is extracted from the time signal (the so-called time window) and the frequency spectrum is calculated using the FFT algorithm. During this process the instrument assumes that the signal in this time window (time data set) is continuously periodical, that is, it is repeated over time, as seen in Figure 7.

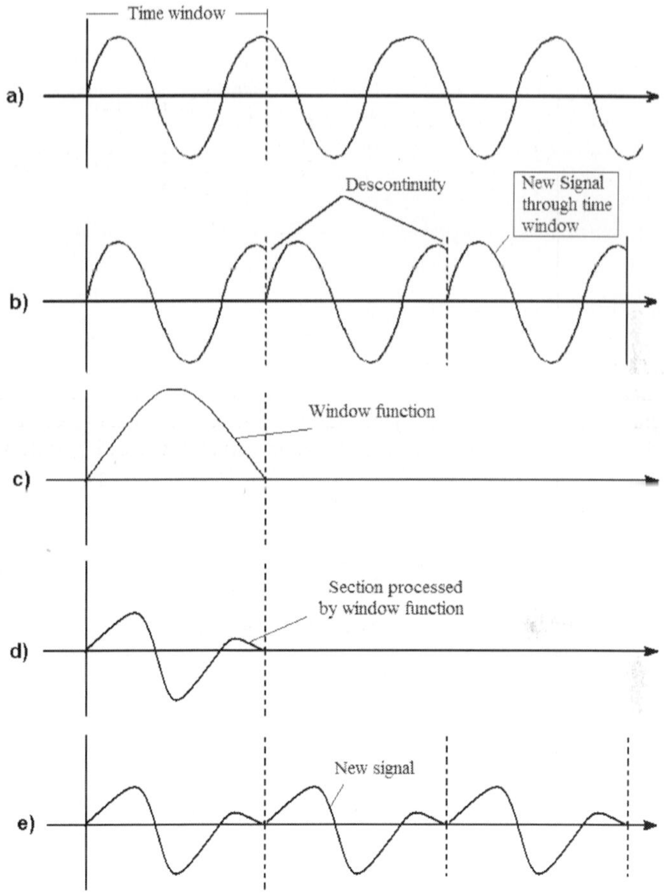

Fig. 7. Obtaining fast Fourier transform.

Depending on the structure and circumstances of the signal, some interruptions in the sequence can occur at the edges of the time window - which will reflect on the visual components of vibration.

These sequence interruptions always occur when the number of periods of the sign in the time window is not an integer. To eliminate such interruptions, a weight function is applied to the signal within the time window. As a rule, this is done so that the values of the signal at the beginning and end of the time window are reduced to zero.

Then, when the time signal on the computer is rebuilt, all the interruptions in the signal sequence are removed.

Evidently, because of this "manipulation", the machine's original vibration signal is distorted. To correct this, the results of the transformation are multiplied by a correction factor so that the exact values of magnitude are maintained, after processing.

3.3 Vibration analysis by global levels

In the case of application for predictive maintenance, the international technical standards, including the ISO, define two criteria for the adoption of a global value. One method assesses the severity of vibration by absolute measurement of non-rotating parts. The other one assesses conditions of the machine by direct measurement of oscillation of the shaft. (Gonçalves et all, 2007).

According to NBR 10082 a rating of acceptable levels of vibration severity for similar machines is established and grouped into classes. Table 3 shows the guidance offered by this standard, where:

Class I - Small machines activated by directly coupled electric motor, maximum power of 15 kW.

Class II – Mid-sized machines, class I type, with power greater than 15 kW, up to 75 kW. Motors or machines rigidly mounted up to 300 KW.

Class III - Large driving machines and other large machines (> 75 kW) with rotating masses mounted on rigid and heavy foundations, which are relatively rigid in the measuring of vibration.

Class IV - Machines of the Class III type, mounted on relatively flexible foundations in the measuring of vibration, for example, a set of turbogenerators.

Range of vibration severity		Avaliation of quality for differents machine classes			
Range	Velocity in limits (mm/s)	Class I	Class II	Class III	Class IV
0.28	0.28	A	A	A	A
0.45	0.45	A	A	A	A
0.71	0.71	A	A	A	A
1.12	1.12	B	A	A	A
1.8	1.8	B	B	A	A
2.8	2.8	C	B	B	A
4.5	4.5	C	C	B	B
7.1	7.1	D	C	C	B
11.2	11.2	D	D	C	C
18	18	D	D	D	C
28	28	D	D	D	D
45	45	D	D	D	D
71	upper 45	D	D	D	D

Table 3. Classification and assessment of machines by vibration severity levels.

Where:

A = Proper conditions;

B = Acceptable for continued operation;

C = Tolerable Limit;

D = Non-permissible.

For rotating machines with rotation speeds in the range of 600 to 12,000 rpm (10 to 200 Hz), ISO norm 2372, VDI Richiline 2056, and in Brazil by NBR 10082, take the value of effective vibration speed, known as rms speed of the signal, as the unit of measure for identifying the severity of vibration (Arato, 2004).
The parameter to be measured is the absolute velocity of vibration on the machine parts, preferably the bearings. In this case, the global value chosen as the unit of measure to indicate the vibration severity, the effective value, or simply RMS speed (V_{ef}) is not represented by a single scale of values. This is due to the great diversity of forms, mass, assembly and operational conditions of the equipment, which results in the RMS speed values for different levels of acceptable severity, (Gonçalves et all , 2007).

3.4 Demodulation

In more complex situations, where there is a combination of more than one source of excitement added to the noise transmitted through the support and foundations of the machines, the obtained spectrum of frequencies can present difficulties in the analysis (Arato, 2004).
In cases like this it is necessary to use other more dedicated techniques, such as the technique of demodulation, which enables identifying noise sources responsible for the excitation of resonant responses in the structure, hence allowing to monitor defects that are responsible for impacts of the repeated excitation type, in addition to others that produce modulator signals, even if the level of energy of the source does not allow a direct identification of its frequency in the general spectrum, as it generates amplitudes of minor significance, which remain hidden in the level of background noise.
Taking into consideration, by generalization, that the modulation in magnitude of a signal is defined as the multiplication of one sign for another, a nonlinear inherent process that creates new frequencies are not present in any of the signals involved. The identification of the noise source associated with the defect requires identifying the frequency of the modulating signal, (Arato, 2004).
The process of identifying the modulating frequency of a modulated signal is known as demodulation, and includes the following steps, (Arato, 2004):

a. Filtering of the signal by band-pass filter for the frequency range identified as modulated;
b. Detection of the modulator signal;
c. Spectral analysis of this detected modulator signal.

For the detection of the modulator signal there are several techniques. Application of the Hilbert transform that can be obtained from X (f), which is the Fourier transform of the filtered signal x(t) , according to the equations below.

$$x_{re}(t) = \mathrm{Re}\left[2\int_{0}^{\infty} X(f) e^{i2\pi ft} df \right] \quad (5)$$

$$x_{im}(t) = \mathrm{Im}\left[2\int_{0}^{\infty} X(f) e^{i2\pi ft} df \right] \quad (6)$$

Obtaining the signals $x_{re}(t)$ and $x_{im}(t)$ from which an analytical signal $z(t) = x_{re}(t) + ix_{im}(t)$ (Bendat(1986) can be constructed, (apud Arato & Silva, 2000), which can be represented by

Equation 7, where A(t) is the envelope and $\phi(t)$ is the instantaneous phase of the signal x(t), according to Equations 8 and 9.

$$z(t) = A(t) \cdot e^{i\phi(t)} \tag{7}$$

$$A(t) = \sqrt{x^2(t) + x_1^2(t)} \tag{8}$$

$$\phi(t) = tg^{-1}\left(\frac{x(t)}{x_1(t)}\right) \tag{9}$$

4. Materials and methods

In this work, to verify the effectiveness of the techniques studied, a reducer of the worn drive type was monitored. For this monitoring a test bench was built, where the reducer, coupled with its entry shaft to an electric motor, by means of an elastic coupling, had a load of an unbalanced mass in its output shaft.

A photograph of the bench is shown below in Figure 8. The electric motor used is a WEG, 220 V, 60 Hz, three-phase power with 0.5 CV power and 1720 rpm.

Fig. 8. Test bench for verification of the studied techniques.

The reducer used was a Macopema, ZM reducer, worn thread, with a reduction of 1:30, 0.53 CV at the entry and 0.31 CV at the output, with oil capacity of 0.25 liters.

<table>
<tr><td colspan="3">Rolling element bearing (www.skf.com)</td><td colspan="2" rowspan="2">Shaft rotation</td><td rowspan="2">Engagement</td></tr>
<tr><td>Model</td><td>6008</td><td>6204</td></tr>
<tr><td>f_{ip}</td><td>197 Hz</td><td>142 Hz</td><td>exit</td><td>entry</td><td rowspan="4">28.67 Hz</td></tr>
<tr><td>f_{ep}</td><td>147 Hz</td><td>87.5 Hz</td><td rowspan="3">0.95 Hz</td><td rowspan="3">28.67 Hz</td></tr>
<tr><td>f_{rp}</td><td>191 Hz</td><td>114 Hz</td></tr>
<tr><td>f_{gp}</td><td>13.2 Hz</td><td>11 Hz</td></tr>
</table>

Table 4. Preferred vibration frequencies of the reducer.

At the output of the reducer a bearing was attached and after the bearing a 7.5 kg mass with a 195 mm arm.

The preferred vibration frequencies of the reducer analyzed were calculated, as illustrated in Table 4.

Where:

f_{ip} = defect frequency of inner race.

f_{ep} = defect frequency of outer race.

f_{rp} = defect frequency of rolling bearing elements.

f_{gp} = defect frequency of cage.

The tests were conducted after a running period of 168 hours, for four weeks, and each week (168 hours) oil samples were collected.

Oil recommended by the manufacturer was used; beyond oil with several percentages of liquid contamination and oil with various percentages of solid contamination.

This work presents results obtained from the first four weeks of testing, with ISO 320 oil that was recommended by the manufacturer of the reducer.

The vibration measurements were collected in the three directions of the reducer. Analyses were performed in time and frequency in order to determine the beginning and severity of the active wear where the sensors were placed for collection of the vibration signs. Figure 9 shows the points along the reducer where the sensors were placed for collection of the vibration signs.

Fig. 9. Collection points of vibration signals.

According to the norms the bearings should be monitored first, thus points 3 and 7 were chosen. Points 2 and 5 represent the other two directions. These points contain all the information provided by points 1, 4, 6 and 8.

The time vibration signals were obtained by measuring the vibration speed of the reducer. For such measures piezoelectric accelerometers, a 4-channel Conditioner/ Amplifier, data acquisition system DaqBooK and a Notebook were used. The sampling frequencies were of 500 Hz, 1 kHz, 5 kHz and 10 kHz, and the corresponding analog filters were of 141 Hz, 281 Hz, 2250 Hz and 4500 kHz. For each frequency 10 samples were taken out of 2048 points.

The time vibration signals obtained were processed using the algorithm FFT (Fast Fourier Transform), and analyzed in the laboratory through the "software" DASYlab.

As the accelerometer measures the vibration speed of the reducer, using a reading indicator, the measured value of greatness is obtained directly, that is, the value of the effective vibration speed for each distinct sampling frequency. The value of the vibration severity however, is obtained when a vibration signal of a sampling frequency of 5000Hz is read, but subjected to a high-passed filter of 10 Hz and a low-passed of 1000 Hz. Both the effective value of the vibration speed and the severity of vibration were obtained using the "software" DASYlab, which contains numerous tools as: reading indicators, filters, etc.
The vibration analysis was achieved by spectral analysis, analysis by demodulation and the values of the effective vibration speed and vibration severity.
For demodulation of the signal it was necessary to use a computational routine on the Matlab platform called DEMOD, created by Arato (2004) and responsible for the calculation of Hilbert transformed.
Only the time signals obtained in points 2 and 5, when subjected to high frequencies of sampling were demodulated, due to the fact they are the only signals demonstrating frequencies that can be resonant. After demodulation the signal was processed to obtain the spectrum of the demodulated signal.
Oil samples were prepared in the rotary particle depositor (RPD), and then examined and photographed using the optical microscope Neophot 21 with adapted light transmitted. In the RPD lamina the particles are arranged in three separate rings, depending on the size of the particle, due to this, it was necessary to capture the images by observing these three rings separately.
Using the automatic monitor of ferrous particles, the PQ index of the samples was obtained. Also the viscosity, water content, the acidity index and atomic absorption of the oil samples were obtained at the end of each test.

5. Results and discussions

5.1 Analysis of lubricant

Initially, the inner elements of the reducer were photographed for a subsequent comparison and verification of wearing.

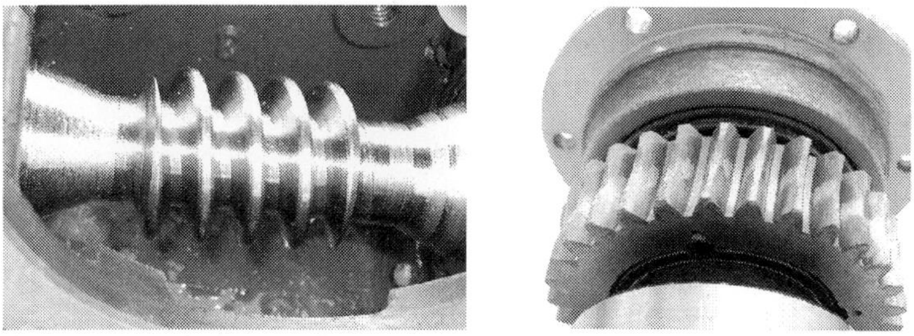

Fig. 10. Inner elements of the reducer. On the left is the worn screw and on the right the crown gear.

Figure 11 shows the photographs after the first week of the experiment.
In the figure above it can be seen severe wear particles (1^A), bronze particles (2^A), laminar particles (3^A) and cutting particles (4^A).
Figure 12 shows the photos obtained after the second week of the experiment.

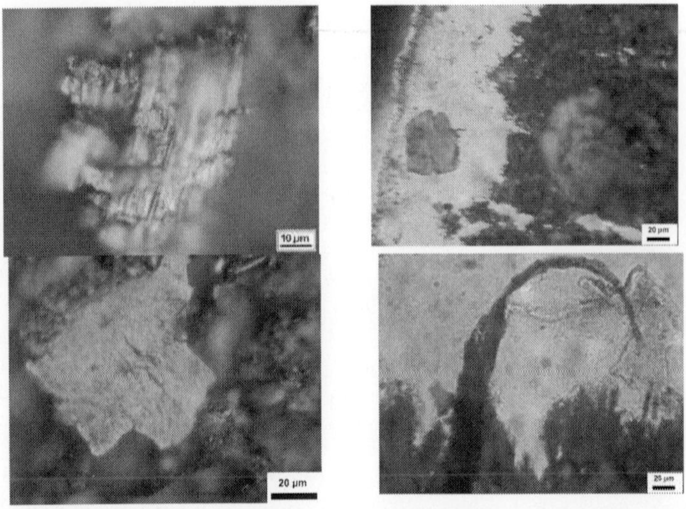

Fig. 11. Wear particles generated in the first week of the first experiment with transmitted and reflected light. First picture: inner ring of the RPD. Second photo: outer ring. Other photos: intermediary ring.

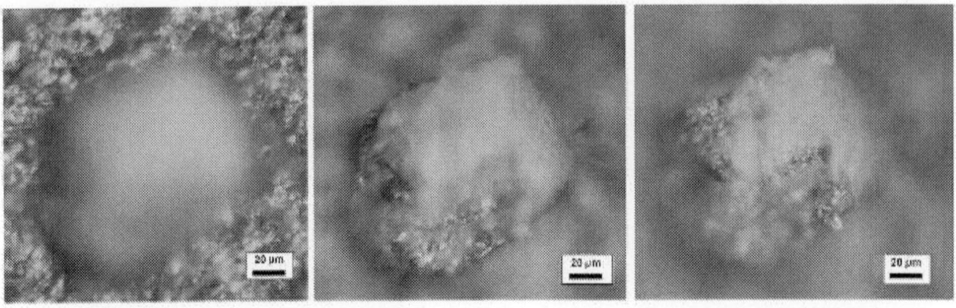

Fig. 12. Wear particles generated in the second week of the first experiment with reflected light and in the inner ring of RPD. The three photos are from the same field of vision but with different focal heights.

By Figure 12 the visual difficulty of the laminar particles can be perceived.
Figure 13 illustrates the photos obtained after the third week of the experiment.
Through Figure 14 the presence of oxide (1^{A}), laminar particles (2^{A}), bronze detained between the ferrous particles (3A) and parts difficult to focus can be seen (4^{A} and 5^{A}). The last two figures also show the need to vary the type of light in the verification of the particles.
Table 5 presents the tests values of TAN, viscosity at 40ºC and 100ºC, after the end of the last week of the first experiment. Table 6 shows the result of the atomic absorption at the end of this experiment. Table 7 shows the figures obtained in the direct ferrography during the experiments, represented by the PQ index of the magnetic particles counter. Figure 15 shows the state of the reducer after the experiments.

Fig. 13. Wear particles generated in the third week of the first experiment with reflected and transmitted light in the intermediary ring of RPD.

Through the figure above the severe wear particles represented by striation (1^{A}) and the particles overlapping the others (2^{A}) can be seen.
In Figure 14 the photos obtained after the fourth week of the experiment are shown.

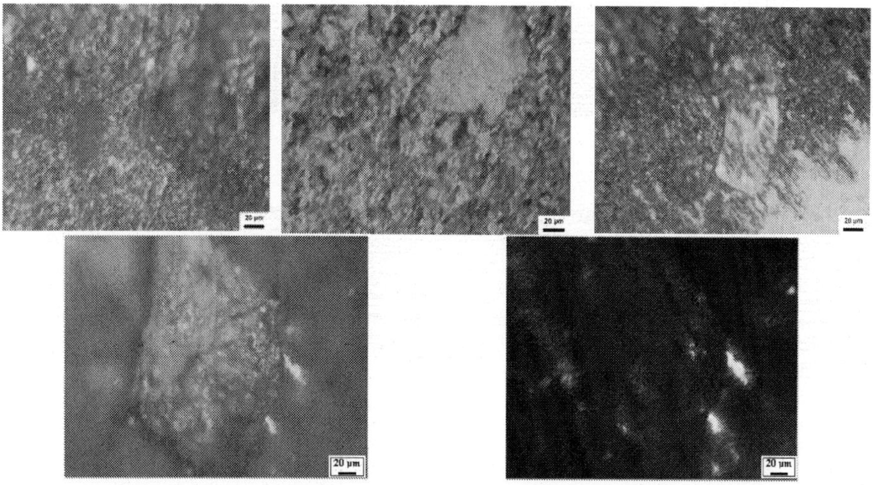

Fig. 14. Wear particles generated in the fourth week of the first experiment The first photo is of the inner ring of RPD observed with transmitted and reflected light. The second photo is of the outer ring with the same lights. The third one is the intermediary ring with the same lights. The last two are also of the intermediary ring; the second to last is with transmitted and reflected light and the last one with only transmitted light.

Water	TAN	Viscosity at 40^{0}C	Viscosity at 100^{0}C
ASTM D 95	ASTM D 664	ASTM D445	ASTM D 445
(%)	mg KOH/g	cSt	cSt
0.00	1.17	295	42

Table 5. Tests conducted on the oil at the end of the experiment.

Atomic absorption						
Cu	Si	Al	Fe	Pb	Cr	Ni
Ppm	ppm	ppm	ppm	ppm	ppm	ppm
580	18	4	142	0	0	7

Table 6. Atomic Absorption conducted on the oil at the end of the experiment. Atomic absorption

Test A			
1st Sample	**2nd Sample**	**3nd Sample**	**4nd Sample**
670	1680	3050	4000

Table 7. PQ Rates obtained on the particle monitor.

Fig. 15. Inner elements of the reducer after the first test.

Through Table 5 it was noticed that the viscosity at 100 ^{0}C increased from allowable range (27.9-33.3), according to the specifications of the new oil, to 42.03. This means an alert since the 10% of the permissible range was exceeded. This may be an indication of oxidation of the oil.
In Table 6 large amounts of Cu was noticed in the bronze of the crown and Fe in the alloy steel from which the pinion is manufactured. The Si is an indicator of external contamination.
In Table 7 the gradual wear of the pinion can be seen. As the samples were not changed during the first experiment, the quantity of metallic particles accumulated during the weeks of this first experiment.

5.2 Analysis of vibrations

Several vibration measures were taken at various points of the reducer. In the following Figures some measures taken at some points are presented. Table 8 shows the effective value and the severity value of the vibration velocities.

	Vibration effective value (mm/s)			**Vibration Severity**
	500Hz	5000Hz	10KHz	NBR 10082
Point 2	0.27	0.71	0.99	0.64
Point 3	0.34	0.43	0.54	0.51
Point 5	0.37	0.44	1.09	0.42
Point 7	0.51	0.64	0.68	0.61

Table 8. Effective value of vibration speeds (mm/s) and severity values of vibration by the NBR 10082 norm, (10 to 1000Hz), at the end of the last week of the first experiment.

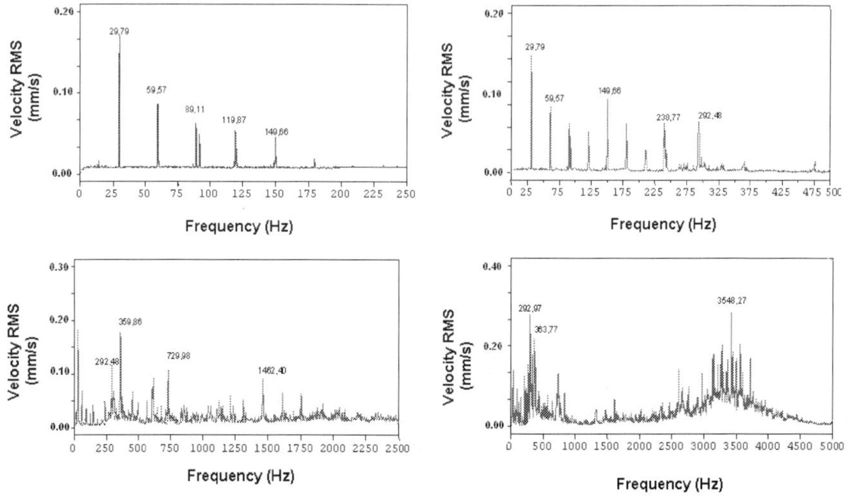

Fig. 16. Vibration spectra obtained at point 2, for sampling frequencies of 500, 1000, 5000 and 10000 Hz with analog filters of 141, 281, 2250 and 4500 Hz respectively, at the last week of the first experiment.

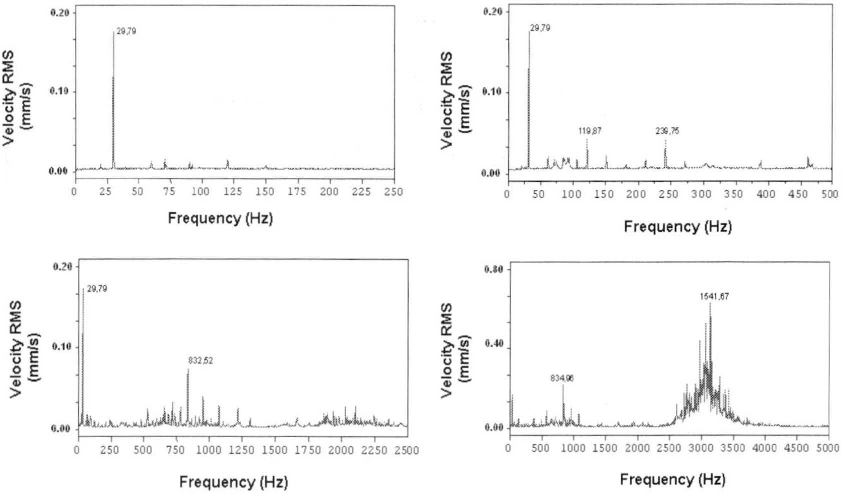

Fig. 17. Vibration spectra obtained at point 5, for sampling frequencies of 500, 1000, 5000 and 10000 Hz with analog filters of 141, 281, 2250 and 4500 Hz, respectively, after the last week of the first experiment.

Figure 18 presents a graph showing the trend of the values obtained in the four weeks of the first test for points 2, 3, 5 and 7.

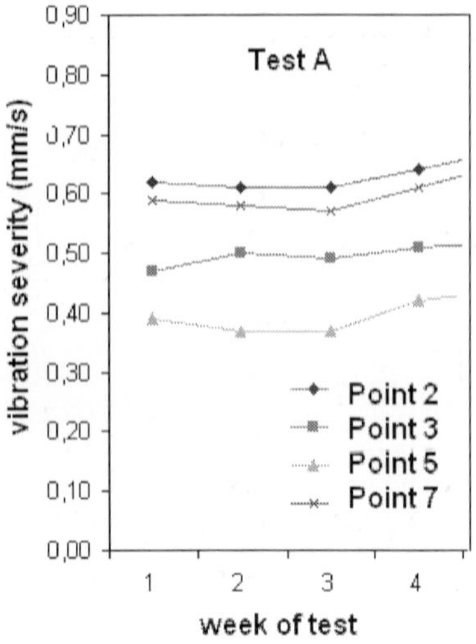

Fig. 18. Evolution of the vibration severity in accordance with the NBR 10082 norm.

Considering that our experiment fits in Class I of the NBR 10082 norm, we found that all the values are within the range considered to be in good conditions, that is, within the range A of Table 3.

Resonant frequencies were found in points 2 and 5 of the reducer. These were demodulated when subjected to a sampling frequency of 10kHz.

Figure 19 shows Point 2 with a cutoff frequency of 250 and 450Hz, where demodulation was performed on the sampling signal of 10 KHz at the end of the first experiment.

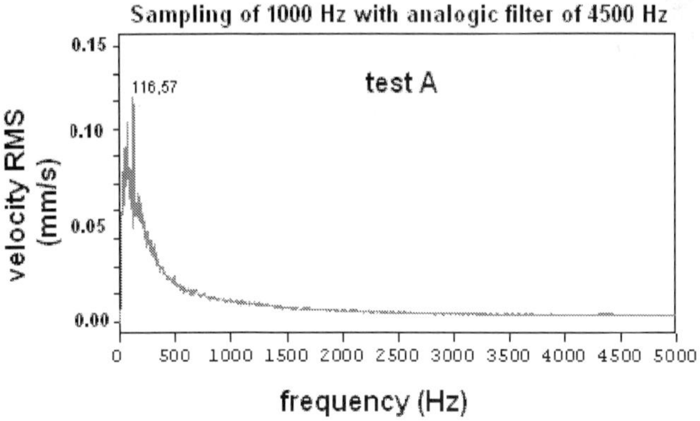

Fig. 19. Filtered time signal around the resonance frequency using a cutoff frequency of 250 Hz and 450 after demodulation by the program DEMOD followed by its spectrum.

Figure 20 shows Point 5 with a cutoff frequency of 2900 and 3300Hz, where demodulation was done on the sampling signal of 10 KHz at the end of the first experiment.

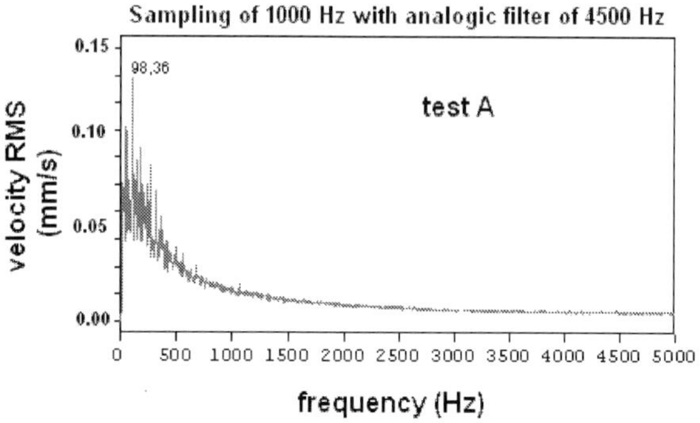

Fig. 20. Filtered time signal around the resonance frequency using a cutoff frequencies of 2900 and 3300 Hz, after demodulation by the program DEMOD followed by its spectrum.

Through demodulation of the signal it was observed that these frequencies are not related to bearing defects.

6. Conclusions

The type of reducer used, worn drive, is difficult to analyze because in this case there is a low rotation shaft (exit of the crown, which in our case is of 0.95 Hz) and the engagement frequency coincides with the rotation shaft frequency, given that the worn screw has a single entry.

In the studied reducer the oil analysis showed a certain advantage, given that since from the beginning this operation demonstrated the reducer's improper operation, while the vibration analysis was difficult to use given that the rotation shaft and engagement coincided.

The experiments also demonstrated that if in a system an abnormal occurrence takes place, for example an extra load during a certain period of time, the subsequent vibration analysis may not perceive what took place if no permanent damage occurred, however the oil analysis will acknowledge that there was such a problem during the operation.

No bearing defects were found in the spectra and neither in the demodulation.

After disassembling the reducer it was found that the axis bearings of the worn screw were loose. The literature shows that the loose bearings are represented by the frequency of the spin axis and its harmonics, which is the same case as the wear. This fact was difficult to check because of the coincidence of such frequencies.

One of the simplest tests, the counting of particles through the automatic monitor of PQA particles, showed a high generation of magnetic particles from the first sample.

The atomic absorption performed at the end of each test also provided important information about the tribological condition of the system, where a large amount of copper particles can be noted from the crown and iron resulting from the pinion.

The reducer's incorrect operation can also be verified by a visual inspection of the particles. Most of the photos demonstrate the presence of many ferrous particles and some copper particles. We also found many wear particles by friction (laminar particles) and severe wear particles by sliding, which are particles with striation.

7. Acknowledgment

For Continuos Funding and support, I thank:
FUNDUNESP - Unesp Development Foundation;
FAPESP- State of Sao Paulo Research Foundation;
CNPq - National Counciul for Scientific and Technological Development;
FINEP - National Fund for Studies and Projects.

8. References

Anderson A, Sweeney A, Willians G.(1999). How to make decisions. In: ___ Quantitative approaches to decision making. 9.ed. South Western: College Publishing, 1999. p.666-671.

Arato, A.(2004). Manutenção preditiva usando análise de vibrações. Barueri: Manole, 2004. 190 p.

ASSOCIAÇÃO BRASILEIRA DE NORMAS TÉCNICAS - ABNT. NBR 10082: vibração mecânica de máquinas com velocidades de operação de (600 a 12000) rpm - bases para especificação e padrões de avaliação, 1987.

Azovtev, A.; Barkov, A.(1998). "Development of Condition Diagnostics in Russia Using Vibration", Vibro Acoustical Systems and Technologies, Inc. (VAST), Saint Petersburg, Russsia, 16p., 1998.

Barraclough T G, Sperring T P, Roylance, B J, Nowell, T.(1999). Generic-based wear debris identification -on the first step towards morphological classification. In: PROCEEDING OF THE INTERNATIONAL CONFERENCE ON CONDITION MONITORING, 1999, Swansea. Proceedings... Oxford: Coxmore Publishing, 1999. p.525-538.

Bendat, Julio & Pierson, Allan G.(1986). "Randon data: Analysis and Measurment Procedures", John Wiley, EUA, 1986.

Diana, G.; Chelif, F.(1999). "Vibration Mechanics", SFM: Société Française des Mécaniciens, Acoustical and Vibratory Surveillance Methods and Diagnostic Techniques, Senlis, France, p.475-486, 10-12 October, 1999, vol 2/2.

Gonçalves A C, Lago D F, Cunha R C. (2007). Vibration and wear Particles Analysis in a test Stand. Industrial Lubrication and Tribology. Vol 59, issue 5, 2007, ISSN: 0036-8792

Gonçalves, A. C. ; Campos-Silva, J.B.(2011). Predictive Maintenance of a Reducer with Contaminated oil Under an Excentrical Load Throug Vibration and Oil Analysis. Journal of the Brazilian Society of Mechanical Sciences and Engineering (Impresso), v. XXXIII, p. 1/1-7, 2011.

Gonçalves, A. C. ; Padovese, L. R.(2010) . Vibration and oil analysis for monitoring problems related to water contamination in rolling bearing. In: First International Brazilian Conference on Tribology - TriboBR 2010, 2010, Rio de Janeiro. First International Brazilian Conference on Tribology - TriboBR 2010 and ITS - IFTo MM 2010 - 2nd International Tribology Symposium of IFToMM, 2010.

Grenn A. (2003). "The study of the failure modes of a spur gear using vibration an particle analysis techniques" JAMES COOK UNIVERSITY School of engineering, 2003

Hutchings, I. M.(1992). Tribology: friction and wear of engineering materials. Londres: Edward, 1992.

Kato K.; Adachi K. Wear mechanisms.(2001) In: BHUSHAN, B. (Ed.). Modern tribology handbook: principles of tribology. Boca Raton: CRC Press, 2001.

Kimura, R. K. ; Gonçalves, A. C.(2009) . Análysis of biodiesel type and contamination level on lubricant properties of diesel stationary enginees. In: 20th International Congress of Mechanical Engineering, 2009, Gramado. 20th International Congress of Mechanical Engineering -COBEM2009, 2009.

Lockwood F E, Dalley R. Lubricant analysis. (1992). In: HENRY, S. D. et al. (Ed.). ASM Handbook: friction, lubrication, and wear technology. Metals Park: The Materials Information Society, 1992. v.18, p.299-312.

Mirshawha, V. (1991). "Manutenção Preditiva: Caminho para Zero Defeitos", São Paulo, Makron, McGraw-Hill, 1991, 317p.

Modern Tribology Handbook, Vol. I ("Principles of Tribology"), Edited by Bharat Bhushan, CRC Press, 2001

Parker.(2000). The handbook of hydraulic filtration. Metamora: Parker, 2000. p 12-19.

POA. (2002). "Practicing Oil Analysis", November - December, 2002, p. 46 - 50.

Rabinowicz, E.(1995). Friction and wear of materials. 2. ed. New York: John Willey and Sons, 1995.

Wang, H.; Williams, K.(1995). "The Vibrational Analysis and Experimental Verification of a Plane Electrical Machine Stator Model", Mechanical Systems and Signal Processing, p.429- 438, 1995.

Beam Structural Modelling in Hydroelastic Analysis of Ultra Large Container Ships

Ivo Senjanović, Nikola Vladimir, Neven Hadžić and Marko Tomić
University of Zagreb, Faculty of Mechanical Engineering and Naval Architecture
Croatia

1. Introduction

Ultra large container ships are very sensitive to the wave load of quartering seas due to considerably reduced torsional stiffness caused by large deck openings. As a result, their natural frequencies can fall into the range of encounter frequencies in an ordinary sea spectrum. Therefore, the wave induced hydroelastic response of large container ships becomes an important issue in structural design. Mathematical hydroelastic model incorporates structural, hydrostatic and hydrodynamic parts (Senjanović et al. 2007, 2008a, 2009b, 2010b). Beam structural model is preferable in the early design stage and for determining global response, while for more detailed analyses 3D FEM model has to be used. The hydroelastic analysis is performed by the modal superposition method, which requires dry natural vibrations of the structure to be determined. For each mode dynamic coefficients (added mass and damping) and wave load are calculated based on velocity potential. The governing equation of ship motion in rough sea specified for the impulsive (slamming) load as a transient problem is solved in time domain. The motion equation is also given for the case of harmonic wave excitation (springing), which is solved in the frequency domain.

In the chapter, methodology of the ship hydroelastic analysis is described, and position and role of the beam structural model is explained. Beam finite element for coupled horizontal and torsional vibrations, that includes warping of ship cross-section, is constructed. Shear influence on both bending and torsion is taken into account. The strip element method is used for determination of normal and shear stress flows, and stiffness moduli, i.e. shear area, torsional modulus, shear inertia modulus (as a novelty), and warping modulus.

In the modelling of large container ships it is important to appropriately account for the contribution of transverse bulkheads to hull stiffness and the behavior of relatively short engine room structure. In the former case, the equivalent torsional modulus is determined by increasing ordinary (St. Venant) value, depending on the ratio of the strain energy of a bulkhead and corresponding hull portion. Equivalent torsional modulus of the engine room structure is also determined utilizing the energy approach. It is assumed that a short closed structure behaves as an open one with the contribution of decks.

Application of the beam structural model for ship hydroelastic analysis is illustrated in case of a very large container ship. Correlation of dry natural vibrations analysis results for the beam model with those for 3D FEM model shows very good agreement. Hydroelastic analysis emphasizes peak values of transfer functions of displacements and sectional forces in resonances, i.e. in the case when the encounter frequency is equal to one of the natural frequencies.

2. Methodology of ship hydroelastic analysis

A structural model, ship and cargo mass distributions and geometrical model of ship surface have to be defined to perform ship hydroelastic analysis. At the beginning, dry natural vibrations have to be calculated, and after that modal hydrostatic stiffness, modal added mass, damping and modal wave load are determined. Finally, wet natural vibrations as well as the transfer functions (RAO) for determining ship structural response to wave excitation are obtained (Senjanović et al. 2008a, 2009b), Fig. 1.

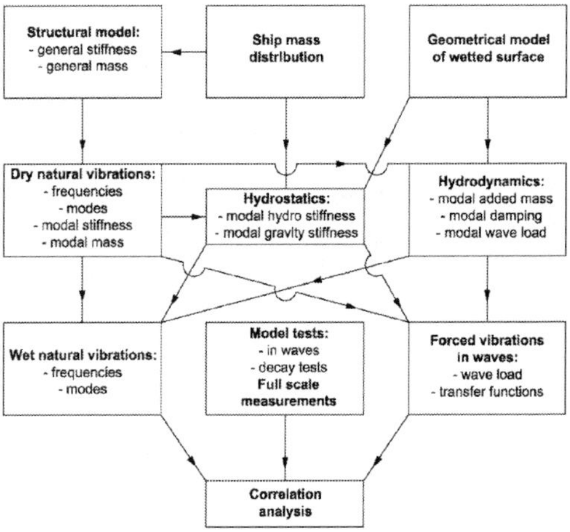

Fig. 1. Methodology of ship hydroelastic analysis

3. General remarks on structural model

A ship hull, as an elastic non-prismatic thin-walled girder, performs longitudinal, vertical, horizontal and torsional vibrations. Since the cross-sectional centre of gravity and centroid, as well as the shear centre positions are not identical, coupled longitudinal and vertical, and horizontal and torsional vibrations occur, respectively. The shear centre in ships with large hatch openings is located below the keel and therefore the coupling of horizontal and torsional vibrations is extremely high. The above problem is rather complex due to geometrical discontinuity of the hull cross-section, Fig. 2.

The accuracy of the solution depends on the reliability of stiffness parameters determination, i.e. of bending, shear, torsional and warping moduli. The finite element method is a powerful tool to solve the above problem in a successful way. One of the first solutions for coupled horizontal and torsional hull vibrations, dealing with the finite element technique, is given in (Kawai, 1973, Senjanović & Grubišić, 1991). Generalised and improved solutions are presented in (Pedersen, 1985, Wu & Ho, 1987). In all these references, the determination of hull stiffness is based on the classical thin-walled girder theory, which doesn't give a satisfactory value for the warping modulus of the open cross-section (Haslum & Tonnessen, 1972, Vlasov, 1961). Apart from that, the fixed values of stiffness moduli are determined, so that the application of the beam theory for hull vibration analysis is limited to a few lowest natural modes only. Otherwise, if the mode dependent stiffness parameters are used the application of the beam theory can be extended up to the tenth natural mode (Senjanović & Fan, 1989, 1992, 1997).

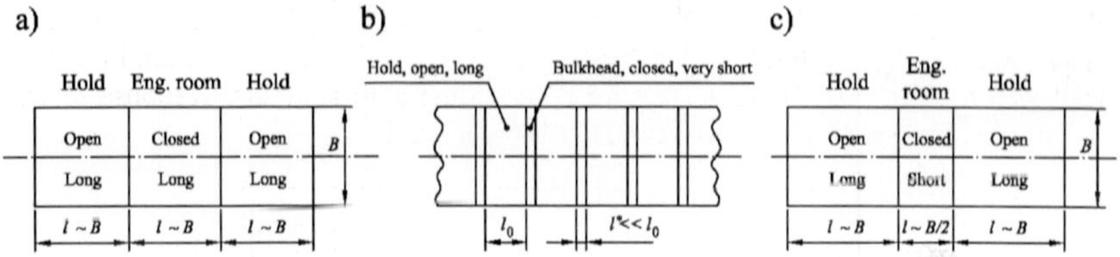

Fig. 2. Discontinuities of ship hull

4. Consistent differential equations of beam vibrations

Referring to the flexural beam theory (Timoshenko & Young, 1955, Senjanović, 1990), the total beam deflection, w, consists of the bending deflection, w_b, and the shear deflection, w_s, while the angle of cross-section rotation depends only on the former, Fig. 3

$$w = w_b + w_s, \quad \varphi = \frac{\partial w_b}{\partial x} \tag{1}$$

The cross-sectional forces are the bending moment and the shear force

$$M = -EI_b \frac{\partial \varphi}{\partial x}, \tag{2}$$

$$Q = GA_s \frac{\partial w_s}{\partial x}, \tag{3}$$

where E and G are the Young's and shear modulus, respectively, while I_b and A_s are the moment of inertia of cross-section and shear area, respectively.
The inertia load consists of the distributed transverse load, q_i, and the bending moment, μ_i, and in the case of coupled horizontal and torsional vibration is specified as

$$q_i = -m\left(\frac{\partial^2 w}{\partial t^2} + c\frac{\partial^2 \psi}{\partial t^2}\right), \tag{4}$$

$$\mu_i = -J_b \frac{\partial^2 \varphi}{\partial t^2}, \tag{5}$$

where m is the distributed mass, J_b is the mass moment of inertia about z-axis, and c is the distance between the centre of gravity and the shear centre, $c = z_G - z_S$, Fig. 4.

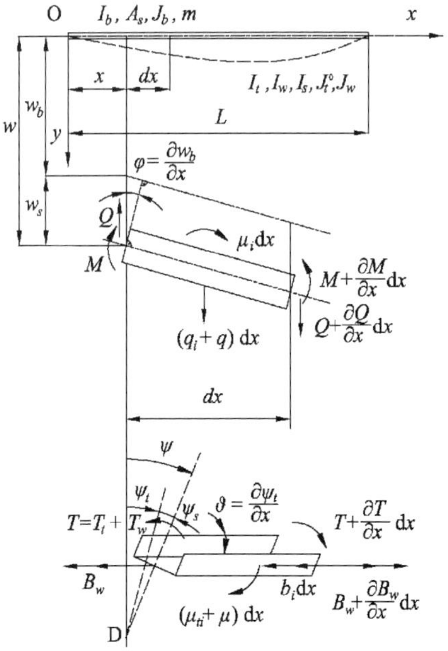

Fig. 3. Beam bending and torsion

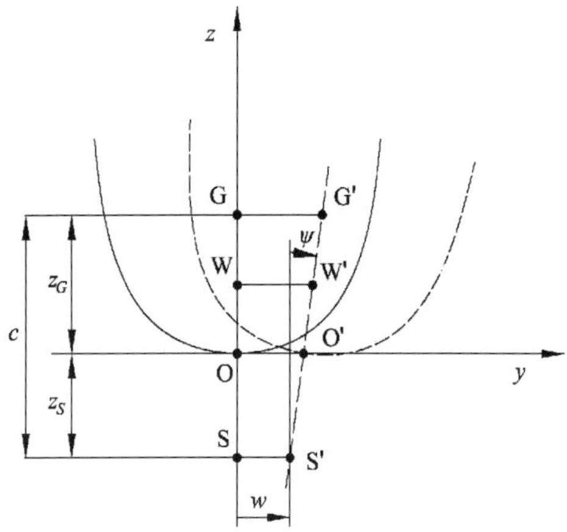

Fig. 4. Cross-section of a thin-walled girder

In a similar way the total twist angle, ψ, consists of the pure twist angle, ψ_t, and the shear contribution, ψ_s, while the second torsional displacement, which causes warping of cross-section, is variation of the pure twist angle, i.e. Fig. 3 (Pavazza, 2005)

$$\psi = \psi_t + \psi_s, \ \vartheta = \frac{\partial \psi_t}{\partial x}. \tag{6}$$

The cross-sectional forces include the pure torsional torque, T_t, warping bimoment, B_w, and additional torque due to restrained warping, T_w

$$T_t = GI_t \vartheta, \tag{7}$$

$$B_w = -EI_w \frac{\partial \vartheta}{\partial x}, \tag{8}$$

$$T_w = GI_s \frac{\partial \psi_s}{\partial x}, \tag{9}$$

where I_t, I_w and I_s are the torsional modulus, warping modulus and shear inertia modulus, respectively.

The inertia load consists of the distributed torque, μ_{ti}, and the bimoment, b_i, presented in the following form:

$$\mu_{ti} = -J_t \frac{\partial^2 \psi}{\partial t^2} - mc \frac{\partial^2 w}{\partial t^2}, \tag{10}$$

$$b_i = -J_w \frac{\partial^2 \vartheta}{\partial t^2}, \tag{11}$$

where J_t is the mass polar moment of inertia about the shear centre, and J_w is the mass bimoment of inertia with respect to the warping centre, Fig. 4.

Considering the equilibrium of a beam differential element, one can write for flexural vibrations

$$\frac{\partial M}{\partial x} = Q + \mu_i, \tag{12}$$

$$\frac{\partial Q}{\partial x} = -q_i - q, \tag{13}$$

and for torsional vibrations (Pavazza, 1991)

$$\frac{\partial B_w}{\partial x} = T_w + b_i, \tag{14}$$

$$\frac{\partial T_t}{\partial x} + \frac{\partial T_w}{\partial x} = -\mu_{ti} - \mu. \tag{15}$$

The above equations can be reduced to two coupled partial differential equations as follows. Substituting Eqs. (2) and (3) into (12) yields

$$\frac{\partial w_s}{\partial x} = -\frac{EI_b}{GA_s}\frac{\partial^2 \varphi}{\partial x^2} + \frac{J_b}{GA_s}\frac{\partial^2 \varphi}{\partial t^2}. \tag{16}$$

By inserting Eqs. (3) and (4) into (13) leads to

$$EI_b \frac{\partial^4 \varphi}{\partial x^4} + m\frac{\partial^2 \varphi}{\partial t^2} - \left(J_b + m\frac{EI_b}{GA_s}\right)\frac{\partial^4 \varphi}{\partial x^2 \partial t^2} + \frac{mJ_b}{GA_s}\frac{\partial^4 \varphi}{\partial t^4} + mc\frac{\partial^3 \psi}{\partial x \partial t^2} = \frac{\partial q}{\partial x}. \tag{17}$$

In a similar way, substituting Eqs. (8) and (9) into (14) yields

$$\frac{\partial \psi_s}{\partial x} = -\frac{EI_w}{GI_s}\frac{\partial^2 \vartheta}{\partial x^2} + \frac{J_w}{GI_s}\frac{\partial^2 \vartheta}{\partial t^2}. \tag{18}$$

By inserting Eqs. (7), (9) and (10) into (15) one finds

$$EI_w \frac{\partial^4 \vartheta}{\partial x^4} - GI_t\frac{\partial^2 \vartheta}{\partial x^2} + J_t\frac{\partial^2 \vartheta}{\partial t^2} - \left(J_w + J_t\frac{EI_w}{GI_s}\right)\frac{\partial^4 \vartheta}{\partial x^2 \partial t^2} + \frac{J_w}{GI_s}\frac{\partial^4 \vartheta}{\partial t^4} + mc\frac{\partial^3 w}{\partial x \partial t^2} = \frac{\partial \mu}{\partial x}. \tag{19}$$

Furthermore, ψ in (17) can be split into $\psi_t + \psi_s$ and the later term can be expressed with (18). Similar substitution can be done for $w = w_b + w_s$ in (19), where w_s is given with (16). Thus, taking into account that $\varphi = \partial w_b / \partial x$ and $\vartheta = \partial \psi_t / \partial x$, Eqs. (17) and (19) after integration per x read

$$\begin{aligned} &EI_b \frac{\partial^4 w_b}{\partial x^4} + m\frac{\partial^2 w_b}{\partial t^2} - \left(J_b + m\frac{EI_b}{GA_s}\right)\frac{\partial^4 w_b}{\partial x^2 \partial t^2} + \frac{mJ_b}{GA_s}\frac{\partial^4 w_b}{\partial t^4} \\ &+mc\left(\frac{\partial^2 \psi_t}{\partial t^2} - \frac{EI_w}{GI_s}\frac{\partial^4 \psi_t}{\partial x^2 \partial t^2} + \frac{J_w}{GI_s}\frac{\partial^4 \psi_t}{\partial t^4}\right) = q \end{aligned} \tag{20}$$

$$\begin{aligned} &EI_w \frac{\partial^4 \psi_t}{\partial x^4} - GI_t\frac{\partial^2 \psi_t}{\partial x^2} + J_t\frac{\partial^2 \psi_t}{\partial t^2} - \left(J_w + J_t\frac{EI_w}{GI_s}\right)\frac{\partial^4 \psi_t}{\partial x^2 \partial t^2} \\ &+\frac{J_w}{GI_s}\frac{\partial^4 \psi_t}{\partial t^4} + mc\left(\frac{\partial^2 w_b}{\partial t^2} - \frac{EI_b}{GA_s}\frac{\partial^4 w_b}{\partial x^2 \partial t^2} + \frac{J_b}{GA_s}\frac{\partial^4 w_b}{\partial t^4}\right) = \mu. \end{aligned} \tag{21}$$

After solving Eqs. (20) and (21) the total deflection and twist angle are obtained by employing (16) and (18), i.e.

$$w = w_b + w_s = w_b - \frac{EI_b}{GA_s}\frac{\partial^2 w_b}{\partial x^2} + \frac{J_b}{GA_s}\frac{\partial^2 w_b}{\partial t^2} + f(t) \tag{22}$$

$$\psi = \psi_t + \psi_s = \psi_t - \frac{EI_w}{GI_s}\frac{\partial^2 \psi_t}{\partial x^2} + \frac{J_w}{GI_s}\frac{\partial^2 \psi_t}{\partial t^2} + g(t) \tag{23}$$

where $f(t)$ and $g(t)$ are integration functions, which depend on initial conditions.
The main purpose of developing differential equations of vibrations (20) and (21) is to get insight into their constitution, position and role of the stiffness and mass parameters, and

coupling, which is realized through the inertia terms. If the pure torque T_t is excluded from the above theoretical consideration, it is obvious that the complete analogy between bending and torsion exists, (Pavazza, 1991).

Application of Eqs. (20) and (21) is limited to prismatic girders. For more complex problems, like ship hull, the finite element method is on disposal.

The shape functions of beam finite element for vibration analysis have to satisfy the following consistency relations for harmonic vibrations obtained from Eqs. (22) and(23), (Senjanović, 1990)

$$w = w_b + w_s = \left(1 - \omega^2 \frac{J_b}{GA_s}\right) w_b - \frac{EI_b}{GA_s} \frac{\mathrm{d}^2 w_b}{\mathrm{d}x^2} \tag{24}$$

$$\psi = \psi_t + \psi_s = \left(1 - \omega^2 \frac{J_w}{GI_s}\right) \psi_t - \frac{EI_w}{GI_s} \frac{\mathrm{d}^2 \psi_t}{\mathrm{d}x^2}. \tag{25}$$

5. Beam finite element

The properties of a finite element for the coupled horizontal and torsional vibration analysis can be derived from the total element energy. It consists of the strain energy, the kinetic energy, the work of the distributed external lateral load, q, and the torque, μ, and the work of the boundary forces. Thus, according to (Senjanović, 1990, Senjanović & Grubišić, 1991),

$$\begin{aligned} E_{tot} &= \frac{1}{2}\int_0^l \left[EI_b \left(\frac{\partial^2 w_b}{\partial x^2}\right)^2 + GA_s \left(\frac{\partial w_s}{\partial x}\right)^2 + EI_w \left(\frac{\partial^2 \psi_t}{\partial x^2}\right)^2 + GI_s \left(\frac{\partial \psi_s}{\partial x}\right)^2 + GI_t \left(\frac{\partial \psi_t}{\partial x}\right)^2 \right] \mathrm{d}x \\ &+ \frac{1}{2}\int_0^l \left[m \left(\frac{\partial w}{\partial t}\right)^2 + J_b \left(\frac{\partial^2 w_b}{\partial x \partial t}\right)^2 + 2mc \frac{\partial w}{\partial t}\frac{\partial \psi}{\partial t} + J_w \left(\frac{\partial^2 \psi_t}{\partial x \partial t}\right)^2 + J_t \left(\frac{\partial \psi}{\partial t}\right)^2 \right] \mathrm{d}x \\ &- \int_0^l (qw + \mu\psi) \mathrm{d}x + (Qw - M\varphi + T\psi + B_w \vartheta)_0^l , \end{aligned} \tag{26}$$

where l is the element length. Since the beam has four displacements, $w, \varphi, \psi, \vartheta$, a two-node finite element has eight degrees of freedom, i.e. four nodal shear-bending and torsion-warping displacements respectively, Fig. 5,

$$\{U\} = \begin{Bmatrix} w(0) \\ \varphi(0) \\ w(l) \\ \varphi(l) \end{Bmatrix}, \quad \{V\} = \begin{Bmatrix} \psi(0) \\ \vartheta(0) \\ \psi(l) \\ \vartheta(l) \end{Bmatrix} \tag{27}$$

Therefore, the basic beam displacements, w_b and ψ_t, can be presented as the third-order polynomials

$$\begin{aligned} & w_b = \langle a_k \rangle \{\xi^k\}, \ \psi_t = \langle d_k \rangle \{\xi^k\}, \ k = 0,1,2,3, \\ & \xi = \frac{x}{l}, \ \langle \dots \rangle = \{ \dots \}^T . \end{aligned} \tag{28}$$

Furthermore, satisfying alternately the unit value for one of the nodal displacement {*U*} and zero values for the remaining displacements, and doing the same for {*V*}, it follows that:

$$w_b = \langle w_{bi} \rangle \{U\},\ w_s = \langle w_{si} \rangle \{U\},\ w = \langle w_i \rangle \{U\},$$
$$\psi_t = \langle \psi_{ti} \rangle \{V\},\ \psi_s = \langle \psi_{si} \rangle \{V\},\ \psi = \langle \psi_i \rangle \{V\},\ i = 1,2,3,4, \tag{29}$$

where w_{bi}, w_{si}, w_i and ψ_{ti}, ψ_{si}, ψ_i are the shape functions specified below by employing relations (24) and (25)

$$w_{bi} = \langle a_{ik} \rangle \{\xi^k\},\ w_{si} = \langle b_{ik} \rangle \{\xi^k\},\ w_i = \langle c_{ik} \rangle \{\xi^k\}$$
$$\psi_{ti} = \langle d_{ik} \rangle \{\xi^k\},\ \psi_{si} = \langle e_{ik} \rangle \{\xi^k\},\ \psi_i = \langle f_{ik} \rangle \{\xi^k\} \tag{30}$$

$$\left[a_{ik}\right] = \frac{1}{a(a+12\beta)} \begin{bmatrix} a+6\beta & 0 & -3a & 2a \\ -4\beta(a+3\beta)l & a(a+12\beta)l & -2a(a+3\beta)l & a^2 l \\ 6\beta & 0 & 3a & -2a \\ -2\beta(a-6\beta)l & 0 & -a(a-6\beta)l & a^2 l \end{bmatrix} \tag{31}$$

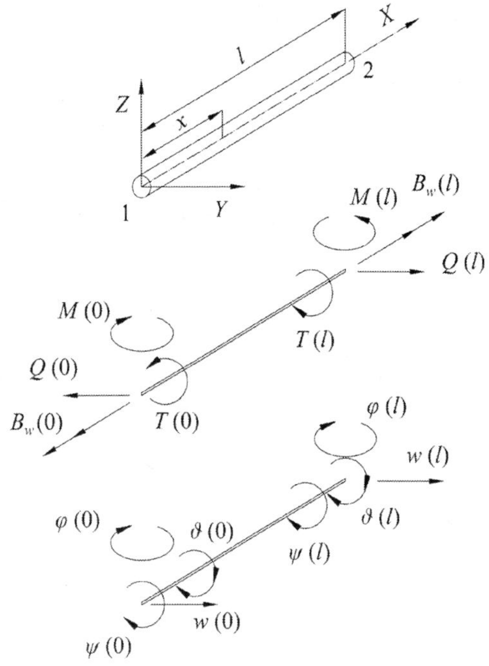

Fig. 5. Beam finite elemet

$$\begin{aligned} b_{i0} &= -(1-a)a_{i0} - 2\beta a_{i2} \\ b_{i1} &= -(1-a)a_{i1} - 6\beta a_{i3} \\ b_{i2} &= -(1-a)a_{i2} \\ b_{i3} &= -(1-a)a_{i3}, \end{aligned} \tag{32}$$

$$\left[c_{ik}\right]=\left[a_{ik}\right]+\left[b_{ik}\right],\quad i=1,2,3,4,\quad k=0,1,2,3 \tag{33}$$

$$\alpha = 1-\omega^2\frac{J_b}{GA_s},\ \beta=\frac{EI_b}{GA_s l^2} \tag{34}$$

Constitution of torsional matrices $\left[d_{ik}\right]$, $\left[e_{ik}\right]$ and $\left[f_{ik}\right]$ is the same as $\left[a_{ik}\right]$, $\left[b_{ik}\right]$ and $\left[c_{ik}\right]$, but parameters a and β have to be exchanged with

$$\eta = 1-\omega^2\frac{J_w}{GI_s},\ \gamma=\frac{EI_w}{GI_s l^2} \tag{35}$$

according to (25). By substituting Eqs. (29) (26) one obtains

$$E_{tot} = \frac{1}{2}\begin{Bmatrix} U \\ V \end{Bmatrix}^T \begin{bmatrix} k_{bs} & 0 \\ 0 & k_{ws}+k_t \end{bmatrix}\begin{Bmatrix} U \\ V \end{Bmatrix} + \frac{1}{2}\begin{Bmatrix} \dot{U} \\ V \end{Bmatrix}^T \begin{bmatrix} m_{sb} & m_{st} \\ m_{ts} & m_{tw} \end{bmatrix}\begin{Bmatrix} \dot{U} \\ V \end{Bmatrix} - \begin{Bmatrix} q \\ \mu \end{Bmatrix}^T \begin{Bmatrix} U \\ V \end{Bmatrix} - \begin{Bmatrix} P \\ R \end{Bmatrix}^T \begin{Bmatrix} U \\ V \end{Bmatrix}, \tag{36}$$

where, assuming constant values of the element properties,

$$\left[k\right]_{bs} = \left[EI_b\int_0^l \frac{\mathrm{d}^2 w_{bi}}{\mathrm{d}x^2}\frac{\mathrm{d}^2 w_{bj}}{\mathrm{d}x^2}\mathrm{d}x + GA_s\int_0^l \frac{\mathrm{d}w_{si}}{\mathrm{d}x}\frac{\mathrm{d}w_{sj}}{\mathrm{d}x}\mathrm{d}x\right] \text{ - bending-shear stiffness matrix,}$$

$$\left[k\right]_{ws} = \left[EI_w\int_0^l \frac{\mathrm{d}^2 \psi_{ti}}{\mathrm{d}x^2}\frac{\mathrm{d}^2 \psi_{tj}}{\mathrm{d}x^2}\mathrm{d}x + GI_s\int_0^l \frac{\mathrm{d}\psi_{si}}{\mathrm{d}x}\frac{\mathrm{d}\psi_{sj}}{\mathrm{d}x}\mathrm{d}x\right] \text{ - warping-shear stiffness matrix,}$$

$$\left[k\right]_{t} = \left[GI_t\int_0^l \frac{\mathrm{d}\psi_{ti}}{\mathrm{d}x}\frac{\mathrm{d}\psi_{tj}}{\mathrm{d}x}\mathrm{d}x\right] \text{ - torsion stiffness matrix,}$$

$$\left[m\right]_{sb} = \left[m\int_0^l w_i w_j \mathrm{d}x + J_b\int_0^l \frac{\mathrm{d}w_{bi}}{\mathrm{d}x}\frac{\mathrm{d}w_{bj}}{\mathrm{d}x}\mathrm{d}x\right] \text{ - shear-bending mass matrix,}$$

$$\left[m\right]_{tw} = \left[J_t\int_0^l \psi_i \psi_j \mathrm{d}x + J_w\int_0^l \frac{\mathrm{d}\psi_{ti}}{\mathrm{d}x}\frac{\mathrm{d}\psi_{tj}}{\mathrm{d}x}\mathrm{d}x\right] \text{ - torsion-warping mass matrix,}$$

$$\left[m\right]_{st} = \left[mc\int_0^l w_i \psi_j \mathrm{d}x\right],\ \left[m\right]_{ts} = \left[m\right]_{st}^T \text{ - shear-torsion mass matrix,}$$

$$\begin{aligned} &\{q\} = \left\{\int_0^l q w_j\,\mathrm{d}x\right\} \text{ - shear load vector,} \\ &\{\mu\} = \left\{\int_0^l \mu \psi_j\,\mathrm{d}x\right\} \text{ - torsion load vector,} \\ &i,j=1,2,3,4. \end{aligned} \tag{37}$$

The vectors $\{P\}$ and $\{R\}$ in Eq. (36) represent the shear-bending and torsion-warping nodal forces, respectively,

$$\{P\}=\begin{Bmatrix} -Q(0) \\ M(0) \\ Q(l) \\ -M(l) \end{Bmatrix}, \quad \{R\}=\begin{Bmatrix} -T(0) \\ B_w(0) \\ T(l) \\ -B_w(l) \end{Bmatrix} \tag{38}$$

The above matrices are specified in Appendix A, as well as the load vectors for linearly distributed loads along the finite element, i.e.

$$q=q_0+q_1\xi, \quad \mu=\mu_0+\mu_1\xi. \tag{39}$$

The total element energy has to be at its minimum. Satisfying the relevant conditions

$$\frac{\partial E_{tot}}{\partial\{U\}}=\{0\}, \quad \frac{\partial E_{tot}}{\partial\{V\}}=\{0\} \tag{40}$$

and employing Lagrange equations of motion, the finite element equation yields

$$\{f\}=[k]\{\delta\}+[m]\{\ddot{\delta}\}-\{f\}_{q\mu}, \tag{41}$$

where

$$\begin{gathered} \{f\}=\begin{Bmatrix} P \\ R \end{Bmatrix}, \quad \{f\}_{q\mu}=\begin{Bmatrix} q \\ \mu \end{Bmatrix}, \quad \{\delta\}=\begin{Bmatrix} U \\ V \end{Bmatrix} \\ [k]=\begin{bmatrix} k_{bs} & 0 \\ 0 & k_{ws}+k_t \end{bmatrix}, \quad [m]=\begin{bmatrix} m_{sb} & m_{st} \\ m_{ts} & m_{tw} \end{bmatrix}. \end{gathered} \tag{42}$$

It is obvious that coupling between the bending and torsion occurs through the mass matrix only, i.e. by the coupling matrices $[m]_{st}$ and $[m]_{ts}$.

6. Contribution of transverse bulkheads to the hull stiffness

This problem for container ships is extensively analyzed in (Senjanović et al., 2008b), where torsional modulus of ship cross-section is increased proportionally to the ratio of bulkhead strain energy and strain energy of corresponding hull portion. The bulkhead is considered as an orthotropic plate with very strong stool (Szilard, 2004). Bulkhead strain energy is determined for the given warping of cross-section as a boundary condition. The warping causes bulkhead screwing and bending. Here, only the review of the final results is presented. Bulkhead deflection (axial displacement) is given by the following formula, Fig. 6:

$$u(y,z)=-y\left\{(z-d)+\left[1-\left(\frac{y}{b}\right)^2\right]\frac{z^2}{H}\left(2-\frac{z}{H}\right)\right\}\psi', \tag{43}$$

where H is the ship height, b is one half of bulkhead breadth, d is the distance of warping centre from double bottom neutral line, y and z are transverse and vertical coordinates, respectively, and ψ' is the variation of twist angle.

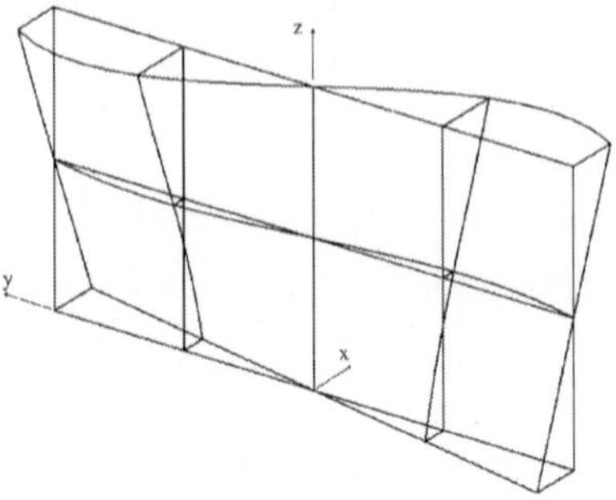

Fig. 6. Shape of bulkhead deformation

The bulkhead grillage strain energy includes vertical and horizontal bending with contraction and torsion (Senjanović et al., 2008b).

$$U_g = \frac{1}{1-\nu^2}\left[\frac{116H^3}{35b}i_y + \frac{32b^3}{105H}i_z + \frac{8Hb}{75}\nu\left(i_y + i_z\right) + \frac{143Hb}{75}(1-\nu)i_t\right]E\psi'^2 \tag{44}$$

where i_y, i_z and i_t are the average moments of inertia of cross-section and torsional modulus per unit breadth, respectively. The stool strain energy is comprised of the bending, shear and torsional contributions

$$U_s = \left[\frac{12h^2 I_{sb}}{b} + 72(1+\nu)\frac{h^2}{b^3}\frac{I_{sb}^2}{A_s} + \frac{9bI_{st}}{10(1+\nu)}\right]E\psi'^2 \tag{45}$$

where I_{sb}, A_s and I_{st} are the moment of inertia of cross-section, shear area and torsional modulus, respectively. Quantity h is the stool distance from the inner bottom, Fig. 7.

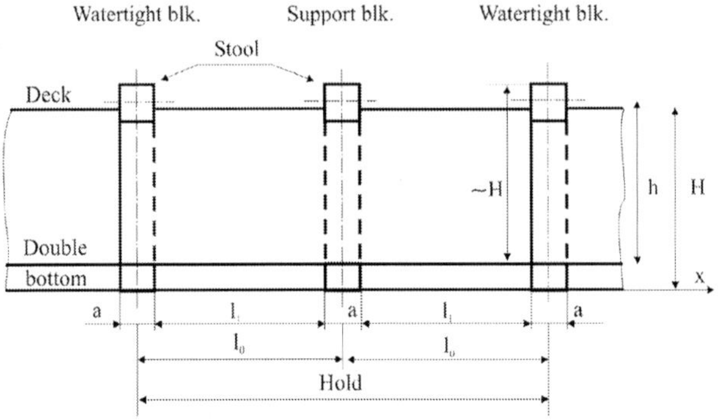

Fig. 7. Longitudinal section of container ship hold

The equivalent torsional modulus yields, Fig. 7

$$I_t^* = \left[1 + \frac{a}{l_1} + \frac{4(1+\nu)C}{I_t l_0}\right] I_t, \quad C = \frac{U_g + U_s}{E\psi'^2}, \tag{46}$$

where a is the web height of bulkhead girders (frame spacing), l_0 is the bulkhead spacing, $l_1 = l_0 - a$ is the net length, and C is the energy coefficient.

7. Contribution of engine room structure to the hull stiffness

Ultra Large Container Ships are characterized by relatively short engine room structure with length of about a half of ship breadth. Its complex deformation is illustrated in a case of a 7800 TEU container ship, Fig. 8. The deck shear deformation is predominant, while hold transverse bulkhead stool is exposed to bending. Due to shortness of the engine room, its transverse bulkheads are skewed but somewhat less pronounced than warping of the hold bulkheads. Warping of the transom is negligible, and that is an important fact when specifying boundary conditions in vibration analysis.

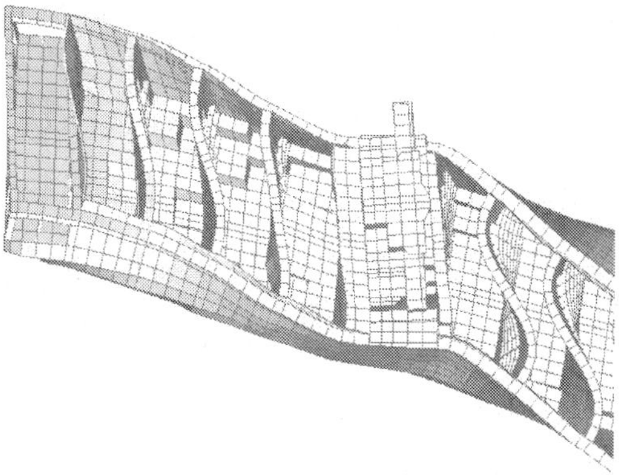

Fig. 8. Deformation of 7800 TEU container ship aft structure

7.1 Stiffness of engine room structure

A short engine room structure can be considered either as a closed segment with relevant stiffness or as an open segment with increased stiffness due to deck contribution (Pedersen, 1985). The latter simulation in fact gives results which agree better with 3D FEM results, than the former one (Pedersen, 1983). Deck contribution to hull stiffness can be determined by energy approach, as it is done in the case of transverse bulkheads (Senjanović et al., 2008b). Such a beam model is consistent at global level of energy balance, and that is sufficient for application in ship hydroelastic analysis, where proper natural frequencies and mode shapes of dry hull are required.

In the case of short engine room, torsion induces distortion of cross-section while hull bending is negligible. Solution of that complex problem is described here by employing the energy balance approach and concept of the effective stiffness due to reason of simplicity. A

closed hull segment is considered as open one with deck influence. For that purpose let us determine deck strain energy. All quantities related to closed and open cross-section are designated by $(.)^{*}$ and $(.)^{\circ}$, respectively
As it can be seen in Fig. 8, the upper deck is exposed to large deformation, while the double bottom in-plane deformation is quite small. The relative axial displacement of the internal upper deck boundaries, with respect to double bottom, is result of their warping

$$U=|U_D|+|U_B|=(|w_D|+|w_B|)\psi_t' \tag{47}$$

It causes deck in-plane (membrane) deformation. The problem can be solved in an approximate analytical way by considering deck as a beam. Its horizontal anti-symmetric deflection consists of pure bending and shear contribution, Fig. 9. The former is assumed in the form

$$u_b=\frac{y}{2b}\left[3-\left(\frac{y}{b}\right)^2\right]U_b, \tag{48}$$

which satisfies relevant boundary conditions: $u_b(0)=0$ and $u_b''(0)=0$, where U_b is the boundary bending deflection. Shear deflection depends on bending deflection

$$u_s=-\frac{EI}{GA}\frac{\mathrm{d}^2u_b}{\mathrm{d}y^2}=2(1+\nu)\left(\frac{a}{b}\right)^2\frac{y}{b}U_b, \tag{49}$$

where the internal deck cross-section area, $A=2at$, its moment of inertia, $I=\frac{2}{3}a^3t$, and the relation $E=2(1+\nu)G$, are taken into account, Fig. 9. Total deflection is obtained by summing up its constitutive parts, Eqs. (48) and (49). Relation between total boundary deflection and the bending boundary deflection reads

$$U=\left[1+2(1+\nu)\left(\frac{a}{b}\right)^2\right]U_b \tag{50}$$

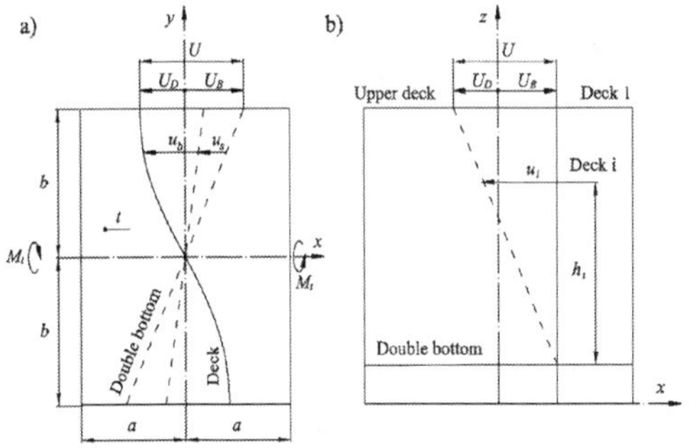

Fig. 9. Deck deformation and double bottom rotation, a)-bird view, b)-lateral view

The total internal deck strain energy consists of the bending and shear contributions

$$E_1 = \frac{1}{2} EI \int_{-b}^{b} \left(\frac{d^2 u_b}{dy^2} \right)^2 dy + \frac{1}{2} GA \int_{-b}^{b} \left(\frac{du_s}{dy} \right)^2 dy \tag{51}$$

By substituting Eqs. (48) and (49) into (51), one finds

$$E_1 = 4(1+\nu) Gt \left(\frac{a}{b} \right)^3 \left[1 + 2(1+\nu) \left(\frac{a}{b} \right)^2 \right] U_b^2 \tag{52}$$

Finally, by taking into account Eqs. (47) and (50), yields

$$E_1 = \frac{4(1+\nu) Gt \left(\frac{a}{b} \right)^3}{1 + 2(1+\nu) \left(\frac{a}{b} \right)^2} \left(|w_D| + |w_B| \right)^2 \psi_t'^2 \tag{53}$$

On the other hand, total energy of the closed hull segment can be obtained by summing up energy of open segment and the deck strain energy, i.e.

$$E_{tot}^* = E_w^\circ + E_t^\circ + E_1 - E_\mu \tag{54}$$

where

$$E_w^\circ = \frac{1}{2} \int_{-a}^{a} B_w^\circ \psi_t'' dx, \quad E_t^\circ = \frac{1}{2} \int_{-a}^{a} T_t^\circ \psi_t' dx, \quad E_\mu = \int_{-a}^{a} \mu_x \psi dx. \tag{55}$$

Within a short span $2a$, constant value of ψ_t' (as for deck) can be assumed, so that second term in Eq. (26) by inserting T_t° from Eqs. (7), leads to

$$E_t^\circ = GI_t^\circ a \psi_t'^2 . \tag{56}$$

E_t° and E_1 in (54) can be unified into one term since both depend on $\psi_t'^2$

$$E_t^\circ + E_1 = Ga\tilde{I}_t \psi_t'^2 \tag{57}$$

where

$$\tilde{I}_t = (1 + C) I_t^\circ, \quad C = \frac{E_1}{E_t^\circ} \tag{58a, b}$$

$\tilde{I}_t$ is the effective torsional modulus which includes both open cross-section and deck effects.

Engine room structure is designed in such a way that the hold double skin continuity is ensured and necessary decks are inserted between the double skins. Strain energy is derived for the first (main) deck and for the others it can be assumed that their strain energy is proportional to the deck plating volume, V, and linearly increasing deformation with the deck distance from inner bottom, h, Fig. 9, since the double bottom is much stiffer than decks. In that way the coefficient C, Eq. (58b), by employing (53) and (56), reads

$$C = \frac{\sum E_i}{E_t} = \frac{4(1+\nu)t_1\left(\frac{a}{b}\right)^3\left(|w_D|+|w_B|\right)^2 k}{\left[1+2(1+\nu)\left(\frac{a}{b}\right)^2\right] I_t^{\circ} a} \tag{59}$$

where

$$k = \sum \frac{V_i}{V_1}\left(\frac{h_i}{h_1}\right)^2 . \tag{60}$$

In the above consideration distortion of cross-sections is not included and that is subject of further investigation.

7.2 Torsion of segmented girder

Let us consider a girder consisted of three segments, Fig. 10. The end segments are open and the middle one is closed, so that the girder is symmetric with respect to the z axis. Each segment is specified in its local coordinate system. The properties of the middle and end segments are designated by $(.)^*$ and $(.)^{\circ}$, respectively. The relevant expressions for displacements and sectional forces are listed below (Senjanović et al., 2009a, 2010a):

$$\begin{aligned} u &= w\frac{\mathrm{d}\psi_t}{\mathrm{d}x} = w\left(\frac{A_1}{l} + A_2 a \mathrm{sh}ax + A_3 a \mathrm{ch}ax + \psi_p'\right), \\ T_t &= GI_t\left(\frac{A_1}{l} + A_2 a \mathrm{sh}ax + A_3 a \mathrm{ch}ax + \psi_p'\right), \\ T_w &= -GI_t\left(A_2 a \mathrm{sh}ax + A_3 a \mathrm{ch}ax\right) - EI_w\psi_p''', \\ T &= GI_t\left(\frac{A_1}{l} + \psi_p'\right) - EI_w\psi_p''', \end{aligned} \tag{61}$$

$$B_w = GI_t\left(A_2 \mathrm{ch}ax + A_3 \mathrm{sh}ax\right) + EI_w\psi_p'',$$

where ψ_p represents particular solution of differential equation and coefficient α yields

$$a = \sqrt{\frac{GI_t}{EI_w}} . \tag{62}$$

The symbols A_i and B_i are used for the integration constants of the closed and open segments. The girder is loaded with torque M_t at the ends, while $\mu_x = 0$. The ends are fixed against warping.

The boundary and compatibility conditions in the considered case, yield

$$\begin{aligned}&\psi^*(a)=\psi^\circ(0),\ \psi_t^{*\prime}(a)=\psi_t^{\circ\prime}(0),\\&T^*(a)=T^\circ(0),\ B_w^*(a)=B_w^\circ(0),\\&u^\circ(l^\circ)=0,\ T^\circ(l^\circ)=M_t.\end{aligned}\tag{63}$$

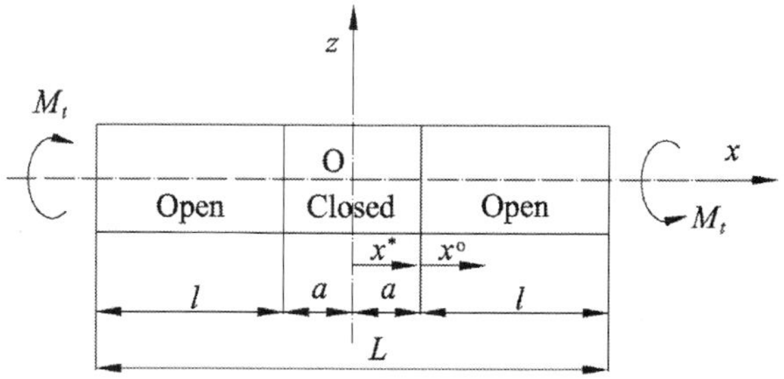

Fig. 10. Torsion of segmented girder

From the third and last conditions (63) one finds

$$A_1=\frac{M_t a}{G\tilde{I}_t},\ B_1=\frac{M_t l^\circ}{GI_t^\circ}.\tag{64}$$

The remaining four conditions (63) lead to the system of algebraic equations (Senjanović et al., 2010a) and its analytical solution reads:

$$A_3=\frac{D_{A3}}{D},\ B_2=\frac{D_{B2}}{D},\ B_3=\frac{D_{B3}}{D},\tag{65}$$

where

$$\begin{aligned}&D_{A3}=\frac{M_t}{G}\left[\left(1-\frac{I_t^\circ}{\tilde{I}_t}\right)\mathrm{ch}\beta l^\circ-1\right],\\&D_{B2}=-\frac{M_t}{G}\mathrm{sh}\alpha a\left[\left(1-\frac{\tilde{I}_t}{I_t^\circ}\right)\mathrm{ch}\beta l^\circ+\frac{\tilde{I}_t}{I_t^\circ}\right],\\&D_{B3}=\frac{M_t}{G}\left[\left(1-\frac{\tilde{I}_t}{I_t^\circ}\right)\mathrm{sh}\alpha a\ \mathrm{sh}\beta l^\circ-\frac{\alpha}{\beta}\mathrm{ch}\alpha a\right],\\&D=I_t^\circ\alpha\mathrm{ch}\alpha a\ \mathrm{ch}\beta l^\circ+\tilde{I}_t\beta\mathrm{sh}\alpha a\ \mathrm{sh}\beta l^\circ.\end{aligned}\tag{66}$$

8. Numerical procedure for vibration analysis

A thin-walled girder is modelled with a set of beam finite elements. Their assemblage in the global coordinate system, performed in the standard way, results in the matrix equation of motion, which may be extended by the damping forces

$$[K]\{\Delta\}+[C]\{\dot{\Delta}\}+[M]\{\ddot{\Delta}\}=\{F(t)\}, \tag{67}$$

where $[K]$, $[C]$ and $[M]$ are the stiffness, damping and mass matrices, respectively; $\{\Delta\},\{\dot{\Delta}\}$ and $\{\ddot{\Delta}\}$ are the displacement, velocity and acceleration vectors, respectively; and $\{F(t)\}$ is the load vector.

In case of natural vibration $\{F(t)\} = \{0\}$ and the influence of damping is rather low for the most of the structures, so that the damping forces may be ignored. Assuming

$$\{\Delta\}=\{\varphi\}e^{i\omega t}, \tag{68}$$

where $\{\varphi\}$ and ω are the mode vector and natural frequency respectively, Eq. (67) leads to the eigenvalue problem

$$([K]-\omega^2[M])\{\varphi\}=\{0\}, \tag{69}$$

which may be solved by employing different numerical methods (Bathe, 1996) The basic one is the determinant search method in which ω is found from the condition

$$|[K]-\omega^2[M]|=0 \tag{70}$$

by an iteration procedure. Afterwards, $\{\varphi\}$ follows from (69) assuming unit value for one element in $\{\varphi\}$.

The forced vibration analysis may be performed by direct integration of Eq. (67), as well as by the modal superposition method. In the latter case the displacement vector is presented in the form

$$\{\Delta\}=[\varphi]\{X\}, \tag{71}$$

where $[\varphi]=[\{\varphi\}]$ is the undamped mode matrix and $\{X\}$ is the generalised displacement vector. Substituting (71) into (67), the modal equation yields

$$[k]\{X\}+[c]\{\dot{X}\}+[m]\{\ddot{X}\}=\{f(t)\}, \tag{72}$$

where

$$\begin{aligned}
&[k]=[\varphi]^T[K][\varphi] - \text{modal stiffness matrix}\\
&[c]=[\varphi]^T[C][\varphi] - \text{modal damping matrix}\\
&[m]=[\varphi]^T[M][\varphi] - \text{modal mass matrix}\\
&\{f(t)\}=[\varphi]^T\{F(t)\} - \text{modal load vector.}
\end{aligned} \tag{73}$$

The matrices $[k]$ and $[m]$ are diagonal, while $[c]$ becomes diagonal only in a special case, for instance if $[C] = a_0 [M] + \beta_0 [K]$, where a_0 and β_0 are coefficients (Senjanović, 1990).

Solving (72) for undamped natural vibration, $[k] = [\omega^2 m]$ is obtained, and by its backward substitution into (72) the final form of the modal equation yields

$$\left[\omega^2\right]\{X\}+2[\omega][\zeta]\{\dot{X}\}+\{\ddot{X}\}=\{\varphi(t)\}, \tag{74}$$

where

$$\begin{aligned} [\omega]&=\left[\sqrt{\frac{k_{ii}}{m_{ii}}}\right] \text{ - natural frequency matrix} \\ [\zeta]&=\left[\frac{c_{ij}}{2\sqrt{(k_{ii}m_{ii})}}\right] \text{ - relative damping matrix} \\ \{\varphi(t)\}&=\left\{\frac{f_i(t)}{m_{ii}}\right\} \text{ - relative load vector.} \end{aligned} \tag{75}$$

If $[\zeta]$ is diagonal, the matrix Eq. (74) is split into a set of uncoupled modal equations.
If vibration excitation is of periodical nature it can be split into harmonics, and the structure response for each of them is determined in the frequency domain. In a case of general or impulsive excitation the vibration problem has to be solved in the time domain.
Several numerical methods are available for this purpose, as for instance the Houbolt, the Newmark and the Wilson θ method (Bathe, 1996), as well as the harmonic acceleration method (Lozina, 1988, Senjanović, 1984).
It is important to point out that all stiffness and mass matrices of the beam finite element (and consequently those of the assembly) are frequency dependent quantities, due to coefficients a and η in the formulation of the shape functions, Eqs. (34) and (35). Therefore, for solving the eigenvalue problem (69) an iteration procedure has to be applied. As a result of frequency dependent matrices, the eigenvectors are not orthogonal. If they are used in the modal superposition method for determining forced response, full modal stiffness and mass matrices are generated. Since the inertia terms are much smaller than the deformation ones in Eqs. (24) and (25), the off-diagonal elements in modal stiffness and mass matrices are very small compared to the diagonal elements and can be neglected.
It is obvious that the usage of the physically consistent non-orthogonal natural modes in the modal superposition method is not practical, especially not in the case of time integration. Therefore, it is preferable to use mathematical orthogonal modes for that purpose. They are created by the static displacement relations yielding from Eqs. (24) and (25) with $\omega = 0$, that leads to $a = \eta = 1$. In that case all finite element matrices, defined with Eqs. (37) and in Appendix A, can be transformed into explicit form, Appendix B.

9. Cross-section properties of thin-walled girder

Geometrical properties of a thin-walled girder include cross-section area A, moment of inertia of cross-section I_b, shear area A_s, torsional modulus I_t, warping modulus I_w and shear inertia modulus I_s. These parameters are determined analytically for a simple cross-section as pure geometrical properties (Haslum & Tonnessen, 1972, Pavazza, 1991, 2005, Vlasov, 1961).
However, determination of cross-section properties for an open multi-cell cross-section, as for instance in case of ship structures, is quite a difficult task. Therefore, the strip element method is applied for solving this statically indetermined problem (Cheung, 1976). That is well-known and widely used theory of thin-walled girders, which is only briefly described

here. Firstly, axial node displacements are calculated due to bending caused by shear force, and due to torsion caused by variation of twist angle. Then, shear stress in bending τ_b, shear stress due to pure torsion τ_t, shear and normal stresses due to restrained warping τ_w and σ_w, respectively, are determined. Based on the equivalence of strain energies induced by sectional forces and calculated stresses, it is possible to specify cross-section properties in the same formulation as presented below. Furthermore, those formulae can be expressed by stress flows, i.e. stresses due to unit sectional forces (Senjanović & Fan, 1992, 1993).

Shear area:

$$A_s = \frac{Q^2}{\int_A \tau_b^2 dA} = \frac{1}{\int_A g_b^2 dA}, \quad g_b = \frac{\tau_b}{Q}. \tag{76}$$

Torsional modulus:

$$I_t = \frac{T_t^2}{\int_A \tau_t^2 dA} = \frac{1}{\int_A g_t^2 dA}, \quad g_t = \frac{\tau_t}{T_t}. \tag{77}$$

Shear inertia modulus:

$$I_s = \frac{T_w^2}{\int_A \tau_w^2 dA} = \frac{1}{\int_A g_w^2 dA}, \quad g_w = \frac{\tau_w}{T_w}. \tag{78}$$

Warping modulus:

$$I_w = \frac{B_w^2}{\int_A \sigma_w^2 dA} = \frac{1}{\int_A f_w^2 dA}, \quad f_w = \frac{\sigma_w}{B_w}; \quad I_w = \int_A w^2 dA. \tag{79}$$

The above quantities are not pure geometrical cross-section properties any more, since they also depend on Poisson's ratio as a physical parameter.

The mass parameters can be expressed with the given mass distribution per unit length, m, and calculated cross-section parameters, i.e.

$$J_b = \frac{m}{A} I_b, \quad J_t^0 = \frac{m}{A} I_p, \quad J_w = \frac{m}{A} I_w. \tag{80}$$

where $I_p = I_{by} + I_{bz}$ is the polar moment of inertia of cross-section.

10. Illustrative numerical examples

For the illustration of the procedure related to engine room effective stiffness determination, 3D FEM analysis of ship-like pontoon has been undertaken. The 3D FEM model is constituted according to 7800 TEU container ship with main dimensions $L_{pp} x B x H = 319x42.8x24.6$ m, and detailed desciption given in (Tomašević, 2007). The complete hydroelastic analysis of the same ship has been performed.

Stiffness properties of ship hull are calculated by program STIFF, based on the theory of thin-walled girders (STIFF, 1990), Fig. 11.

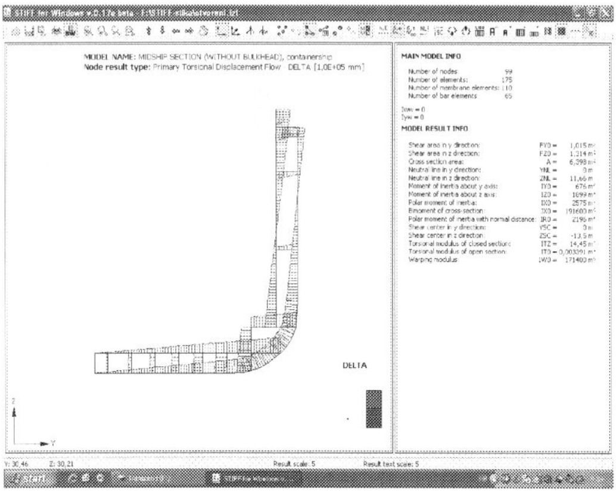

Fig. 11. Program STIFF – warping of ship cross-section

Influence of the transverse bulkheads is taken into account by using the equivalent torsional modulus for the open cross-sections instead of the actual values, i.e. $I_t^* = 2.4I_t$. This value is applied for all ship-cross sections as the first approximation.

10.1 Analysis of ship-like segmented pontoon

Torsion of the segmented pontoon of the length $L = 300$ m, with effective parameters is considered. Torsional moment $M_t = 40570$ kNm is imposed at the pontoon ends. The pontoon is considered free in the space and the problem is solved analytically according to the formulae given in Section 4. The following values of the basic parameters are used: $a = 10.1$ m, $b = 19.17$ m, $t_1 = 0.01645$ m, $w_D = -221$ m^2, $w_B = 267$ m^2, $I_t^\circ = 14.45$ m^4, $k = 1.894$. As a result $C = 22.42$, Eq. (59), and accordingly $\tilde{I}_t = 338.4$ m^4, Eq. (58a), are obtained. Since $\tilde{I}_t = 0.36 I_t^*$, effect of the short engine room structure on its torsional stiffness is obvious.

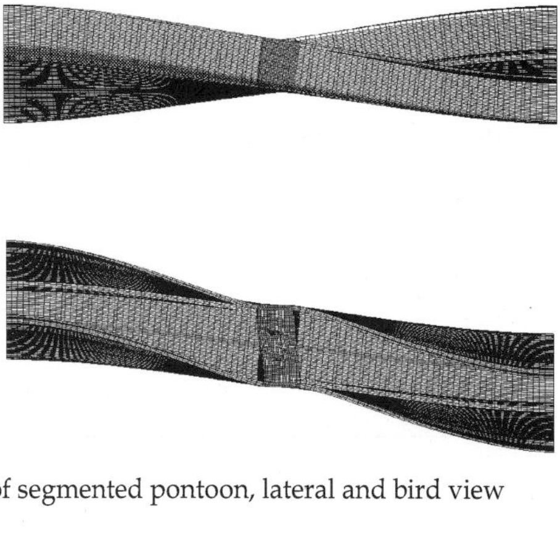

Fig. 12. Deformation of segmented pontoon, lateral and bird view

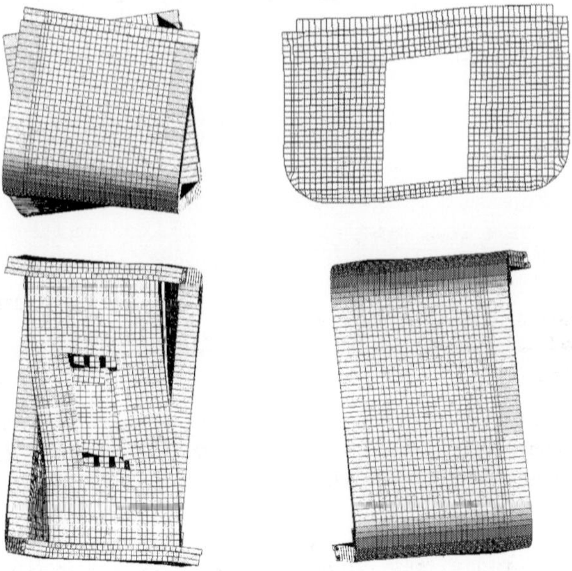

Fig. 13. Lateral, axial, bird and fish views on deformed engine room superelement

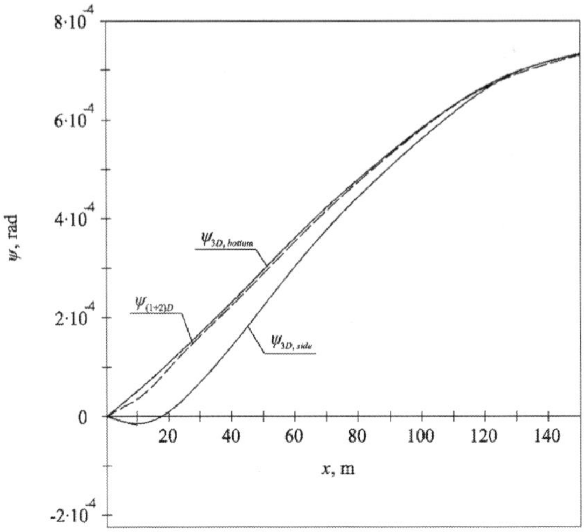

Fig. 14. Twist angles of segmented pontoon

The 3D FEM model of segmented pontoon is made by commercial software package SESAM and consists of 20 open and 1 closed (engine room) superelement. The pontoon ends are closed with transverse bulkheads. The shell finite elements are used. The pontoons are loaded at their ends with the vertical distributed forces in the opposite directions, generating total torque M_t = 40570 kNm. The midship section is fixed against transverse and vertical displacements, and the pontoon ends are constrained against axial displacements (warping). Lateral and bird view on the deformed segmented pontoon is shown in Fig. 12, where the influence of more rigid engine room structure is evident. Detailed view on this pontoon portion is presented in Fig. 13. It is apparent that segment of very stiff double bottom and sides rotate as a "rigid body", while decks and transverse bulkheads are exposed to shear deformation. This deformation causes the distortion of the cross-section, Fig. 13.

Twist angles of the analytical beam solution and that of 3D FEM analysis for the pontoon bottom are compared in Fig. 14. As it can be noticed, there are some small discrepancies between $\psi_{(1+2)D}$ and $\psi_{3D,bottom}$, which are reduced to a negligible value at the pontoon ends

Fig. 14 also shows twist angle of side structure and the difference $\delta = \psi_{3D,\text{bottom}} - \psi_{3D,\text{side}}$ represents distortion angle of cross-section which is highly pronounced. As it is mentioned before, the problem will be further investigated.

10.2 Validation of 1D FEM model

The reliability of 1D FEM analysis is verified by 3D FEM analysis of the considered ship. For this purpose, the light weight loading condition of dry ship with displacement Δ=33692 t is taken into account. The equivalent torsional stiffness of the engine room structure, as well as equivalent stiffness of fore and aft peaks is not taken into account in this example for the time being. However, it will be done in the next step of investigation. The lateral and bird view of the first dominantly torsional and second dominantly horizontal mode of the wetted surface, determined by 1D model, is shown in Fig. 15.

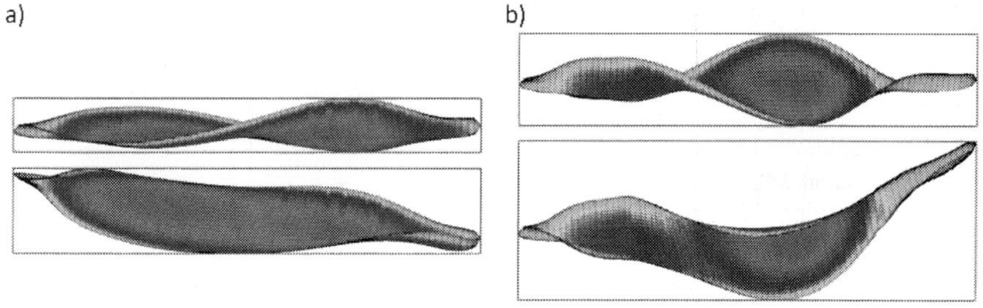

Fig. 15. The first and second mode, lateral and bird view, light weight, 1D model

The first and second 3D dry coupled natural modes of the complete ship structure are shown in Fig. 16. They are similar to that of 1D analysis for the wetted surface. Warping of the transverse bulkheads, which increases the hull torsional stiffness, is evident.

The first four corresponding natural frequencies obtained by 1D and 3D analyses are compared in Table 1.

Mode no.	Vert.		Horiz. + tors.		Mode no.
	1D	3D	1D	3D	
1	7.35	7.33	4.17	4.15	1(H0 + T1)
2	15.00	14.95	7.34	7.40	2(H1 + T2)
3	24.04	22.99	12.22	12.09	3(H2 + T3)
4	35.08	34.21	15.02	16.22	4(H3 + T4)

Table 1. Dry natural frequencies, light weight, ω_i [rad/s]

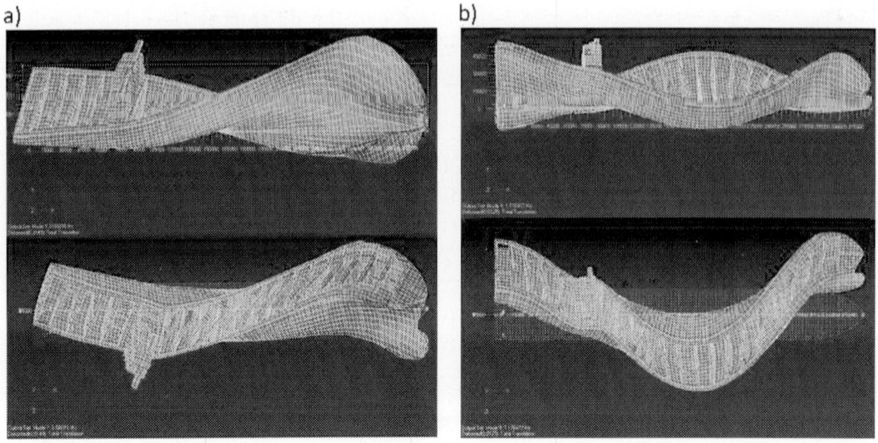

Fig. 16. The first and second mode, lateral and bird view, light weight, 3D model

Quite good agreement is achieved. Values of natural frequencies for higher modes are more difficult to correlate, since strong coupling between global hull modes and local substructure modes of 3D analysis occurs.

10.3 Hydroelastic response of large container ship

Transfer functions of torsional moment and horizontal bending moment at the midship section, obtained using 1D structural model, are shown in Figs. 17 and 18, respectively. The angle of 180° is related to head sea. They are compared to the rigid body ones determined by program HYDROSTAR. Very good agreement is obtained in the lower frequency domain, where the ship behaves as a rigid body, while large discrepancies occur at the resonances of the elastic modes, as expected.

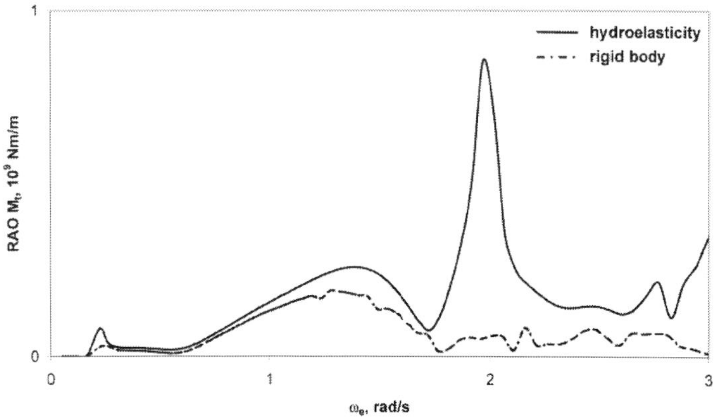

Fig. 17. Transfer function of torsional moment, χ=120°, U=25 kn, x=155.75 m from AP

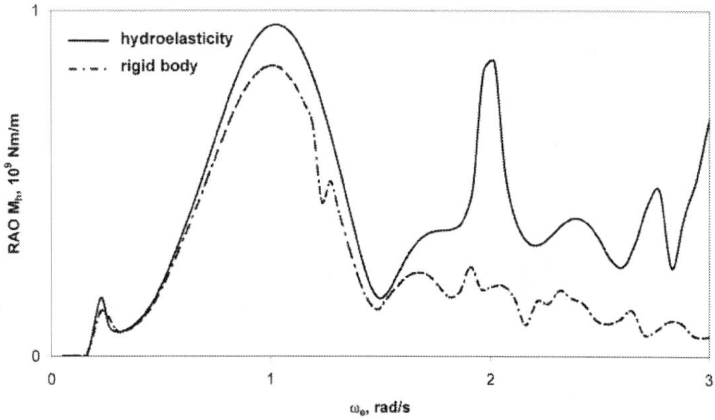

Fig. 18. Transfer function of horizontal bending moment, χ=120°, U=25 kn, x=155.75 m from AP

11. Conclusion

Ultra large container ships are quite elastic and especially sensitive to torsion due to large deck openings. The wave induced response of such ships should be determined by using mathematical hydroelastic models which are consisted of structural, hydrostatic and hydrodynamic parts.

In this chapter the methodology of ship hydroelastic analysis is briefly described, and the role of structural model is discussed. After that, full detail description of the sophisticated beam structural model, which takes shear influence on torsion, as well as contribution of transverse bulkheads and engine room structure to the hull stiffness, is given. Numerical procedure for vibration analysis is also described and determination of ship cross-section properties is explained. The developed theories are illustrated through the numerical examples which include analysis of torsional response of a ship-like segmented pontoon, free vibration analysis of a large container ship and comparison with the results obtained using 3D FEM model, and complete global hydroelastic analysis of a container ship.

It is shown that the used sophisticated beam model of ship hull, based on the advanced thin-walled girder theory with included shear influence on torsion and a proper contribution of transverse bulkheads and engine room structure to its stiffness, is a reasonable choice for determining wave load effects. However, based on the experience, stress concentration in hatch corners calculated directly by the beam model is underestimated. This problem can be overcome by applying substructure approach, i.e. 3D FEM model of substructure with imposed boundary conditions from beam response. In any case, 3D FEM model of complete ship is preferable from the viewpoint of determining stress concentration. Concerning further improvements of the beam model, the distortion induced by torsion is of interest.

The illustrative numerical example of the 7800 TEU container ship shows that the developed hydroelasticity theory, utilizing sophisticated 1D FEM structural model and 3D hydrodynamic model, is an efficient tool for application in ship hydroelastic analyses. The obtained results point out that the transfer functions of hull sectional forces in case of resonant vibration (springing) are much higher than in resonant ship motion.

12. Acknowledgment

This investigation is carried out within the EU FP7 Project TULCS (*Tools for Ultra Large Container Ships*) and the project of Croatian Ministry of Science, Education and Sports *Load and Response of Ship Structures*.

13. Appendix A – consistent finite element properties (frequency dependent formulation)

The stiffness and mass matrices, Eqs. (37), are expressed with one or two integrals, which can be classified in three different types. For general notation of shape functions

$$g_i = \langle g_{ik} \rangle \{ \xi^k \}, \quad i = 1,2,3,4; \quad k = 0,1,2,3, \tag{A1}$$

where q_{ik} are coefficients and $\xi = x/l$, one finds the solutions of integrals in the following form:

$$\begin{aligned} I_{ij}^{(0)}\left(g_{ik}, g_{jk}\right) &= \int_0^l g_i g_j \mathrm{d}x = \langle g_{ik} \rangle \int_0^l \{\xi^k\}\langle \xi^k \rangle \mathrm{d}x \{g_{jk}\} \\ &= l\Big[g_{i0}g_{j0} + \frac{1}{2}\left(g_{i1}g_{j0} + g_{i0}g_{j1}\right) + \frac{1}{3}\left(g_{i2}g_{j0} + g_{i1}g_{j1} + g_{i0}g_{j2}\right) \\ &+ \frac{1}{4}\left(g_{i0}g_{j3} + g_{i1}g_{j2} + g_{i2}g_{j1} + g_{i3}g_{j0}\right) + \frac{1}{5}\left(g_{i1}g_{j3} + g_{i2}g_{j2} + g_{i3}g_{j1}\right) \\ &+ \frac{1}{6}\left(g_{i2}g_{j3} + g_{i3}g_{j2}\right) + \frac{1}{7} g_{i3}g_{j3} \Big] \end{aligned} \tag{A2}$$

$$I_{ij}^{(1)}\left(g_{ik},g_{jk}\right)=\int_0^l \frac{\mathrm{d}g_i}{\mathrm{d}x}\frac{\mathrm{d}g_j}{\mathrm{d}x}\mathrm{d}x=\left\langle g_{ik}\right\rangle\int_0^l\left\{k\xi^{k-1}\right\}\left\langle k\xi^{k-1}\right\rangle\mathrm{d}x\left\{g_{jk}\right\}$$
$$=\frac{1}{l}\Big[g_{i1}g_{j1}+g_{i1}g_{j2}+g_{i2}g_{j1}+g_{i1}g_{j3}+g_{i3}g_{j1}$$
$$+\frac{4}{3}g_{i2}g_{j2}+\frac{3}{2}\left(g_{i2}g_{j3}+g_{i3}g_{j2}\right)+\frac{9}{5}g_{i3}g_{j3}\Big] \quad \text{(A3)}$$

$$I_{ij}^{(2)}\left(g_{ik},g_{jk}\right)=\int_0^l \frac{\mathrm{d}^2g_i}{\mathrm{d}x^2}\frac{\mathrm{d}^2g_j}{\mathrm{d}x^2}\mathrm{d}x=\left\langle g_{ik}\right\rangle\int_0^l\left\{k(k-1)\xi^{k-2}\right\}\left\langle k(k-1)\xi^{k-2}\right\rangle\mathrm{d}x\left\{g_{jk}\right\}$$
$$=\frac{4}{l^3}\left[g_{i2}g_{j2}+\frac{3}{2}\left(g_{i2}g_{j3}+g_{i3}g_{j2}\right)+3g_{i3}g_{j3}\right]. \quad \text{(A4)}$$

Thus, the finite element properties can be written in the following systematic way suitable for coding.
Stiffness matrices

$$\begin{aligned}
\left[k\right]_{bs}&=EI_b\left[I_{ij}^{(2)}\left(a_{ik},a_{jk}\right)\right]+GA_s\left[I_{ij}^{(1)}\left(b_{ik},b_{jk}\right)\right]\\
\left[k\right]_{ws}&=EI_w\left[I_{ij}^{(2)}\left(d_{ik},d_{jk}\right)\right]+GI_s\left[I_{ij}^{(1)}\left(e_{ik},e_{jk}\right)\right]\\
\left[k\right]_{t}&=GI_t\left[I_{ij}^{(1)}\left(d_{ik},d_{jk}\right)\right]
\end{aligned} \quad \text{(A5)}$$

Mass matrices

$$\begin{aligned}
\left[m\right]_{sb}&=m\left[I_{ij}^{(0)}\left(c_{ik},c_{jk}\right)\right]+J_b\left[I_{ij}^{(1)}\left(a_{ik},a_{jk}\right)\right]\\
\left[m\right]_{tw}&=J_t\left[I_{ij}^{(0)}\left(f_{ik},f_{jk}\right)\right]+J_w\left[I_{ij}^{(1)}\left(d_{ik},d_{jk}\right)\right]\\
\left[m\right]_{st}&=m_c\left[I_{ij}^{(0)}\left(c_{ik},f_{jk}\right)\right],\ \left[m\right]_{ts}=\left[m\right]_{st}^T
\end{aligned} \quad \text{(A6)}$$

Load vectors

$$\begin{aligned}
\{q\}&=l\left\{q_0\left(c_{i0}+\frac{1}{2}c_{i1}+\frac{1}{3}c_{i2}+\frac{1}{4}c_{i3}\right)+q_1\left(\frac{1}{2}c_{i0}+\frac{1}{3}c_{i1}+\frac{1}{4}c_{i2}+\frac{1}{5}c_{i3}\right)\right\}\\
\{\mu\}&=l\left\{\mu_0\left(f_{i0}+\frac{1}{2}f_{i1}+\frac{1}{3}f_{i2}+\frac{1}{4}f_{i3}\right)+\mu_1\left(\frac{1}{2}f_{i0}+\frac{1}{3}f_{i1}+\frac{1}{4}f_{i2}+\frac{1}{5}f_{i3}\right)\right\}
\end{aligned} \quad \text{(A7)}$$

14. Appendix B – simplified finite element properties, from appendix A (frequency independent formulation)

Stiffness matrices:

$$\left[k\right]_{bs}=\frac{2EI_b}{(1+12\beta)l^3}\begin{bmatrix}6 & 3l & -6 & 3l\\ & 2(1+3\beta)l^2 & -3l & (1-6\beta)l^2\\ & & 6 & -3l\\ Sym. & & & 2(1+3\beta)l^2\end{bmatrix} \quad \text{(B1)}$$

$$[k]_{ws} = \frac{2EI_w}{(1+12\gamma)l^3}\begin{bmatrix} 6 & 3l & -6 & 3l \\ & 2(1+3\gamma)l^2 & -3l & (1-6\gamma)l^2 \\ & & 6 & -3l \\ Sym. & & & 2(1+3\gamma)l^2 \end{bmatrix} \quad \text{(B2)}$$

$$[k]_t = \frac{GI_t}{30(1+12\gamma)^2 l}\begin{bmatrix} 36 & 3(1-60\gamma)l & -36 & 3(1-60\gamma)l \\ & 4(1+15\gamma+360\gamma^2)l^2 & -3(1-60\gamma)l & -(1+60\gamma-720\gamma^2)l^2 \\ & & 36 & -3(1-60\gamma)l \\ Sym. & & & 4(1+15\gamma+360\gamma^2)l^2 \end{bmatrix} \quad \text{(B3)}$$

Mass matrices:

$$[m]_{sb} = [m]_s + [m]_b \quad \text{(B4)}$$

$$[m]_s = \frac{ml}{420(1+12\beta)^2}\begin{bmatrix} 156+3528\beta+20160\beta^2 & (22+462\beta+2520\beta^2)l & 54+1512\beta+10080\beta^2 & -(13+378\beta+2520\beta^2)l \\ & (4+84\beta+504\beta^2)l^2 & (13+378\beta+2520\beta^2)l & -(3+84\beta+504\beta^2)l^2 \\ & & 156+3528\beta+20160\beta^2 & -(22+462\beta+2520\beta^2)l \\ Sym. & & & (4+84\beta+504\beta^2)l^2 \end{bmatrix} \quad \text{(B5)}$$

$$[m]_b = \frac{J_b}{30(1+12\beta)^2 l}\begin{bmatrix} 36 & (3-180\beta)l & -36 & (3-180\beta)l \\ & (4+60\beta+1440\beta^2)l^2 & (-3+180\beta)l & -(1+60\beta-720\beta^2)l^2 \\ & & 36 & (-3+180\beta)l \\ Sym. & & & (4+60\beta+1440\beta^2)l^2 \end{bmatrix} \quad \text{(B6)}$$

$$[m]_{tw} = [m]_t + [m]_w \quad \text{(B7)}$$

$$[m]_t = \frac{J_t l}{420(1+12\gamma)^2}\begin{bmatrix} 156+3528\gamma+20160\gamma^2 & (22+462\gamma+2520\gamma^2)l & 54+1512\gamma+10080\gamma^2 & -(13+378\gamma+2520\gamma^2)l \\ & (4+84\gamma+504\gamma^2)l^2 & (13+378\gamma+2520\gamma^2)l & -(3+84\gamma+504\gamma^2)l^2 \\ & & 156+3528\gamma+20160\gamma^2 & -(22+462\gamma+2520\gamma^2)l \\ Sym. & & & (4+84\gamma+504\gamma^2)l^2 \end{bmatrix} \quad \text{(B8)}$$

$$[m]_w = \frac{J_w}{30(1+12\gamma)^2 l}\begin{bmatrix} 36 & (3-180\gamma)l & -36 & (3-180\gamma)l \\ & (4+60\gamma+1440\gamma^2)l^2 & (-3+180\gamma)l & -(1+60\gamma-720\gamma^2)l^2 \\ & & 36 & (-3+180\gamma)l \\ Sym. & & & (4+60\gamma+1440\gamma^2)l^2 \end{bmatrix} \quad \text{(B9)}$$

$$[m]_{st} = \frac{mlc}{420(1+12\beta)(1+12\gamma)} \times \begin{bmatrix} 156+1764\beta+1764\gamma+20160\beta\gamma & (22+252\beta+210\gamma+2520\beta\gamma)l & 54+756\beta+756\gamma+10080\beta\gamma & -(13+168\beta+210\gamma+2520\beta\gamma)l \\ (22+210\beta+252\gamma+2520\beta\gamma)l & (4+42\beta+42\gamma+504\beta\gamma)l^2 & (13+210\beta+168\gamma+2520\beta\gamma)l & -(3+42\beta+42\gamma+504\beta\gamma)l^2 \\ 54+756\beta+756\gamma+10080\beta\gamma & (13+168\beta+210\gamma+2520\beta\gamma)l & 156+1764\beta+1764\gamma+20160\beta\gamma & -(22+252\beta+210\gamma+2520\beta\gamma)l \\ -(13+210\beta+168\gamma+2520\beta\gamma)l & -(3+42\beta+42\gamma+504\beta\gamma)l^2 & -(22+210\beta+252\gamma+2520\beta\gamma)l & (4+42\beta+42\gamma+504\beta\gamma)l^2 \end{bmatrix} \tag{B10}$$

$$[m]_{ts} = [m]_{st}^T \tag{B11}$$

Load vectors:

$$\{q\} = \frac{q_0 l}{12}\begin{Bmatrix} 6 \\ l \\ 6 \\ -l \end{Bmatrix} + \frac{q_1 l}{60(1+12\beta)}\begin{Bmatrix} 9+120\beta \\ (2+30\beta)l \\ 21+240\beta \\ -(3+30\beta)l \end{Bmatrix} \tag{B12}$$

$$\{\mu\} = \frac{\mu_0 l}{12}\begin{Bmatrix} 6 \\ l \\ 6 \\ -l \end{Bmatrix} + \frac{\mu_1 l}{60(1+12\gamma)}\begin{Bmatrix} 9+120\gamma \\ (2+30\gamma)l \\ 21+240\gamma \\ -(3+30\gamma)l \end{Bmatrix} \tag{B13}$$

Stiffness ratios:

$$\beta = \frac{EI_b}{GA_s l^2}, \; \gamma = \frac{EI_w}{GI_s l^2}. \tag{B14}$$

15. References

Bathe, KJ. (1996). *Finite Element Procedures,* Prentice Hall

Cheung, YK. (1976). *Finite Strip Method in Structural Analysis,* Pergamon Press

Haslum, K. & Tonnessen, A. (1972). An Analysis of Torsion in Ship Hull, *European Shipbuilding,* No.5/6, pp. 67-89

Kawai, T. (1973). The Application of Finite Element Method to Ship Structures, *Computers & Structures,* Vol.3, No.5, pp. 1175-1194, ISSN 0045-7949

Lozina, Ž. (1988). A Comparison of Harmonic Acceleration Method with the Other Commonly Used Methods for Calculation of Dynamic Transient Response, *Computers & Structures,* Vol.29, No.2, pp. 227-240, ISSN 0045-7949

Pavazza, R. (1991). *Bending and Torsion of Thin-Walled Beams of Open Section on Elastic Foundation,* Ph.D. Thesis. University of Zagreb, (in Croatian)

Pavazza, R. (2005). Torsion of Thin-Walled Beams of Open Cross-Sections with Influence of Shear, *International Journal of Mechanical Sciences,* Vol.47, No.7, pp. 1099-1122, ISSN 0020-7403

Pedersen, PT. (1983). A Beam Model for the Torsional-Bending Response of Ships Hulls, *RINA Transactions,* Vol.31, pp. 171-182

Pedersen, PT. (1985). Torsional Response of Container Ships, *Journal of Ship Research*, Vol.29, pp. 194-205, ISSN 1542-0604

Senjanović, I. (1984). Harmonic Acceleration Method for Dynamic Structural Analysis, *Computers & Structures*, Vol.18, No.1, pp. 71-80, ISSN 0045-7949

Senjanović, I. (1990). *Ship Vibrations*, Part II, University of Zagreb, (in Croatian)

Senjanović, I. & Fan, Y. (1989). A Higher-Order Flexural Beam Theory, *Computers & Structures*, Vol.32, No.5, pp. 973-986, ISSN 0045-7949

Senjanović, I. & Fan, Y. (1992). A Higher-Order Theory of Thin-Walled Girders with Application to Ship Structures, *Computers & Structures*, Vol.43, No.1, pp. 31-52, ISSN 0045-7949

Senjanović, I. & Fan, Y. (1993). A Finite Element Formulation of Initial Ship Cross-Section Properties, *Brodogradnja*, Vol.41, No.1, pp. 27-36, ISSN 0007-215X

Senjanović, I. & Fan, Y. (1997). A Higher-Order Torsional Beam Theory, *International Journal for Engineering Modelling*, Vol.32, No.1-4, pp. 25-40, ISSN 1330-1365

Senjanović, I. & Grubišić, R. (1991). Coupled Horizontal and Torsional Vibration of a Ship Hull with Large Hatch Openings, *Computers & Structures*, Vol.41, No.2, pp. 213-226, ISSN 0045-7949

Senjanović, I., Malenica, Š., Tomašević, S. & Rudan, S. (2007). Methodology of Ship Hydroelastic Investigation, *Brodogradnja*, Vol.58, No.2, pp. 133-145, ISSN 0007-215X

Senjanović, I. , Tomašević, S., Tomić, M., Rudan, S., Vladimir N. & Malenica, Š. (2008a). Hydroelasticity of Very Large Container Ships, *Proceedings of International Conference on Design and Operation of Container Ships*, pp. 51-70, RINA, London

Senjanović, I., Tomašević, S., Rudan, S. & Senjanović, T. (2008b). Role of Transverse Bulkheads in Hull Stiffness of Large Container Ships, *Engineering Structures*, Vol.30, No.9, pp. 2492-2509, ISSN 0141-0296

Senjanović, I., Tomašević, S. & Vladimir, N. (2009a). An Advanced Theory of Thin-Walled Girders with Application to Ship Vibrations, *Marine Structures*, Vol.22, No.3, pp. 387-437, ISSN 0951-8339

Senjanović, I. , Tomašević, S., Vladimir N. & Malenica, Š. (2009b). Numerical Procedure for Ship Hydroelastic Analysis, *Proceedings of International Conference on Computational Methods in Marine Engineering*, pp. 259-264, CIMNE, Barcelona

Senjanović, I., Vladimir, N. & Tomić, M. (2010a). The Contribution of the Engine Room Structure to the Hull Stiffness of Large Container Ships, *International Shipbuilding Progress*, Vol.57, No.1-2, pp. 65-85, ISSN 0020-868X

Senjanović, I. , Tomašević, S., Vladimir N. Tomić, M. & Malenica, Š. (2010b). Application of an Advanced Beam Theory to Ship Hydroelastic Analysis, *Proceedings of International Workshop on Advanced Ship Design for Pollution Prevention*, pp. 31-42, Taylor & Francis, London

STIFF (1990). *User's Manual*, University of Zagreb

Szilard, R. (2004). *Theories and Applications of Plate Analysis*, John Wiley & Sons, New York

Timoshenko, S. & Young, DH. (1955). *Vibrations Problems in Engineering*, D. Van Nostrand

Tomašević, S. (2007). *Hydroelastic Model of Dynamic Response of Container Ship in Waves*, Ph.D. Thesis. University of Zagreb, (in Croatian)

Vlasov, VZ. (1961). *Thin-Walled Elastic Beams*, Israel Program for Scientific Translation, Jerusalem

Wu, YS. & Ho, CS. (1987). Analysis of Wave Induced Horizontal and Torsion Coupled Vibrations of Ship Hull. *Journal of Ship Research,* Vol.31, No.4, pp. 235-252, ISSN 1542-0604